Physics Research and Technology

nova
science publishers
www.novapublishers.com

Physics Research and Technology

Theory of Fluids
Marian Apostol, PhD
2024. ISBN: 979-8-89113-464-5 (Hardcover)
2024. ISBN: 979-8-89113-474-4 (eBook)

Photonic Crystal Fibers (PCFs): Advances in Research and Applications
Mário F. S. Ferreira, PhD (Editor)
2024. ISBN: 979-8-88697-618-2 (Hardcover)
2024. ISBN: 979-8-89113-331-0 (eBook)

Ferrite Materials and Technologies
Ravi Panwar, PhD and Dharmendra Singh (Editors)
2023. ISBN: 979-8-89113-086-9 (Hardcover)

Fields of Particles with Spin, Theory and Applications
Alina Ivashkevich, PhD, Nina Krylova, PhD, Elena Ovsiyuk, PhD,
Vasiliy Kisel, PhD, Vladimir Balan, PhD and Viktor Red'kov, PhD
2023. ISBN: 979-8-89113-012-8 (Hardcover)
2023. ISBN: 979-8-89113-113-2 (eBook)

75 Years of the Pion
Igor Strakovsky, PhD, Alexey A. Petrov, PhD
and Nicolai Popov, PhD (Editors)
2023. ISBN: 979-8-88697-917-6 (Hardcover)
2023. ISBN: 979-8-89113-164-4 (eBook)

Conjugate Construction of Quantum Optics:
From Foundations to Applications
Jeffrey Zheng, PhD (Editor)
2023. ISBN: 979-8-88697-651-9 (Hardcover)

More information about this series can be found at
https://novapublishers.com/product-category/series/physics-research-and-technology/

Marian Apostol

Theory of Fluids

nova science publishers

www.novapublishers.com

Library of Congress Cataloging-in-Publication Data

Names: Apostol, Marian (Professor of Theoretical Physics), editor.
Title: Theory of fluids / Marian Apostol, PhD (author), Professor, Institute for Physics and Nuclear Engineering, Magurele, Romania.
Description: New York : Nova Science Publishers, [2024] | Series: Physics research and technology | Includes bibliographical references and index.
Identifiers: LCCN 2024000933 (print) | LCCN 2024000934 (ebook) | ISBN 9798891134645 (hardcover) | ISBN 9798891134744 (adobe pdf)
Subjects: LCSH: Fluids. | Fluids--Mathematics.
Classification: LCC QC145.2 .A66 2024 (print) | LCC QC145.2 (ebook) | DDC 532.001--dc23/eng/20240129
LC record available at https://lccn.loc.gov/2024000933
LC ebook record available at https://lccn.loc.gov/2024000934

Published by Nova Science Publishers, Inc. † New York

Contents

1 Introduction

Brief history. The theory of fluids, as all physical theories, began with simple empirical observations. Empirical observations are disparate and of limited validity. In order to inter-relate them and to give them a sense, a meaning, physical and mathematical concepts have been imagined, with their unifying methods. Mathematics and theoretical physics appeared this way. The concepts have themselves a limited validity, and the problem of their reconciling appears on this occasion. Bernoulli claimed that gases are made of atoms, which move around, collide and put pressure on the walls of the vessel. This atomistic conception of matter was to be reconciled with the concept of continuous media, which is very obvious at first glance for fluids, solids and condensed matter in general. Euler found it very convenient to write down Newton's equation of motion for a flow of fluid with the position as variable, at the price of introducing non-linear terms in velocity, which raise many difficulties and unknowns. Euler's equation was the beginning of the theory of fluids. A simpler form of motion in fluids, discovered first by Poisson and Stokes, was called waves. Meantime, the things were complicated by heat, which may appear in the motion of the fluids. The harmonic analysis invented by Fourier for the heat transfer is a milestone in mathematics and theoretical physics. On this occasion the thermoconduction was incorporated in the theory of fluids. Finally, the other source of heat, arising from viscosity, was added to Euler's equation by Navier and Stokes, following the work of Cauchy. The motion of the fluids became more complicated. Moreover, another particularity of the flow of the fluids was discovered by Helmholtz, who noticed that the rotations of the flow, called vortices, can be viewed to a large extent as distinct entities. So attractive was the vortex, that Thomson (Lord Kelvin) attempted to construct the whole world by vortices, reviving an older idea of Descartes. Slowly, the theory of fluids got formally a unitary aspect, and many applications have been worked out by Strutt (Lord Rayleigh) and many others. Despite of a satisfactory conceptual frame and many technical insigths, a full understanding of the flow of the fluids is still lacking, especially related to the friction near boundaries and the resulting turbulence; not to mention discontinuities, instabilities, undeterminacies and incompatibilities.

The whole development of the theory of fluids has been collected and systemized, on various occasions, by Lamb in his *Hydrodynamics*, Kirchhoff in his *Vorlesungen der Mathematische Physik*, Sommerfeld in *Mechanik der deformierbaren Medien* and by Landau and Lifshitz in their sixth volume of the *Course of Theoretical Physics*, which, partially, we have closely followed in the present book. Apart from many minor points, the book includes original contributions concerning the completeness of the equations of fluid mechanics, the solutions of the Navier-Stokes equations, equilibrium with thermal gradient in gravitational field, the entropic sound, the dynamics of vortices, the kinetic-molecular drag force, the origin of turbulence, the singular turbulent wind, the turbulent motion, the linearized equations of fluid mechanics, thermal waves and a non-linear equation describing clouds and wreaths of smoke; a short chapter includes an original investigation of explosions.

Fluid dynamics. Fluids are continuous bodies which flow. They have not a definite shape, like liquids or gases. The flow is the internal motion of fluids. Fluids are viewed as continuous, homogeneous and isotropic macroscopic matter. A point of fluid (a particle) is a volume of fluid which is much smaller than the dimensions of the fluid, but much larger than the distances between the fluid molecules. We owe this fundamental concept of continuous media to Euler.

The motion of fluids is defined by the velocity field $\boldsymbol{v}(\boldsymbol{r},t)$, where $\boldsymbol{r}$ is the position of a fluid point and t denotes the time. The velocity obeys the Navier-Stokes equation of motion

$$\begin{gathered}\frac{d\boldsymbol{v}}{dt}=\frac{\partial\boldsymbol{v}}{\partial t}+(\boldsymbol{v}grad)\boldsymbol{v}=\\=-\frac{1}{\rho}gradp+\frac{\eta}{\rho}\Delta\boldsymbol{v}+\frac{1}{\rho}(\frac{1}{3}\eta+\zeta)grad\,div\boldsymbol{v}+\boldsymbol{f}\ ,\end{gathered}\tag{1.1}$$

where ρ is the mass density, p is the pressure, η is the first viscosity coefficient, ζ is the second viscosity coefficient and $\boldsymbol{f}$ is an external force per unit mass. The viscosity coefficients η and ζ are constant and uniform in fluid (and both positive). Equation (1.1) is the application of Newton's law of motion to fluids, including the internal friction (viscosity).

Equation (1.1) includes, apart from the empirical viscosity coefficients, two other parameters: the density ρ and the pressure p; they are functions of position and time. Fluids flow in external conditions, and a full characterization of the flow requires the knowledge of these conditions. In general, a fluid in motion is not in thermodynamic equilibrium, because the viscosity generates heat (it accounts for an internal friction) and the spatial and temporal dependence implies heat transfer from point to point in fluid (heat conduction, thermoconduction). Consequently, rigorously speaking, we could not use equations of states for determining the pressure, leaving aside the variation of the density for compressible fluids. The five unknowns in equation (1.1), namely the three velocity components, the density and the pressure, satisfy three Navier-Stokes equations (1.1) and the continuity equation

$$\frac{\partial \rho}{\partial t} + div(\rho \boldsymbol{v}) = 0 \ , \tag{1.2}$$

which ensures the mass conservation; we have a determined problem (the completeness of the problem of the fluid motion) by giving the pressure. This remark was suggested by Euler and Lagrange.

The equations of motion ensure the momentum conservation and the conservation of the angular momentum. If the fluid is compressible we need to take into account the internal energy of the fluid (as well as its enthalpy), in order to write down the energy conservation:

$$\begin{aligned} &\tfrac{\partial}{\partial t}\left(\tfrac{1}{2}\rho v^2 + \rho\varepsilon\right) = \\ &= -\partial_j\left[\rho v_j\left(\tfrac{1}{2}v^2 + w\right) - v_i\sigma'_{ij}\right] - \sigma'_{ij}\partial_j v_i + \rho T\tfrac{ds}{dt} \ , \end{aligned} \tag{1.3}$$

where ε is the internal energy per unit mass, $w = \varepsilon + p/\rho$ is the enthalpy per unit mass ($1/\rho$ is the volume of the unit mass), s is the entropy per unit mass, T is the temperature and the viscosity tensor σ'_{ij} is given by

$$\sigma'_{ij} = \eta\left(\partial_i v_j + \partial_j v_i - \frac{2}{3}\delta_{ij}\partial_k v_k\right) + \zeta\delta_{ij}\partial_k v_k \ . \tag{1.4}$$

In equation (1.3) the *lhs* is the variation in time of the energy density; $\rho\boldsymbol{v}\left(\frac{1}{2}v^2+w\right)$ is the energy flux density (energy flow, energy current) due to the mass transfer (including also the work done by the pressure forces); while $-v_i\sigma'_{ij}$ is the energy flux density arising from momentum transfer in collisions, occurring by collisions. The last two terms are related to the dissipated heat $-div\boldsymbol{q}$, where $\boldsymbol{q}=-\kappa gradT$ is the heat flux density and κ is the thermal conductivity (at constant pressure):

$$\begin{aligned}\rho T\tfrac{ds}{dt}&=\rho T\left(\tfrac{\partial s}{\partial t}+(\boldsymbol{v}grad)s\right)=\\&=\sigma'_{ij}\partial_j v_i+div(\kappa gradT)\ .\end{aligned}\tag{1.5}$$

This equation shows the increase in time of the entropy, by thermal conduction and heat production. We can use $ds=(\partial s/\partial T)_p dT$, where $c_p=T(\partial s/\partial T)_p$ is the specific heat per unit mass at constant pressure (the volume of the unit mass, *i.e.* the density, is not constant for compressible fluids); we get

$$\rho c_p\left[\frac{\partial T}{\partial t}+(\boldsymbol{v}grad)T\right]=\sigma'_{ij}\partial_j v_i+div(\kappa gradT)\ ,\tag{1.6}$$

which is a generalizaton of Fourier's equation of heat transfer. Equation (1.5) is the equation of heat transfer. $\sigma'_{ij}\partial_j v_i$ is the source term for heat; it accounts for heat production by internal friction; it is a second-order term in velocity, which contributes positive changes to entropy, as for irreversible processes. $div(\kappa gradT)$ is the thermal conduction (heat diffusion); it brings also positive contributions to the entropy; part of the heat production is thermally conducted (dissipated) by this term; the other part is transported by the term on the left in equation (1.5). It is worth noting that s refers to a fluid particle which is not a closed thermodynamic ensemble, nor one in equilibrium. The total entropy of the fluid increases by second-order contributions in velocity and variations of temperature, arising from viscosity and thermal conduction. In the equation of energy conservation the thermodynamic quantities obey the well-known equilibrium relations $dw=d\varepsilon+dp/\rho$ and $d\varepsilon=-pd(1/\rho)+Tds$. The solution of the flow equations, energy conservation and heat transfer included, brings small contributions to the thermodynamic quantities for small velocities (small variations of the velocity); *e.g.*, much smaller than the sound velocity. For instance, the entropy is changed by second-order contributions, according to equation (1.5) (pressure is changed by first-order terms, in general); such changes correspond to small deviations from thermodynamic equilibrium.

The equation of heat conduction (equation (1.6)), the equation of continuity (1.2) and the Navier-Stokes equation (1.1) are coupled equations for temperature, velocity and density, which are determined (five equations and five unknowns).

A further complexity occurs for a fluid mixture, where the fluid equations must be supplemented by the conservation of the mixture components.

Equation (1.1) should be solved with boundary (and initial) conditions. At the solid surfaces edging the fluid the velocity must be zero (the fluid adheres to solid surfaces).

For high values of the viscosity coefficients in equation (1.1) the velocity is small and the non-linear term in velocity may be neglected. Great simplifications are usually obtained in solving equation (1.1) for steady flow and incompressible fluids.

A special situation occurs for vanishing viscosity. According to Feynman a vanishing viscosity makes water dry, while the real water is wet, *i.e.* viscid. For zero viscosity and thermoconductivity the flow proceeds with constant entropy; it is adiabatic. If the entropy is constant and uniform in fluid, the adiabatic flow is called isentropic. Under these circumstances the fluids are called ideal fluids, and the equation of motion of ideal fluids is Euler's equation

$$\frac{\partial \boldsymbol{v}}{\partial t} + (\boldsymbol{v} grad)\boldsymbol{v} = -\frac{1}{\rho} grad p + \boldsymbol{f} \ . \tag{1.7}$$

The continuity equation is added to Euler's equation and the pressure should be given as function of density (equation of state). Then, the problem is determined (the entropy is constant and uniform, the ideal flow is isentropic). Equation (1.7) has some particularities. First (besides being non-linear), it includes only first-order spatial derivatives. The vanishing of the normal-to-surface component of the velocity (not the tangential component) may act as a boundary condition. This leaves room for friction of the fluid with the edging solid surfaces, with the solid bodies immersed in fluid, past which fluid flows, or even with other (immiscible) fluids which the fluid may be in contact with (in the later case the normal velocity for the two fluids must be the same). In general, solutions of equations (1.1) or (1.7) may be unstable with respect to small perturbations. The instability arises from perturbations provided by interaction with the edging surfaces. Such instabilities consist of an irregular motion, fluctuating in time and space, with curling velocities, called eddies, and rotating velocities, called vortices. Such a motion is called a turbulent motion, and such a fluid exhibits turbulence. We may imagine the ideal Euler's equation as the Navier-Stokes equation (1.1) in the limit of vanishing viscosity; then, for small viscosity, we may see that the velocities have a high spatial variations, by necessity irregular. Indeed, the turbulence is observed below a small, finite, critical viscosity, and it is ubiquitous. Turbulent flow is the rule, while the regular, streamlined (laminar) flow is exception. The conditions for the occurrence of the turbulence are not known; they depend not only on viscosity, but also on the values of the velocity and on the extent of solid surfaces generating the friction; all these three elements are embodied in the Reynolds number; it seems that a critical value of the Reynolds number should be exceeded, in order for turbulence to occur. Usual simplifications in treating turbulent flows are uniform density (incompressible fluid) and steady mean flow (independent of time).

Few examples of elements of a turbulent flow are known. One is the Helmholtz vortex, which is a model of cyclon. However, two or more interacting vortices are unstable, having the tendency to merge into one, bigger vortex, or to go away from one another. Another is the turbulent boundary layer in a flow over a flat surface. A limiting case of fully developed turbulence consists of a statistical ensemble of small vortices, which exhibits a constant rate of producing turbulence energy in time and a velocity correlation function which decreases like the two-thirds power of distance; this statistical turbulence is known as Kolmogorov turbulence. It shows that turbulent flows have little to do with equations of mechanical motion; attempting to solve equations of fluid motion for turbulent flows is, perhaps, like looking for solution in the wrong place. In fact, at least the transient, partially developed turbulence in limited spatial regions may have not a solution in terms of differential equations. Heisenberg thought that God may know something about the constant, universal speed of light in vacuum, but not much about turbulence.

The Navier-Stokes equation (1.1) and the continuity equation (1.2) lead to the momentum conservation

$$\frac{\partial(\rho v_i)}{\partial t} = -\partial_i p - \partial_j(\rho v_i v_j) + \eta\Delta v_i + \left(\frac{1}{3}\eta + \zeta\right)\partial_i div\boldsymbol{v} \qquad (1.8)$$

($p\delta_{ij} + \rho v_i v_j$ is the tensor of the momentum flux density); for a turbulent regime the velocity is resolved into a mean component $\overline{\mathbf{v}}$ and a fluctuating component $\mathbf{v}'$; equation (1.8) leads to the Reynolds equation

$$0 = \overline{\frac{\partial(\rho v_i)}{\partial t}} = \partial_i\overline{p} - \partial_j(\rho\overline{v}_i\overline{v}_j) - \partial_j(\rho\overline{v_i'v_j'}) + \\ + \eta\Delta\overline{v}_i + \left(\tfrac{1}{3}\eta + \zeta\right)\partial_i div\overline{\boldsymbol{v}} \ , \qquad (1.9)$$

where $\overline{p}$ is the mean pressure. Apart from mean terms occurring in the Navier-Stokes and continuity equations, we can see the occurrence of an additional term given by the Reynolds stress tensor $\rho\overline{v_i'v_j'}$: the fluctuations generate an additional force but, at the same time, make the Reynolds equation an "open" equation (the Reynolds tensor has six components). This latter point shows that turbulence is governed by other laws than the laws of the mechanical motion; very likely indefinite to a large extent. It is worth noting the special nature of the objects called velocity fluctuations; they satisfy conditions of the type $\overline{v_i'} = 0$, $\overline{\partial_j v_i'} = 0$, but $\overline{v_i'v_j'} \neq 0$. We note the continuity equations $\partial_i\overline{v_i} = 0$ and $\partial_i v_i' = 0$ (for incompressible fluids); the average is taken with respect to time. Also, a slow time dependence can be allowed after averaging over fast-varying time components (or ensemble averages), leading to meteorological equations.[1] In this context, various flow equations can be constructed for various quantities, in a similar manner as equations (1.8) or (1.9) (turbulence included); "closure" assumptions are needed for such "open" equations.

Description of the book. Euler endowed the fluid particles with thermodynamical quantities like pressure, internal energy, entropy, enthalpy, all in infinitesimal volumes. Any such function $f(t, \boldsymbol{r})$ depends

[1] R. B. Stull, *An Introduction to Boundary Layer Meteorology*, Kluwer, Dordrecht (1988).

on the time t and position $\boldsymbol{r}$. However, the position may itself be a function $\boldsymbol{r}(t)$ of time, as in the Lagrange formulation. There should exist an equilibrium state for fluids, both mechanical and thermodynamical, which does not depend on the time. The flow of the fluids is a small deviation from this equilibrium state. Such a deviation is represented by the total derivative $df/dt = \partial f/\partial t + \boldsymbol{v}gradf$, where $\boldsymbol{v} = d\boldsymbol{r}/dt$ is the velocity. It follows that the flow is represented by the velocity $\boldsymbol{v}(t, \boldsymbol{r})$, which is the unknown function in the equation of motion. At first sight, the separation of the unknown $\boldsymbol{v}$ from the variables t and $\boldsymbol{r}$ looks like a meaningful representation of the problem of the fluid flow. However, the action of an external field, given by a potential $\varphi(\boldsymbol{r})$, depends on the position of the particle with respect to the equilibrium state, *i.e.* it depends on $\varphi(\boldsymbol{r}_0+\boldsymbol{u})-\varphi(\boldsymbol{r}_0) = \boldsymbol{u}grad\varphi(\boldsymbol{r}_0)$, where $\boldsymbol{u}(t, \boldsymbol{r})$ is the displacement of the particle from its equilibrium position $\boldsymbol{r}_0$. Obviously, $\boldsymbol{v} = d\boldsymbol{u}/dt$. This circumstance introduces a new unknown, the displacement $\boldsymbol{u}$, which may complicate the problem.

This problem appears in potential flow, surface gravitational waves and, especially, in the so-called Oberbeck-Boussinesq free convection, where the equilibrium should be resolved first and independently of the flow. The displacement $\boldsymbol{u}$ is useful in compressible flows, where it helps in transforming the continuity equation into an identity. We are thus led to a simplified form of the equations of the fluid motion. These equations are appropriate not only for sound waves, but also for more complicated motions, because the displacement refers to the relative positions of the particles, and, as such, it should not necessarily be small, it is sufficient for it to have small variations.[2]

The motion of the fluids is subjected to three equations for the three components of the velocity (Euler equations, or Navier-Stokes equations). However, these equations include two other unknowns: density and pressure. The pressure is a function of density and entropy (or temperature), and the continuity equation is a fourth equation. The missing fifth equation is the heat transfer equation for entropy (or temperature). Although suggested originally by Euler and Lagrange, this fifth equation has been fully understood only after the develop-

[2] M. Apostol, "Non-inertial electromagnetic effects in matter. Gyromagnetic effect", Solid State Commun. **152** 1567 (2012).

ment of Thermodynamics. It implies a special observation related to the adiabatic processes. Part of the entropy (temperature) is compensated by local dilatations/compressions, as in adiabatic processes, while the remaining part is conducted and dissipated. Of course, such a mechanism implies small deviations from equilibrium. This way, the equations of the fluid mechanics are complete.

The real fluids are viscid. Usually, the viscosity is small and it has small effects inside the fluid, where it may be neglected. However, it has important effects on the boundaries of the fluids, where it produces the boundary layer. In estimating such effects it is important to note that the boundaries are finite, while the usual mathematical methods assume infinite boundaries. This may lead to a misuse of mathematical techniques of treating such problems. In this respect, for instance, the laminar boundary layer (Prandtl equations) and the free convection (Oberbeck-Boussinesq equations) may acquire much simpler solutions, at least qualitatively.

For large velocities it is favourable to transfer the energy to particles collisions, in order to reduce the mechanical energy of the flow. This energy is usually taken and dissipated by viscosity. If the viscosity is small, or we are close to a boundary, where the collision motion of the particles is restricted, the heat transfer is not possible anymore, and surfaces of large discontinuities may appear, which are unstable. The fluid motion becomes irregular and undetermined. This is the phenomenon of turbulence. In turbulent motion the velocity fluctuates, and we may have a homogeneous turbulence with statistically averaged correlation functions of velocity. The equations of motion acquire several undetermined parameters; they are known as Reynolds equations. Few physical constraints hold, which reduce the number of free parameters (the so-called closure equations). This is why the resolution of such equations may contain many arbitrary assumptions. In one fortunate case, where additional physical assumptions are valid, we get a singular inverse wind in a boundary layer, with a discontinuous cusp in velocity and an abrupt change of direction.

Apart from the classical theory of fluids described briefly above, special situations exist in fluids, which require a distinct treatment. For instance, if thermoconduction is high, the adiabatic processes may not take place for local equilibrium, and a special sound, called entropic sound, may appear, propagating with the mean thermal velocity. For dilute solvents non-linear diffusion effects may take place, which suggest clouds or wreaths of smoke. The interaction at the surface of two fluids, or between a fluid and a solid, generates surface phenomena, like capillarity. The superfluid state introduces not only two distinct fluids, but an additional equation of motion for the superfluid component, resulting from the quantum-mechanical character of the superfluidity. A special subject is related to the vorticial motion in ideal fluids, which exhibits an independent character, unfortunately unstable; but it may be a meaningful representation of the turbulence. An average over fluctuating, singular vortices leaves us with a gas of singularities (or a solution of vortices in the background fluid).

Fluids may flow with relativistic velocities. The relativistic fluid dynamics is not included in this book. If large velocities appear (relativistic or not), with large, localized variations, we may have shock waves. This subject is not included in this book. Instead, (non-relativistic) explosions are presented.

General problem of fluid mechanics. The flow of fluids should be determined with respect to an equilibrium state of the fluids. The equilibrium should be both global mechanical equilibrium and, at least local, thermodynamical equilibrium. The flow implies small variations with respect to these equilibria. A relativistic local thermodynamical equilibrium may involve inconsistencies, such that a relativistic flow is rather a flow of a whole volume of fluid, which amounts in fact to a moving reference frame. As such, it may not be viewed as a proper fluid flow, but rather as an application of the relativistic transformations, though the thermodynamic equilibrium is relativistic in the rest frames of the fluid particles.[3] For a proper fluid flow the non-relativistic thermodynamic equilibrium is more appropriate.

Discontinuities may appear in fluids, which propagate. These are shock waves; they are not a proper flow of fluid, but rather a propagation of shocks.

Interactions may propagate in fluids, through collisions, as waves of density and pressure. This is the sound, which, again, may be viewed

[3] W. Pauli, *Theory of Relativity*, Pergamon, London (1958); L. Landau and E. Lifshitz, *Course of Theoretical Physics*, vol. 6 (*Fluid Mechanics*), Elsevier, Oxford (1989).

not as a proper fluid flow, but rather as propagation of interaction. In order to preserve the local non-relativistic thermodynamic equilibrium the flow velocity should be smaller than the sound velocity, which is of the order of the mean thermal velocity. Even so, we need to distinguish between a small-velocity regime and a high-velocity regime. The fluid mechanics is a scale problem.[4]

The Navier-Stokes equation may include an internal pressure, an external pressure and an external force. The relative variations of the internal pressure are small, such that they may only be relevant for the small-velocity regime. This is the realm of a proper fluid flow. In the high-velocity regime the term quadratic in velocity in the Navier-Stokes equation is important, and we have instabilities and turbulence.

A particular situation arises from the small capillarity effects generated by adherence forces. Another special situation relates to the superfluid flow of quantum fluids.

Comments on the theory of fluids. The theory of fluids is one of the most sophisticated subject in Theoretical Physics. Apart from its overwhelming relevance in practical endeavours, like engineering, environment, climate, navigation and air flights, it is a basic subject in mathematics and theoretical physics, due to its intricate physical hypotheses and equations. It suffices only to recall that the basic equation of fluid mechanics, the Euler equation, is non-linear. Most of the mathematical techniques of solving differential equations, like Fourier expansion, potential theory, wave equation, Green functions, were prompted by the theory of fluid mechanics. A special point of great relevance in this respect is the interplay between ideal fluids and viscous fluids. It is well known that one of the Millennium Problems, still unsolved, is the regular solution of the Navier-Stokes equations for real fluids, especially in the limit of small viscosity. Related to this problem, the ubiquitous turbulent behavior, so often encountered in atmosphere, climate, or in modern microfluidics, remains unapproachable, to a large extent.

The difficulties arise from the fact that a fluid flows, *i.e.* its particles have each its own motion and at the same time they are contiguous, *i.e.*, their motion should conform to the motion of their neighbors.

[4]Probably this is why Landau gave to fluid mechanics "a part of his soul", as Lifshitz says, *loc. cit.*, Preface to the 2nd English edition.

It is not seldom that ruptures may appear in fluid flows, as in shock waves, which require a completely different treatment. The difficulty described above is incorporated in the starting approach due to Euler, which, on one side, is very elegant and convenient, and, on the other, entails, probably insurmountable, difficulties.

Euler endowed the fluid particles with thermodynamical quantities like pressure, internal energy, entropy, enthalpy, all in infinitesimal volumes. Any such function depends on the time and position, where the position is itself a function of time. There should exist an equilibrium state for fluids, both mechanical and thermodynamical, which does not depend on the time. The flow of the fluids is a small deviation from this equilibrium state. Such a deviation is represented by the total derivative, which includes the velocity. It follows that the flow is represented by the velocity, which is the unknown function in the equation of motion. At first sight, the unknown velocity is separated from the variables time and position, which looks like a meaningful representation of the problem of the fluid flow. However, the action of an external field, given by a potential, depends on the position of the particle with respect to the equilibrium state, *i.e.* it depends on the displacement of the particle from its equilibrium position. This circumstance introduces a new unknown, the displacement of the particle, which may complicate the problem.

This problem appears in potential flow, surface gravitational waves and, especially, in the so-called Oberbeck-Boussinesq free convection, where the equilibrium should be resolved first and independently of the flow. The displacement is useful in compressible flows, where it helps in transforming the continuity equation in an identity. We are thus led to a simplified form of the equations of the fluid motion. These equations are appropriate not only for waves, but also for more complicated motions, because the displacement refers to the relative positions of the particles, and, as such, it should not be necessarily small, it is sufficient for it to have small variations.

The motion of the fluids is subjected to three equations for the three components of the velocity (Euler equations, or Navier-Stokes equations). However, these equations include two other unknowns: density and pressure. The pressure is a function of density and entropy (or temperature), and the continuity equation is a fourth equation. The missing fifth equation is the heat transfer equation for entropy (or temperature). Although suggested originally by Euler and Lagrange, this fifth equation has been fully understood only after the development of Thermodynamics. It implies a special observation related to the adiabatic processes. Part of the entropy (temperature) is compensated by local dilatations/compressions, as in adiabatic processes, while the remaining part is conducted and dissipated. Of course, such a mechanism implies small deviations from equilibrium. This way, the equations of the fluid mechanics are complete.

The real fluid are viscid. Usually, the viscosity is small and it has small effects inside the fluid, where it may be neglected. However, it has important effects on the boundaries of the fluids, where it produces the boundary layer. In estimating such effects it is important to note that the boundaries are finite, while the usual mathematical methods assume infinite boundaries. This may lead occasionally to a misuse of mathematical techniques of treating such problems. In this respect, for instance, the laminar boundary layer (Prandtl equations) and the free convection (Oberbeck-Boussinesq equations) may acquire much simpler, qualitative solutions.

In analyzing the fluid motion is of utmost importance to keep in mind that the fluid is a continuous medium. The Euler equation and the Navier-Stokes equation have a non-linear term (the inertial, convective, advective term). In a continuous medium this term may produce special effects. If the velocity is very large at one point, in comparison with the velocity at the adjacent points, the fluid cannot flow, continuously, at that point; it either displaces the adjacent fluid points, or changes direction, or returns, etc. In such a way the turbulence may appear. Large variations of the velocity over short distances may appear due to small obstacles, or by vorticial motion. A large velocity at one point may be dissipated as heat, but the heat dissipation is a volume effect, and the volume of the large velocity is too small. The dissipation cannot occur easily, and the energy is transferred to adjacent fluid elements by molecular collisions. On this occasion, the fluid domains (points) lose their identity and the continuous fluid motion is destroyed in favour of the irregular molecular motion. Of course, such a motion transfer may be returned to the constituents of the original fluid element, such that the velocity has a fluctuating, swirling, and singular, character.

All usual fluid flows produce small variations of density and temperature, such that we may view the fluids as incompressible and isentropic. However, we may have fairly large variations of density and temperature from outside, with important effects. The variations of the density are produced by molecular collisions, which ensure the thermodynamic equilibrium; they imply variations of the volume, usually at constant entropy, *i.e.* at equilibrium. For an ideal gas the change in pressure is $V(\partial p/\partial V) = -NT/V$ (at constant temperature), where V is the volume, T is the temperature and N is the number of molecules. The square of the propagation velocity of such a disturbance is $\simeq NT/\rho V = T/M$, where ρ is the density and M is the molecular mass. The adiabatic process modifies slightly this law. This is the sound velocity. We can see that the sound velocity is of the order of the mean thermal velocity, as expected. Similar orders of magnitude hold usually for real gases, liquids, or other fluids. Therefore, for fluid velocity greater than the sound velocity we may destroy the thermodynamic equilibrium, and the consistency of the fluid domains (points). This does not mean of course that we cannot have bodies flying with higher velocity than the sound velocity through the fluid. But in that case it is not a fluid flow, it is a body motion. Discontinuities may propagate in that case through fluids, which are shock waves.

The molecular collisions dissipate heat in volume, through viscosity; and conduct heat through surfaces, by thermoconductivity. Although the viscosity and the thermoconductivity imply small coefficients, they may bring important effects for large temperature gradients. In general, the viscosity has a rather passive role, of dissipating heat; while the thermoconductivity transfers heat, and, as such, it plays an important role in the free convection or in the entropic (thermal) sound.

For large velocities it is favorable to transfer the energy by particle collisions, in order to reduce the mechanical energy of the flow. This energy is usually taken and dissipated by viscosity. If the viscosity is small, or we are close to a boundary, where the collision motion of the particles is restricted, a full heat transfer is not possible anymore, and surfaces of large discontinuities may appear, which are unstable. The fluid motion becomes irregular and undetermined. This is the phenomenon of turbulence. In turbulent motion the velocity fluctuates, and we may have a homogeneous turbulence with statistically averaged correlation functions of velocity. The equations of motion acquire many undetermined parameters; they are known as Reynolds equations. Few physical constraints hold, which would reduce the number of free parameters (the so-called closure equations). This is why the resolution of such equations may contain many arbitrary assumptions. In one fortunate case, where additional physical assumptions are valid, we get a singular inverse wind in a boundary layer, with a discontinuous cusp in velocity and an abrupt change of direction.

The turbulent motion occurs for large variations of the velocity over small distances. The irregular motion implies fluctuating, swirling, often singular velocities, like swirls, eddies, vortices. The Reynolds equations imply a coupling between small eddies in turbulent motion and a mean flow. But there may exist regions of homogeneous and isotropic turbulence, with singular velocities in a set of discrete points. In general, such vortices do not satisfy the Navier-Stokes equations, not even when averaged over fluctuations. But for a homogeneous and isotropic turbulence the average is such as to lead to zero mean velocity and zero transport (inertial, non-linear) terms; in this case the Navier-Stokes equations are satisfied trivially. We are left with a set of pointlike singularities, which obey Newton's law of motion. They behave as a classical gas, or a solution of vortices in the background fluid, in quasi equilibrium.

Let us put, from outside, a layer of fluid in rotation. The layer is incompressible and continuous. A friction force occurs between this layer and the adjacent layers, due to viscosity. Let us assume that the external force is compensated by the friction forces. Consequently, the angular momentum of the layer is conserved; its density $L = \rho v r$ is constant, where v is the velocity of the layer and r is the radius from the origin. Therefore, $v \sim 1/r$. The friction force, acting upon the adjacent layers, transmits the rotation, such that we have a distribution of velocity $v \sim 1/r$ in the fluid. This is precisely the vortex $\boldsymbol{v} = \boldsymbol{\omega} \times grad \ln r$. Since it is singular in the origin, the velocity suffers there a large variation over small distances, such that we may expect turbulence near $\boldsymbol{r} = 0$.

Apart from the classical theory of fluids, special situations exist in fluids, which require a distinct treatment. For instance, if thermo-conduction is high, the adiabatic processes may not take place for local equilibrium, and a special sound, called entropic sound, may appear, propagating with the mean thermal velocity. For dilute solvents non-linear diffusion effects may take place, which suggest clouds or wreaths of smoke. The superfluid state introduces not only two distinct fluids, but an additional equation of motion for the superfluid component, resulting from the quantum-mechanical character of the superfluidity. A special subject is related to the vorticial motion in ideal fluids, which exhibits an independent character, unfortunately unstable.

If large velocities appear, with large, localized variations, we may have shock waves. Also, the relativistic fluid motion, implies a problematic local equilibrium. These subjects are not included in the present book.

2 Ideal Fluids

2.1 Fluids

Fluids are homogeneous and isotropic continuous media. They consist of "macroscopically small, but microscopically large" domains, *i.e.* domains which are much smaller than the dimension of the fluid, but much larger than the molecular dimension. These domains are viewed as fluid points, particles, or fluid elements. A minimal size of the domains, below which the molecular structure appears, can be estimated.[1] The physical quantities are spatial averages over domains and time averages over the molecular motion. In thermal equilibrium these averages are thermal averages. The fundamental concept of small domains as points in fluid, and in continuous media in general, was introduced by Euler.[2]

As for any other continuous medium, the fluid particles are sufficiently large to be endowed with thermodynamic properties, which means that they may be in thermodynamic equilibrium. Moreover, the fluid particles may suffer thermodynamic processes. The local thermodynamic equilibrium may vary from point to point, so the fluid may not be in global equilibrium. Also, the local equilibrium may vary in time, so we may have local non-equilibrium. The variations of the local thermodynamic equilibrium should be small and slow in comparison with the variations at the molecular level, such that the identity of the particles be preserved. Usually, the thermodynamic equilibrium is attained very quickly, in comparison with other processes. In this book we consider classical non-relativistic fluids.

---

[1] M. Apostol, *Statistical Physics*, Cambridge Scholars Publishing, Newcastle upon Tyne (2021).

[2] L. Euler, "Principes generaux du mouvement des fluides", Mem. de l'Academie des Sciences de Berlin **11** 274 (1757).

Fluids flow. Therefore, we have a field of velocities in fluids, $\boldsymbol{v}(t, \boldsymbol{r})$, depending on the time t and position $\boldsymbol{r}$. The flow implies gradients of pressure, temperature, density, etc, so the flow is a non-equilibrium process. The fluid particles flow in time (Lagrange approach), but it is more convenient to follow the change in time and position of the velocity (Euler approach). Consequently, the change in time of any quantity is given by the total time derivative, which is equal to the partial time derivative plus the transport contribution,

$$\frac{d}{dt} = \frac{\partial}{\partial t} + \boldsymbol{v}grad \ . \tag{2.1}$$

Here a word of caution should be inserted. The position $\boldsymbol{r}$ is viewed both as a variable and a function, $\frac{d\boldsymbol{r}}{dt} = \boldsymbol{v}$. Also, we note that we have two scales of motion in this equation. One is the change of position of the particles, which takes place over a distance Δl in time Δt, such that the velocity is $v = \frac{\Delta l}{\Delta t}$. Another is the change of the velocity over distance l in time τ, which gives rise to another velocity $u = \frac{l}{\tau}$ (the terms $\frac{\partial}{\partial t}$ and $grad$ in equation (2.1)); this is velocity of a wave propagation. If we apply the above rule of differentiation to the position $\boldsymbol{r}$, we get $d\boldsymbol{r}/dt = \partial\boldsymbol{r}/\partial t + (\boldsymbol{v}grad)\boldsymbol{r}$, *i.e.* $\boldsymbol{v} = \partial\boldsymbol{r}/\partial t + \boldsymbol{v}$, which shows indeed that $\boldsymbol{r}$ and t are independent variables. Also, we note that $\boldsymbol{v}grad$ is the time variation along the flow, such that we may write the above rule as $d/dt = \partial/\partial t + (\partial/\partial t)_f$, where the suffix f indicates the differentiation along the flow path.

If $\boldsymbol{v} = 0$ (or $\boldsymbol{v} = const$) fluids are in mechanical equilibrium (at rest, possibly by changing the reference frame). But they may be out of global thermodynamic equilibrium, though locally they are in thermodynamic equilibrium. Then we may have kinetic transport in fluids[3]. For $\boldsymbol{v} \neq 0$ we have flow and macroscopic transport.

To each fluid point a mass density ρ is assigned; it is the (mean) number of molecules in the domain, multiplied by the molecular mass and divided by the volume of the domain. The matter is conserved, *i.e.*

$$\frac{d}{dt}\int \rho dV = 0 \tag{2.2}$$

[3]M. Apostol, *Physical Kinetics*, Cambridge Scholars Publishing, Newcastle upon Tyne (2020).

for any volume V. The change in time in equation (2.2) comes not only from $\frac{\partial\rho}{\partial t}$, but also from the change of the volume, *i.e.*

$$\frac{d}{dt}\int \rho dV = \int \frac{\partial \rho}{\partial t} dV + \oint \rho \boldsymbol{v} d\boldsymbol{f} \ , \tag{2.3}$$

where $d\boldsymbol{f}$ is the surface enclosing the small volume dV. We get the continuity equation

$$\frac{\partial \rho}{\partial t} + div(\rho \boldsymbol{v}) = 0 \ . \tag{2.4}$$

In many usual cases the fluids are incompressible ($\rho = const$), such that this equation reduces to $div\boldsymbol{v} = 0$. We note that $\rho\boldsymbol{v}$ is the mass flux density and, at the same time, it is the momentum of unit volume.[4]

Fluids flow. In their motion the fluid particles collide with each other, and a friction force may appear; also, a transfer of momentum. This is the viscosity. It develops heat, and transports momentum. The mechanical motion is dissipated through heat. Also, heat is developed by compression and dilatation in compressible fluid, which may be compensated by internal motion. By collisions and interactions the heat is transported. This is the thermoconduction. A heat flux density $\boldsymbol{q}$ may exist in fluids (heat per unit cross-sectional area and per unit time) and a local entropy s per unit mass. While viscosity generates heat through collisions in the whole volume of the fluid particles, the thermoconduction generates heat at the surface of the fluid particles.

Also, the particles have an internal energy ε (per unit mass). In general, a thermal non-equilibrium may generate macroscopic currents in fluids, called convection currents, which tend to establish the thermal equilibrium; the mechanical equilibrium may be unstable in that case.

If we neglect the viscosity and the thermoconduction, *i.e.* if there is no heat, the flow is adiabatic. Then we should have the entropy conservation

$$\frac{ds}{dt} = \frac{\partial s}{\partial t} + \boldsymbol{v} grads = 0 \ . \tag{2.5}$$

[4]Since $\frac{d}{dt}\int \boldsymbol{r}\rho dV = -\int \boldsymbol{r} div(\rho\boldsymbol{v})dV = \int \rho\boldsymbol{v} dV$, where the first integral determines the position of the centre of mass.

By combining this equation with the continuity equation (2.4), we get indeed the entropy conservation per unit volume

$$\frac{\partial(\rho s)}{\partial t} + div(\rho \boldsymbol{v} s) = 0 \ . \tag{2.6}$$

If s is constant in time and uniform in space, the flow is called isentropic. If we leave aside the viscosity and the thermoconduction, and the flow is adiabatic, the fluid is called an ideal fluid. Usually, the flow in ideal fluids is isentropic.

If heat transport is present, the variation of the entropy is

$$\frac{d}{dt}\int s dV = \int \frac{ds}{dt} dV + \oint s \boldsymbol{w} d\boldsymbol{f}, \tag{2.7}$$

where $\boldsymbol{w}$ is the mean molecular velocity by which heat is conducted. We note its presence at the surface $d\boldsymbol{f}$ enclosing the volume dV. We get the conservation equation

$$\frac{ds}{dt} = \frac{\partial s}{\partial t} + \boldsymbol{v} grads = -div(s\boldsymbol{w}) \ . \tag{2.8}$$

Multiplying this equation by ρT, where T is the temperature, we get

$$\rho T \frac{ds}{dt} = \rho T \left(\frac{\partial s}{\partial t} + \boldsymbol{v} grads \right) = -div \boldsymbol{q} \ , \tag{2.9}$$

where $\boldsymbol{q} = \rho T s \boldsymbol{w}$ is the heat flux density. On the other hand, the heat flux density is $\boldsymbol{q} = -\kappa gradT$, where κ (> 0) is the thermoconductivity. Indeed, the heat flows from higher to lower temperatures. This is Fourier's law.[5] We note that the molecular velocity disappears from the heat flux density, by the thermoconductivity κ. Equation (2.9) becomes

$$\rho T \frac{ds}{dt} = \rho T \left(\frac{\partial s}{\partial t} + \boldsymbol{v} grads \right) = div(\kappa gradT) \ . \tag{2.10}$$

This is the equation of heat flow, or heat transfer, transport. If viscosity is present, an additional term appears on the right in this equation.

[5] J. Fourier, *Theorie Analytique de la Chaleur*, Firmin Didot, Paris (1822).

We note that the presence of the thermoconductivity, *i.e.* the presence of a molecular velocity $\boldsymbol{w}$ by which heat is conducted, involves collisions, besides interaction, which would involve the presence of the viscosity. Usually we consider κ constant (and positive). Of course, an external source of heat can be included in equation (2.10).

The time variation of the total entropy is given by

$$\begin{gathered}\frac{d}{dt}\int \rho s dV = -\int div(\rho \boldsymbol{v} s) dV + \\ + \int \frac{div(\kappa gradT)}{T} dV = \int \frac{\kappa (gradT)^2}{T^2} dV\end{gathered} \tag{2.11}$$

(where the continuity equation is used). We can see that the entropy increases in time, by thermoconduction, as expected. This justifies Fourier's law. Also, we note that the variation of the total entropy is a second-order effect in the small temperature gradients.

2.2 Euler equation

(Ideal) fluids cannot sustain shear stress: they flow. In fluids there exists only compressional stress, *i.e.* a pressure p, acting on and outwards the surface of each domain (this is known as Pascal's law[6]). Therefore, the surrounding domains act upon each domain with a pressure $-p$ and a force $-gradp$ per unit volume. It follows, according to Newton' law of motion, that the equation of motion is

$$\rho \frac{d\boldsymbol{v}}{dt} = \rho \frac{\partial \boldsymbol{v}}{\partial t} + \rho(\boldsymbol{v} grad)\boldsymbol{v} = -gradp + \boldsymbol{F} \ , \tag{2.12}$$

where $\boldsymbol{F}$ is another force per unit volume, acting in fluid. The force $\boldsymbol{F}$ may include an external pressure gradient and viscosity forces. We leave aside the viscosity, as for ideal fluids. The pressure p in equation (2.12) is the internal pressure. An external force may be the gravitation; for a uniform gravitational acceleration g it reads $F = -\rho g$, acting vertically. Some external forces arise from a potential φ, through $\boldsymbol{F} = -\rho grad\varphi$. For the gravitational force $\varphi = gz$, where z is the

[6] B. Pascal, *Recit de la grande experience de l'equilibre des liqueurs*, Paris (1648).

vertical coordinate. External forces may be viewed as generating gradients of external pressure. Equation (2.12) is Euler's equation.[7] Also, it is known as the equation of ideal fluids. In mechanical equilibrium $-gradp + \boldsymbol{F} = 0$, or $gradp + \rho grad\varphi = 0$. This equation is known as the equation of hydrostatics (mechanical equilibrium is also called hydrostatic equilibrium). We note that, in general, the equation of hydrostatics has no solution, since $\rho grad\varphi$ cannot be a gradient, if ρ is a function of position. If the external force derives from a potential, $\boldsymbol{F} = -\rho grad\varphi$, it is given by $\delta\varphi = \boldsymbol{u} grad\varphi$, where $\boldsymbol{u}$ is the local displacement, and $grad\varphi = grad(\boldsymbol{u} grad\varphi)$, where the derivatives do not affect $grad\varphi$.

Let us suppose that a homogeneous body with density ρ, cross-sectional area S and vertical side l is immersed with a length $h < l$ in a fluid with density ρ_0 and a free surface in gravitational field. The body is acted upwards by a force $F = \rho_0 g S h$, which is the weight of the displaced fluid. This is Archimedes' buoyancy force.[8] In equilibrium it is compensated by the weight $\rho g S l$ of the body, such that $h = (\rho/\rho_0)l$ for $\rho < \rho_0$. The body floats in the fluid. For $\rho = \rho_0$ the body is in equilibrium at any depth. For $\rho > \rho_0$ the body sinks. The centre of mass is placed above the centre of the buoyancy force.

Let us give a vertical motion to the body by a small distance δh. The Archimedes' force changes by $\rho_0 g S \delta h$, such that the equation of motion is $\rho S l \delta\ddot{h} = -\rho_0 g S \delta h$. The body executes oscillations with frequency $\sqrt{\rho_0 g/\rho l} = \sqrt{g/h}$. Similarly, let us condider a rotation of the vertical axis of the body by a small angle θ. The centre of buoyancy is displaced horizontally by a distance of the order $a\theta$, for a symmetric body, where a ($\simeq l$) is the dimension of the cross-section; therefore, we have a torque $-Fa\theta$ and an angular momentum $I\dot{\theta}$, where I is the moment of inertia. The body oscillates with a frequency of the order $\sqrt{Fa/I}$. Since the forces remain in equilibrium, $F = mg$, where m is the mass of the body. On the other hand, the moment of inertia is of the order ma^2, such that the frequency can be estimated as $\omega = \sqrt{g/a}$.[9]

[7] L. Euler, "Principes generaux du mouvement des fluides", Mem. de l'Academie des Sciences de Berlin **11** 274 (1757).

[8] Archimedes of Syracuse, *On Floating Bodies*, Syracuse (250 BC).

[9] More details about such matters, and related others, in R. Fitzpatrick, *Theo-*

It is worth noting that we should first solve the equilibrium and thereafter consider the variations of the pressure and the external force with respect to equilibrium, which determine the velocity. This is known as the Boussinesq observation, or "approximation".[10] Also, caution should be exerted in solving the Euler equation in an infinite domain, in order to have quantities which vanish at infinity, or have well-defined values.

Usually, the pressure gradient and the external force are sufficiently small, in comparison with internal pressure, such that they determine small variations of the velocity, in comparison with the internal molecular velocity. Under these conditions the fluid flow is a macroscopic motion. This is in contrast to the kinetic transport which appears at the molecular level. Large disturbances may also exist in fluids, which give rise to shocks.

Equation (2.12) should be solved with initial conditions and boundary conditions. The boundary condition is the vanishing of the normal velocity. The fluid without viscosity may flow along its boundaries. A liquid without viscosity does not wet surfaces; von Neumann called such a water dry water.[11] We note the non-linear term in velocity in Euler equation, which is called the advection (or convective, or inertial) term. Let us suppose that the velocity increases at some position from zero to v in a short time Δt over a short distance Δx. Also, it may suffer variations in a short time τ over short distances l; the two terms in the total time derivatives are of relative order of magnitude $1/\tau$ and $v/l = \Delta x/l\Delta t$; for the same duration $\Delta t = \tau$ their orders are 1 and $\Delta x/l$; obviously, $\Delta x \ll l$, while both tending to zero, in order for the velocity to exist. Therefore, the inertial term brings a small contribution, and it may often be neglected. It is a second-order effect in v, of the order $\rho v^2 = \Delta p$, where Δp is the pressure variation; while the first term in the total time derivative is of order $lv/\tau = uv$, where $u = l/\tau \gg v$ and $uv \gg v^2$.

---

retical Fluid Mechanics, IOP Publishing, Bristol (2017) and A. Sommerfeld, *Vorlesungen ueber Theoretische Physik*, Band II (*Mechanik der Deformierbaren Medien*), Akad. Verlagsgesellschaft, Leipzig (1970).

[10] J. Boussinesq, *Theorie Analytique de la Chaleur*, Gauthiers-Villars, Paris (1903).

[11] Quoted by R. P. Feynman, *The Feynman Lectures on Physics*, vol. 2, Addison-Wesley, Oxnard, California (1964).

Obviously, the velocity u is the velocity of propagation of the interaction; it is a wave velocity. Indeed, the fluids can sustain sound waves, or other waves, generated by external forces. The change in pressure Δp is much smaller than the internal pressure. The latter is of the order $p = \rho v_{th}^2$, where v_{th} is the mean thermal velocity. Therefore, we have $v \ll v_{th}$; but u may be of the same order of magnitude as v_{th}, or larger. An estimate of the thermal velocity is $v_{th} = \sqrt{3T/M}$, where T is the temperature and M is the molecular mass (for ideal gases). For $T = 300K$ and $M = 10^{-22}g$ we get $v_{th} = 10^4 cm/s$ ($1K = 1.38 \times 10^{-16} erg$, electron mass $m = 10^{-27}g$). In liquids v_{th} is higher, due to interactions.

We may neglect the viscosity (and the thermoconduction) in ideal fluids, but we may not neglect it in flow past boundaries, *e.g.* past solid bodies. This discontinuity in flow may give rise to boundary layers and turbulent disruptions (turbulence).

A special type of flow is the molecular transport.[12] It proceeds with molecular velocities, over distances of the order of the mean freepath. Since, usually, the latter is much longer than the mean separation distance between the molecules, we may neglect the advection term in that case. In condensed matter the transport is effected by (long-wavelength) elementary excitations. The molecules in a gas may be viewed as sui generis elementary excitations. However, important contributions may arise sometimes in condensed matter from the local advection term.[13]

2.3 Completeness

The solution of the continuity equation can be viewed as a relationship, denoted, say, by C, between density and velocity, and written as $C(\rho, \boldsymbol{v}) = 0$. Similarly, the solution of the Euler equation can be written as $E(\rho, \boldsymbol{v}, p; \boldsymbol{F}) = 0$. Thus, we have two equations (one vectorial) and three unknowns (ρ, $\boldsymbol{v}$ and p): the problem is not determined. In general, the presence of time derivatives, spatial gradients and ve-

[12] M. Apostol, *Physical Kinetics*, Cambridge Scholars Publishing, Newcastle upon Tyne (2020).

[13] See, for instance, M. Apostol, *loc. cit.*

locity makes the flow a non-determined, non-equilibrium process. We can assume small variations of all the quantities, which, for consistency, should vary slightly both in time and space. We can expand all the quantities in first-order series of powers of these variations. The coefficients of these expansions may be taken as empirical parameters, though, in some cases we may use those derived from the corresponding quantities in local thermodynamic equilibrium. For example, we may view $\rho = 1/V$ as a "thermodynamic" variable, where V is the volume of unit mass; also, temperature T, or pressure p, or entropy s, etc, may be viewed as another "thermodynamic" variable. However, in general we need two (independent) variables, a circumstance which introduces an additional unknown; so we need an additional equation. For instance, we may write the change in pressure as

$$\delta p = \left(\frac{\partial p}{\partial \rho}\right)_T \delta\rho + \left(\frac{\partial p}{\partial T}\right)_V \delta T \ . \tag{2.13}$$

It is easy to see that

$$\left(\frac{\partial p}{\partial \rho}\right)_T = \frac{K}{\rho} \ , \ \frac{1}{K} = -\frac{1}{V}\left(\frac{\partial V}{\partial p}\right)_T \ , \tag{2.14}$$

where K is the isothermal modulus of compression (bulk modulus; $1/K$ is called compressibility). Similarly, $\alpha = (\partial p/\partial T)_V$ is the thermal pressure coefficient; in equilibrium it is related to the thermal expansion coefficient $\beta = \frac{1}{V}(\partial V/\partial T)_p$ (dilatation coefficient, see Appendix[14]) by

$$\alpha = K\beta \ . \tag{2.15}$$

Therefore, in Euler's equation we may use

$$\delta p = \frac{K}{\rho}\delta\rho + \alpha\delta T \tag{2.16}$$

and

$$grad p = \frac{K}{\rho} grad\rho + \alpha grad T \ . \tag{2.17}$$

[14]L. Landau and E. Lifshitz, *Course of Theoretical Physics*, vol. 5 (*Statistical Physics*), Elsevier, Oxford (1980).

We view the coefficients K, α, β as (positive) constants. We can derive them either from an equation of state, or view them as empirical parameters. We can see that the temperature appears explicitly in Euler's equation. Instead of δT we may choose δs, which, for ideal fluids, is zero ($\delta s = 0$). Also, equation (2.13) can be viewed as an expansion of $\delta\rho$ in terms of the variables p and T, and so on. We need to choose convenient variables, which are independent and lead to simple results.

A similar expansion of the entropy gives

$$\delta s = \left(\frac{\partial s}{\partial \rho}\right)_T \delta\rho + \left(\frac{\partial s}{\partial T}\right)_V \delta T = -\frac{\alpha}{\rho^2}\delta\rho + \frac{c_v}{T}\delta T \ , \tag{2.18}$$

where c_v is the specific heat per unit mass at constant volume; here we use

$$\left(\frac{\partial s}{\partial V}\right)_T = -\frac{\partial^2 F}{\partial V \partial T} = \left(\frac{\partial p}{\partial T}\right)_V \ , \tag{2.19}$$

where F is the free energy.

First we note that for ideal fluids the flow is adiabatic, so $\delta s = 0$. We extract δT from equation (2.18) and insert it in equation (2.16), which becomes

$$\delta p = \frac{K}{\rho}\left(1 + \frac{K\beta^2 T}{\rho c_v}\right)\delta\rho \tag{2.20}$$

(the connection between $\delta\rho$ and δT given by equation (2.18) is often viewed as an expression of the buoyancy effect). Here we use the thermodynamic relations

$$c_p - c_v = \frac{K\beta^2 T}{\rho} \ , \ K_{ad} = \frac{c_p}{c_v}K \ , \tag{2.21}$$

where c_p is the specific heat per unit mass at constant pressure and K_{ad} is the adiabatic modulus of compression (see Appendix[15]). We get

$$\delta p = \frac{K_{ad}}{\rho}\delta\rho \ , \tag{2.22}$$

a result which would have been obtained from the beginning, by choosing the entropy as a variable. We note that the temperature

[15] L. Landau and E. Lifshitz, *loc. cit.*

change disappears from the Euler equation. The temperature variations in adiabatic flow are compensated by dilatation/compression processes. This is due to the fact that the thermal equilibrium is established quite quickly in comparison with other processes (flow, dilatation/compression). The Euler equation and the continuity equation have now two unknowns (ρ and $\boldsymbol{v}$), so the problem of fluid mechanics is complete for ideal fluids.

In general, with equation (2.18), the heat transfer equation (2.10) reads

$$\rho T\frac{ds}{dt} = -\frac{\alpha T}{\rho}\frac{d\rho}{dt} + \rho c_v\frac{dT}{dt} = div(\kappa gradT) \ . \tag{2.23}$$

According to the above discussion, part of the temperature in this equation is compensated by the variation of the density (buoyancy effect), so we are left with

$$\rho c_v\frac{dT}{dt} = div(\kappa gradT) \ . \tag{2.24}$$

This is another form of the Boussinesq procedure.[16] It leads in Euler equation to the replacement of the isothermal modulus K by the adiabatic modulus K_{ad}, such that equation (2.16) becomes

$$gradp = \frac{K_{ad}}{\rho}grad\rho + \alpha gradT \ . \tag{2.25}$$

In equation (2.21) the expansion coefficient squared (β^2) may be viewed as if it is a second-order factor, so it is neglected. Therefore, we replace c_v in equation (2.24) by c_p. Finally, we get the heat transfer equation

$$\rho c_p\frac{dT}{dt} = \rho c_p\left(\frac{\partial T}{\partial t} + \boldsymbol{v}gradT\right) = div(\kappa gradT) \ . \tag{2.26}$$

Indeed, this equation is expected, because, apart from the $\boldsymbol{v}$-term, it is an independent equation, which should be taken with coefficients at equilibrium, *i.e.* for constant pressure (c_p). Usually, we view κ (>0) constant and at constant pressure. The Euler equation becomes

$$\rho\left[\frac{\partial \boldsymbol{v}}{\partial t} + (\boldsymbol{v}grad)\boldsymbol{v}\right] = -\frac{K_{ad}}{\rho}grad\rho - \alpha gradT + \boldsymbol{F} \ ; \tag{2.27}$$

[16]J. Boussinesq, *Theorie Analytique de la Chaleur*, Gauthiers-Villars, Paris (1903).

the continuity equation

$$\frac{\partial\rho}{\partial t}+grad(\boldsymbol{v}\rho)=0 \qquad (2.28)$$

must be added, so we have three equations ((2.26) to (2.28)) and three unknowns ($\boldsymbol{v}$, ρ, T); the problem of fluid mechanics is complete. Of course, an external source of heat can be included in equation (2.26). In these approximations higher-order effects are included (which should be disregarded), through the assumption $c_p\simeq c_v$. We should retain only first-order leading terms in solutions. As regards the parameters entering these equations, in general we should view them as empirical parameters. However, within our approximation, the equation of state of the fluid may be used for deriving them. For example, for illustrative purposes we can use the equation of state of the ideal gases $p=\frac{\rho}{M}T$, where M is the molecular mass. The completeness holds also when viscosity is included. As we shall see in the Viscous Fluids chapter the Euler equation and the heat transfer equation are then modified by the viscosity contributions (Euler's equation becomes the Navier-Stokes equation).

It is worth noting that the problem is also complete by using equation (2.23) instead of equation (2.26), assuming or not that part of density and temperature changes are mutually compensated. However, the compensation phenomenon takes place usually, even if only partially, in which case it is, in general, undetermined. Also, if the temperature variation is sufficiently rapid, the equilibrium cannot be reached, the compensation does not occur, and equation (2.23) should be used.

We can take as variables p and T. Then, the density is a function of p and T, and it introduces a T-dependence in Euler's equation. We have

$$\delta\rho=\frac{\rho}{K}\delta p-\beta\rho\delta T\ ,$$
$$\delta s=-\frac{\beta}{\rho}\delta p+\frac{c_p}{T}\delta T\ , \qquad (2.29)$$

where the Gibbs free energy, given by $d\Phi=Vdp-sdT$, is used. Part of the temperature in the second equation is compensated by the pressure variation, which makes $\delta s=\frac{c_p}{T}\delta T$, *i.e.* the usual form of the heat transfer equation (2.26), and changes the first equation in

$$\delta\rho=\frac{\rho}{K_{ad}}\delta p-\rho\beta\delta T\ . \qquad (2.30)$$

By making use of the thermodynamic relations, this equation can be cast in the form

$$\delta p = \frac{K_{ad}}{\rho}\delta\rho + \frac{c_p}{c_v}\alpha\delta T \ . \tag{2.31}$$

This equation differs from equation (2.25) by the ratio c_p/c_v, which, according to the above discussion, is sometimes replaced by unity. Thus, we can see that we can recover the results obtained above. For ideal gases the difference $c_p - c_v$ is of the same order of magnitude as $c_{p,v}$, but for liquids this difference is much smaller than unity.[17]

We should note that we have three non-linear equations, which are homogeneous, except for the term of external forces. Consequently, we may expect instabilities, incompatibilities and indeterminacies, even in the absence of the external forces. Local thermodynamic equilibrium generates kinetic transport, which is distinct from a macroscopic flow. We expect that for sufficiently slow and slight flows the local thermodynamic equilibrium and the macroscopic transport are preserved, such that the equations are compatible, though additional conditions may be needed for stability. An example and a counterexample are provided by convection.

Finally we note that the problem of completeness of the equations of fluid mechanics is trivial in many particular cases, where specific approximations are made.

2.4 Momentum and energy conservation

By using the continuity equation we get immediately from equation (2.12) the momentum conservation

$$\frac{\partial}{\partial t}(\rho v_i) + \partial_j(p\delta_{ij} + \rho v_i v_j) = 0 \ ; \tag{2.32}$$

$\Pi_{ij} = p\delta_{ij} + \rho v_i v_j$ is called the tensor of the momentum flux density; if an external potential φ is included, the momentum conservation

[17] M. Apostol, "A model for the thermodynamics of simple liquids", Physica **B403** 3946 (2008); *Physical Kinetics*, Cambridge Scholars Publishing, Newcastle upon Tyne (2020).

reads

$$\frac{\partial}{\partial t}(\rho v_i) + \partial_j (p\delta_{ij} + \rho v_i v_j) = -\rho \partial_i \varphi \ . \tag{2.33}$$

The Euler equation can also be written as

$$\frac{\partial v_i}{\partial t} + v_j \partial_j v_i = -\frac{\partial_i p}{\rho} - \partial_i \varphi \ . \tag{2.34}$$

We notice that the enthalpy per unit mass w is such that $dw = Vdp + Tds$, where $V = 1/\rho$ is the volume per unit mass and T is the temperature; therefore,

$$dw = \frac{dp}{\rho} + Tds \ . \tag{2.35}$$

For an adiabatic flow $ds = 0$, such that $dw = dp/\rho$ and equation (2.34) becomes

$$\frac{\partial v_i}{\partial t} + v_j \partial_j v_i = -\partial_i w - \partial_i \varphi \ . \tag{2.36}$$

Let us assume a steady flow, *i.e.* a flow with a constant velocity, $\partial v_i / \partial t = 0$. Then, the velocities are distributed in space as tangents to lines, fixed in space. These are called streamlines, or lines of motion. They do not intersect with one another. Consequently, equation (2.36) reads

$$v \partial_l v + \partial_l (w + \varphi) = 0 \ , \tag{2.37}$$

where l denotes the coordinate along a streamline and v is the velocity along that streamline; from equation (2.37) we get

$$\frac{1}{2} v^2 + w + \varphi = const \ , \tag{2.38}$$

where the *const* depends on the streamline. On the other hand $w = \varepsilon + p/\rho$, where ε is the internal energy per unit mass. In general, ε is a function of density and entropy. For the adiabatic flow the entropy is constant. If ε does not depend on density, as for ideal gases, or the fluid is incompressible, equation (2.38) becomes

$$\frac{1}{2} v^2 + \frac{p}{\rho} + \varphi = const \ ; \tag{2.39}$$

for example, for an incompressible fluid we may write

$$\frac{1}{2}\rho v^2 + p + \rho\varphi = const \ . \tag{2.40}$$

This is known as Bernoulli's law.[18] It is an expression of the energy conservation (integral of motion). In uniform gravitational field

$$\frac{1}{2}\rho v^2 + p + \rho g z = const \ , \tag{2.41}$$

where z is the vertical coordinate and g is the gravitational acceleration. This is Torricelli's law.[19] We note that in equations (2.39) to (2.41) the external force may look as a pression, since $-gradp$ is similar to $-\rho grad\varphi$.

2.5 Vena contracta

If a fluid flows horizontally from a large cross-section with velocity zero and pressure p_0 (actually an external force) through a small cross-section S with velocity v and pressure p, then

$$p_0 = \frac{1}{2}\rho v^2 + p \ , \tag{2.42}$$

according to Bernoulli's law. We can see that the pressure p may become negative, for high velocities, *i.e.* the flow tube gets thinner, and contracts itself. This phenomenon is known as vena contracta.[20] Indeed, from the momentum conservation given by equation (2.32) we get

$$p_0 - p = \rho v^2 \tag{2.43}$$

along the streamlines of the efflux, which is in contradiction with energy conservation given by equation (2.42). Actually, the momentum

[18] D. Bernoulli, *Hydrodynamica*, Argentorati, Strassburg (1738).

[19] E. Torricelli, *De Motu Gravium Naturaliter Accelerato*, Firenze (1643).

[20] H. Lamb, *Hydrodynamics*, Cambridge University Press, London (1879), and References therein.

conservation (equation (2.32)) requires the multiplication by the area of the cross-sections, such that we should write

$$(p_0 - p)S = \rho v^2 S' \ , \tag{2.44}$$

where S is the area of the cross-section of the tube (hole) and S' is the area of the fluid jet. From equations (2.42) and (2.44) we get the ratio of the two areas $S' = \frac{1}{2}S$, which is indeed fairly well verified by experiment.

Related to Torricelli's law, the velocity of a liquid flowing through a small hole opened at depth h in a liquid placed in gravitational field is $v = \sqrt{2gh}$. If the jet is re-oriented vertically, it should go upwards to the same height h; which it does not, in practice. Indeed, the energy is lost by dissipation due to friction.

2.6 Compressible fluids

In compressible fluids we have a local displacement $\boldsymbol{u}(t, \boldsymbol{r})$ which produces a change of volume $\frac{\delta V}{V} = div\boldsymbol{u}$ (dilatation or compression); consequently, a change of pressure occurs, given by $\delta p = -\lambda div\boldsymbol{u}$, *i.e.* $\delta p = -\lambda \frac{\delta V}{V}$. We can see that the coefficient $\lambda = -V(\partial p/\partial V)$ is the modulus of compression K. The change in the free energy per unit volume ($V = 1$; at constant temperature) is

$$dF = -\delta p d(\delta V) = \lambda div\boldsymbol{u} d(div\boldsymbol{u}) \ , \tag{2.45}$$

such that the corresponding free energy per unit volume is

$$F = \frac{1}{2}\lambda(div\boldsymbol{u})^2 = \frac{1}{2}\lambda(\delta V)^2 \ . \tag{2.46}$$

We can see that the coefficient λ corresponds to a Lame coefficient from Elasticity.[21]

The coefficient λ in equation (2.46) is isothermal. But dilatations and compressions generate a change of temperature δT. A corresponding term should be added to the free energy, which becomes

$$F' = \frac{1}{2}\lambda(div\boldsymbol{u})^2 - \lambda\beta\delta T div\boldsymbol{u} = \frac{1}{2}\lambda(\delta V)^2 - \lambda\beta\delta T\delta V \ ; \tag{2.47}$$

[21] L. Landau and E. Lifshitz, *Course of Theoretical Physics*, vol. 7 (*Theory of Elasticity*), Elsevier, Oxford (1986).

indeed, the pressure becomes

$$p' = -\lambda\delta V + \lambda\beta\delta T \ , \tag{2.48}$$

whence we see that the coefficient β is the coefficient of thermal expansion at constant pressure ($p' = 0$), $\beta = \frac{1}{V}(\partial V/\partial T)_p$ ($V = 1$).

On the other hand, from equation (2.47), the change of entropy per unit volume is

$$\delta S = \lambda\beta div\boldsymbol{u} \ . \tag{2.49}$$

For an adiabatic transformation δS should be equal to $-C_v\delta T/T$, C_v being the specific heat per unit volume. Therefore, we have $\delta T = -(\lambda\beta T/C_v)div\boldsymbol{u}$ and the pressure becomes

$$p' = -\lambda div\boldsymbol{u} - (\lambda^2\beta^2 T/C_v)div\boldsymbol{u} = -\lambda_{ad}div\boldsymbol{u} \ , \tag{2.50}$$

where we introduce the adiabatic coefficient of compression

$$\lambda_{ad} = \lambda\left(1 + \lambda\beta^2 T/C_v\right) \ . \tag{2.51}$$

This equation can also be written as

$$\frac{1}{\lambda_{ad}} \simeq \frac{1}{\lambda} - \frac{\beta^2 T}{C_p} \ , \tag{2.52}$$

for $\lambda\beta^2 T/C_p \ll 1$, where we use the thermodynamic relation $C_p - C_v = \lambda\beta^2 T$ between the specific heats; indeed, this is the thermodynamic relation for the compression coefficients (see Appendix[22]).

In a compression (dilatation) the volume entropy ($\lambda\beta div\boldsymbol{u}$) decreases (increases), but the molecular disorder and the temperature increase (decrease; $C_v\delta T/T$), such that the total entropy remains constant. This happens because the thermal equilibrium is restored quickly. This "buoyancy effect" is used above for the completeness of the equations of fluid mechanics.

Turning back to the pressure $p = -\lambda div\boldsymbol{u}$, this time with the adiabatic λ, the force per unit volume is $\lambda grad\, div\boldsymbol{u}$. Leaving aside the non-linear term in velocity, and limiting ourselves to small variations, Euler's equation becomes the longitudinal (compression) Navier-Cauchy

[22] L. Landau and E. Lifshitz, *Course of Theoretical Physics*, vol. 5 (*Statistical Physics*), Elsevier, Oxford (1980).

equation[23]

$$\rho\ddot{u} - \lambda grad\, div\boldsymbol{u} = \boldsymbol{F} \ , \tag{2.53}$$

where the small changes in ρ are neglected (linear elasticity). The flow of the ideal fluid is adiabatic as is the motion (of the sound) in elastic solids.

2.7 Energy conservation

By using the continuity equation, the Euler equation and the enthalpy (equation (2.35)), we have successively

$$\begin{aligned} &\frac{\partial}{\partial t}\left(\frac{1}{2}\rho v^2\right) = \frac{1}{2}\frac{\partial\rho}{\partial t}v^2 + \rho\boldsymbol{v}\frac{\partial\boldsymbol{v}}{\partial t} = \\ &= -\frac{1}{2}v^2 div(\rho\boldsymbol{v}) - \boldsymbol{v}gradp - \rho\boldsymbol{v}(\boldsymbol{v}grad)\boldsymbol{v} = \\ &= -\frac{1}{2}v^2 div(\rho\boldsymbol{v}) - \boldsymbol{v}gradp - \rho\boldsymbol{v}grad\left(\frac{1}{2}v^2\right) = \\ &= -\frac{1}{2}v^2 div(\rho\boldsymbol{v}) - \rho\boldsymbol{v}grad\left(\frac{1}{2}v^2 + w\right) + \rho T\boldsymbol{v}grads \end{aligned} \tag{2.54}$$

(in the absence of the external force $\boldsymbol{F}$). On the other hand, we have

$$\begin{aligned} &d(\rho\varepsilon) = \varepsilon d\rho + \rho d\varepsilon = wd\rho + \rho Tds \ , \\ &\frac{\partial(\rho\varepsilon)}{\partial t} = w\frac{\partial\rho}{\partial t} + \rho T\frac{\partial s}{\partial t} = -wdiv(\rho\boldsymbol{v}) + \rho T\frac{\partial s}{\partial t} \ . \end{aligned} \tag{2.55}$$

By adding equations (2.54) and (2.55), we get the law of energy conservation

$$\begin{aligned} &\frac{\partial}{\partial t}\left(\frac{1}{2}\rho v^2 + \rho\varepsilon\right) = -div\left[\rho\boldsymbol{v}\left(\frac{1}{2}v^2 + w\right)\right] + \\ &+\rho T\left(\frac{\partial s}{\partial t} + \boldsymbol{v}grads\right) \ . \end{aligned} \tag{2.56}$$

The quantity $\frac{1}{2}\rho v^2 + \rho\varepsilon$ is the energy per unit volume (energy density), consisting of the kinetic energy and the internal energy. The quantity

[23] C. L. Navier, Mem. Acad. Sci. t. 7 (1827) (read in Paris Academy in 1821); A. L. Cauchy, "Sur les equations qui expriment les conditions d'equilibre ou les lois du mouvement interieur d'un corps, solide, elastique ou non-elastique", included in *Exercises de Mathematique* (1828) (communicated to the Paris Academy in 1822).

$\rho \mathbf{v} \left(\frac{1}{2}v^2 + w\right)$ is the energy flux density and the work done by pressure per unit time; indeed,

$$\begin{aligned} &\int dV div\left[\rho \boldsymbol{v}\left(\tfrac{1}{2}v^2 + w\right)\right] = \\ &= \oint d\boldsymbol{f} \rho \boldsymbol{v}\left(\tfrac{1}{2}v^2 + w\right) = \\ &= \oint d\boldsymbol{f} \rho \boldsymbol{v}\left(\tfrac{1}{2}v^2 + \varepsilon\right) + \oint d\boldsymbol{f} \boldsymbol{v} p \ . \end{aligned} \tag{2.57}$$

For adiabatic flow the entropy term in equation (2.56) disappears. For an ideal, incompressible fluid the internal energy is constant, $\varepsilon = const$, $\rho\varepsilon = const$ and $w = p/\rho + \varepsilon$; in this case the energy conservation includes only the kinetic energy and the work of the pressure forces.

2.8 Fluid thermodynamics

The equation of state of a monoatomic ideal gas with N atoms in volume V at temperature T is given by the pressure $p = NT/V$; since the mass is $m = NM$, where M is the atomic mass, we get the equation of state

$$p = \frac{\rho}{M}T \ , \tag{2.58}$$

where ρ is the density. Similarly, the energy and the enthalpy per unit mass are

$$\varepsilon = \frac{3}{2M}T \ , \ w = \frac{5}{2M}T \ . \tag{2.59}$$

The free energy per unit mass is

$$F = -\frac{1}{M}T \ln\left[\frac{eM}{\rho}\left(\frac{MT}{2\pi\hbar^2}\right)^{3/2}\right] \ , \tag{2.60}$$

where $\hbar \simeq 10^{-27} erg \cdot s$ is Planck's constant and $e \simeq 2.72$. The entropy per unit mass is

$$s = \frac{1}{M} \ln\left[\frac{e^{5/2}M}{\rho}\left(\frac{MT}{2\pi\hbar^2}\right)^{3/2}\right] \ ; \tag{2.61}$$

these equations are valid for $\hbar^2/Ma^2 \ll T$, where $a = (V/N)^{1/3}$ is the mean inter-atomic separation distance.[24]

The "normal conditions" are $T = 300K$ ($1K = 1.38 \times 10^{-16}erg$) and $p = 10^6 dyn/cm^2$ (separation distance $a \simeq 35$Å, 1Å $= 10^{-8}cm$). In normal conditions the atmospheric air has the density $\rho \simeq 10^{-3}g/cm^3$. We can take $M \simeq 10^{-22}g$ for air and the thermal expansion coefficient $\beta = \frac{1}{V}\left(\frac{\partial V}{\partial T}\right)_p \simeq 10^{13}/erg$, the thermoconductivity $\kappa \simeq 10^{19}/cm \cdot s$, the specific heat per unit mass at constant pressure $c \simeq 10^{23}/g$ and the bulk modulus of compression $K = -V\left(\frac{\partial p}{\partial V}\right) \simeq 10^6 dyn/cm^2$ (with slight variations for constant volume, or constant temperature, constant entropy, etc); $\alpha = K\beta$ is much larger than β. The mean thermal velocity of an atom is $v_{th} = \sqrt{3T/M} \simeq 10^4 cm/s$.

There is no simple equation of state for liquids.[25] The mean thermal velocity is of the same order of magnitude as in gases. For water we can take $M \simeq 10^{-23}g$, $\rho \simeq 1g/cm^3$, $\beta \simeq 10^{12}/erg$, $\kappa \simeq 4 \times 10^{20}/cm \cdot s$, $c \simeq 3 \times 10^{23}/g$ and $K \simeq 2 \times 10^{10} dyn/cm^2$.

2.9 Convection

Fluids that are not in thermodynamic equilibrium may develop macroscopic currents, which tend to establish the thermal equilibrium. This is a convective motion, where the motion is unstable (mechanical equilibrium is destroyed).

Let $V(p, s)$ and $V(p', s')$ be two adjacent volumes of fluid per unit mass, separated by distance dz. If the former volume does not displace the volume $V(p', s')$, and, therefore, the mechanical equilibrium is stable, we need to have the inequality

$$V(p', s) < V(p', s') \ , \tag{2.62}$$

[24] L. Landau and E. Lifshitz, *Course of Theoretical Physics*, vol. 5 (*Statistical Physics*), Elsevier, Oxford (1980).

[25] M. Apostol, "A model for the thermodynamics of simple liquids", Physica **B403** 3946 (2008); *Physical Kinetics*, Cambridge Scholars Publishing, Newcastle upon Tyne (2020).

or

$$\left(\frac{\partial V}{\partial s}\right)_p \frac{ds}{dz} > 0 \ . \tag{2.63}$$

Since $(\partial V/\partial s)_p = \frac{T}{c_p}(\partial V/\partial T)_p > 0$, this inequality becomes

$$\frac{ds}{dz} > 0 \ . \tag{2.64}$$

On the other hand we have

$$\frac{ds}{dz} = \left(\frac{\partial s}{\partial T}\right)_p \frac{dT}{dz} + \left(\frac{\partial s}{\partial p}\right)_T \frac{dp}{dz} > 0 \ . \tag{2.65}$$

The partial derivatives are given by $(\partial s/\partial T)_p = c_p/T$ and $(\partial s/\partial p)_T = -\partial^2\Phi/\partial p\partial T = -(\partial V/\partial T)_p$, where Φ is the Gibbs free energy ($d\Phi = Vdp - sdT$, $V = 1/\rho$). Equation (2.65) becomes

$$\frac{c_p}{T}\frac{dT}{dz} - \frac{\beta}{\rho}\frac{dp}{dz} > 0 \ , \tag{2.66}$$

or

$$-\frac{dT}{dz} < -\frac{\beta T}{\rho c_p}\frac{dp}{dz} \ , \tag{2.67}$$

where $\beta = \frac{1}{V}\left(\frac{\partial V}{\partial T}\right)_p$ is the thermal expansion coefficient (see Appendix[26]). In uniform gravitational field with gravitational acceleration g the pressure gradient is $dp/dz = -\rho g$, such that

$$-\frac{dT}{dz} < \frac{g\beta T}{c_p} \ ; \tag{2.68}$$

for an ideal gas $\beta T = 1$, and the condition of stable equilibrium (absence of convection) is

$$-\frac{dT}{dz} < \frac{g}{c_p} \ . \tag{2.69}$$

We note that a positive thermal gradient is always stable. For ideal gases $p = \frac{\rho}{M}T$, and $\beta = 1/T$. For constant density $dp/dz = \frac{\rho}{M}(dT/dz)$.

[26]L. Landau and E. Lifshitz, *Course of Theoretical Physics*, vol. 6 (*Fluid Mechanics*), Pergamon, Oxford (1987).

For monoatomic ideal gases $Mc_p = \frac{5}{2}$, such that the equilibrium is achieved for $dT/dz > 0$. For ideal gases ($\beta T = 1$) the inequality (2.67) reads also $-dT/dz < \frac{1}{2c_p}(dv^2/dz)$ for a flow with velocity v along the z-axis (Bernoulli's law); or, $d(v^2/2 + c_p T)/dz > 0$. A temperature imbalance ΔT generates a velocity of the order of $v \simeq \sqrt{\Delta T/M}$, which is much smaller than the thermal velocity.

By using the numerical data given above, we get from inequality (2.68) $-dT/dz \lesssim 3K/km$ for air and $-dT/dz \lesssim 1K/100m$ for water. We can see that water is more prone to be unstable against thermal gradients (as other liquids too), and to develop convection currents, as in Benard cells.[27]

2.10 Free convection

Let us consider the atmospheric air in equilibrium in uniform gravitational field. The pressure is

$$p = const - \rho g z \; . \tag{2.70}$$

The hydrostatic pressure $\delta p = -\rho g z$ is very small in comparison with the atmospheric pressure. The density suffers a variation

$$\delta\rho = \frac{\rho}{K}\delta p = -\frac{\rho^2 g z}{K} \tag{2.71}$$

with respect to the equilibrium density ρ_0, where K is the adiabatic bulk modulus of compression. Since this variation is very small, we may neglect it, and write the continuity equation as

$$div\boldsymbol{v} = 0 \; . \tag{2.72}$$

Let us assume that a small temperature imbalance δT is present; we write it as $\delta T = T - T_0$, where T_0 is the equilibrium temperature.

[27] H. Benard, "Les tourbillons cellulaires dans une nappe liquide", Revue Generale des Sciences Pures and Appliquees **11** 1261, 1309 (1900); Lord Rayleigh, "On convection currents in a horizontal layer of fluid, when the higher temperature is on the under side", Phil. Mag. **32** 529 (1916); J. R. A. Pearson, "On convection cells induced by surface tension", J. Fluid Mech. **4** 489 (1958).

Within the same approximation the equation of heat conduction is

$$\rho_0 c_p \left(\frac{\partial T}{\partial t} + \boldsymbol{v} gradT \right) = \kappa \Delta T \ . \tag{2.73}$$

However, the thermal gradient generates a change in density, given by

$$\delta\rho = \left(\frac{\partial \rho}{\partial T} \right)_p \delta T = -\rho_0 \beta \delta T \ , \tag{2.74}$$

and a pressure variation δp; both should be included in Euler's equation

$$\begin{gathered} \rho_0 \left[\frac{\partial \boldsymbol{v}}{\partial t} + (\boldsymbol{v} grad)\boldsymbol{v} \right] = \\ = -\frac{K}{\rho_0} grad\rho - \rho_0 \boldsymbol{g} - grad\delta p + \rho_0 \boldsymbol{g} \beta \delta T \ . \end{gathered} \tag{2.75}$$

We note the change of sign in the buoyancy force $-\boldsymbol{g}\delta\rho$, due to the reaction of the adjacent elements of fluids (see the next section). The first two terms on the right give $grad\rho = -\rho_0^2 \boldsymbol{g}/K$, which is neglected (equation (2.71)). This is the Boussinesq approximation.[28] Equations (2.72), (2.73) and (2.75) are known as the equations of convection in uniform gravitational field; this convection is also called free convection.[29] We note that condition (2.74) is the condition of constant pressure given by equation (2.16). The above equations are further discussed in the Thermal Conduction chapter.

2.11 Equilibrium with thermal gradient

Convection currents have been suggested in viscous fluids with vertical thermal gradients in uniform gravitational field,[30] exhibiting an unstable, even chaotical, behaviour.[31] Also, unstable, stratified fluid

[28] J. Boussinesq, *Theorie Analytique de la Chaleur*, Gauthiers-Villars, Paris (1903).

[29] A. Oberbeck, "Ueber die Waermeleitung der Flussigkeiten bei Beruecksichtigung der Stroemungen infolge von Temperaturdifferenzen", Ann. Physik **243** 271 (1879).

[30] Lord Rayleigh, "On convection currents in a horizontal layer of fluid, when the higher temperature is on the under side", Phil. Mag. **32** 529 (1916).

[31] E. N. Lorenz, "Deterministic nonperiodic flow", J. Atmosph. Sci. **20** 130 (1963); see also D. Turcotte, *Fractals and Chaos in Geology and Geophysics*, Cambridge University Press, NY (1997) and D. . Turcotte and G. Schubert, *Geodynamics*, Cambridge University Press, NY (2002).

layers are expected in these conditions in turbulent fluids flowing along surfaces.[32] We investigate below the conditions of stable equilibrium (absence of free convection) for atmospheric air and water.[33]

We consider a fluid with mass density ρ_0 in a half-space $(x, y, z > 0)$, in mechanical equilibrium (at rest) and in thermodynamic equilibrium at temperature T_0. We apply a uniform gravitational acceleration g and a uniform temperature $T_1 > T_0$ on the surface $z = 0$. We are interested in the changes brought about under these conditions in the state of the fluid. This is a problem different than the one with a given temperature gradient, treated above. In the present problem the thermal gradient is established by the equilibrium. We assume that the thermal gradient is small, such that we expect a small, vertical velocity v and a small variation $\Delta\rho$ in density, both functions of z and independent of time. We limit ourselves to the leading contributions, which is the Boussinesq approximation.[34] The continuity equation reads $\frac{dv}{dz} = 0$, which, with the boundary condition $v\mid_{z=0}= 0$, gives a vanishing velocity $v = 0$. Therefore, within this approximation we have a fluid in mechanical equilibrium. As we shall see shortly, the gravitational effect of the density variation is compensated by the thermal gradient (buoyancy force).

Indeed, the Euler equation reduces to the equilibrium equation

$$0 = -\frac{dp}{dz} - \rho g \ , \tag{2.76}$$

where p is the pressure (depending only on z) and ρ denotes the density. In thermal equilibrium the equation of state of the fluid gives the pressure $p(\rho_0, T_0)$ as a function of density and temperature. The thermal gradient on the surface $z = 0$ generates variations in density

[32]G. L. Mellor, "Analytic prediction of the properties of stratified planetary surface layers", J. Atmosph. Sci. **30** 1061 (1973); G. E. Willis and J. W. Deardorff, "A laboratory model of the unstable planetary boundary layer", J. Atmosph. Sci. **31** 1297 (1974); L. Liu, S. N. Gadde and R. J. A. M. Stevens, "Universal wind profile for conventionally neutral atmospheric boundary layers", Phys. Rev. Lett. **126** 104502 (2021).

[33]M. Apostol, "Equilibrium of a fluid with thermal gradient in uniform gravitational field", J. Theor. Phys. **339** (2022).

[34]J. Boussinesq, *loc. cit.*

and temperature, such that equation (2.76) becomes

$$\left(\frac{\partial p}{\partial \rho}\right)_T \frac{d\rho}{dz} + \left(\frac{\partial p}{\partial T}\right)_V \frac{dT}{dz} + g\Delta\rho + \rho_0 g = 0 \ , \tag{2.77}$$

where $(\partial p/\partial T)_V$ is the thermal pressure coefficient (at constant volume V) and $\Delta\rho = \rho - \rho_0$ is the density variation. The coefficient $\left(\frac{\partial p}{\partial \rho}\right)_T = K/\rho$ is related to the isothermal modulus of compression K (bulk modulus, $1/K = -\frac{1}{V}(\partial V/\partial p)_T$). The density variation is caused by the temperature variation, such that we can write separately the equation of hydrostatic equilibrium

$$\left(\frac{\partial p}{\partial \rho}\right)_T \frac{d\rho}{dz} + \rho_0 g = 0 \tag{2.78}$$

and the equation for local thermodynamic equilibrium (buoyancy force)

$$\left(\frac{\partial p}{\partial T}\right)_V \frac{dT}{dz} + g\Delta\rho = 0 \ . \tag{2.79}$$

It is easy to see from equation (2.78) that the density decreases with increasing altitude. Therefore, the change in density is given by $\Delta\rho = \rho_0\beta(T - T_0)$, where $\beta = \frac{1}{V}(\partial V/\partial T)_p$ is the dilatation coefficient. From equations (2.79) and (2.78) we get the temperature and the density distributions, respectively. The dilatation coefficients are related by (see Appendix[35])

$$\left(\frac{\partial p}{\partial T}\right)_V = K\beta \ , \tag{2.80}$$

such that equation (2.79) becomes

$$K\frac{dT}{dz} + g\rho_0(T - T_0) = 0 \ ; \tag{2.81}$$

the solution is

$$T = \Delta T e^{-g\rho_0 z/K} + T_0 \ , \tag{2.82}$$

[35] L. Landau and E. Lifshitz, *Course of Theoretical Physics*, vol. 5 (*Statistical Physics*), Elsevier, Oxford (1980).

where $\Delta T = T_1 - T_0$. From equation (2.78) and from $\Delta\rho = \rho_0\beta(T - T_0)$ we get the density

$$\rho = \rho_0\left(1 + \beta\Delta T e^{-g\rho_0 z/K}\right) \; ; \tag{2.83}$$

at $z = 0$ the density is

$$\rho_{z=0} = \rho_0\left(1 + \beta\Delta T\right) \; , \tag{2.84}$$

as expected.

It is worth noting the meaning of the above scheme of decoupling the equilibrium equations (2.78) and (2.79). Under a small increase of temperature $T - T_0$ the element of fluid increases its volume by dilatation; the volume variation is $\Delta V = V_0\beta(T - T_0)$, where V_0 is the initial volume. Consequently, the density is diminished by $\Delta\rho = -\rho_0\beta(T - T_0)$. As a consequence of the volume increase, the fluid element acts upon its adjacent fluid elements with a force density $-g\rho_0\beta(T - T_0)$. It follows that the adjacent fluid elements react upon the fluid element with a force $g\rho_0\beta(T - T_0)$. This is the force which occurs in equation (2.81); it is the buoyancy force. It looks like the density change $\Delta\rho = -\rho_0\beta(T - T_0)$ has a changed sign. The buoyancy force is an additional force besides the gravitational force $\rho_0 g$, which occurs in equation (2.78). We cannot use ρg for this force, because $\Delta\rho = -\rho_0\beta(T - T_0)$ in $\rho = \rho_0 + \Delta\rho$ is compensated by the action of the adjacent fluid elements. The general equation of equilibrium $dp/dz + \rho g = 0$ may be misleading in this respect, since it includes only the direct action of the gravitational field upon an isolated fluid element, and leaves aside the reaction of the adjacent fluid elements. The fluid element is not isolated, it is in contact with the adjacent fluid elements. Although the fluid element may obey the equation of state of an ideal gas $p = \rho T/M$, where M is the molecular mass, we cannot use $dp/dz + Mgp/T = 0$, which would lead to the barometric formula $p = p_0 e^{-\int_0^z \frac{Mg}{T}dz}$, with pressure p_0 at $z = 0$, because this equation lacks the reaction of the adjacent fluid elements. A similar problem occurs in the equilibrium of the dilute solutions in gravitational field.[36]

It remains to check the heat transfer. The change δs of the small entropy variation s per unit mass has two origins: one arises from the

[36]L. Landau and E. Lifshitz, *loc.cit.*

total time variation, another from the change in the volume V. This change is given by

$$\delta \int dV s = \int dV \frac{ds}{dt} \delta t + \int s \delta dV = \int dV \frac{ds}{dt} \delta t + \oint s \boldsymbol{w} d\boldsymbol{f} \delta t \ , \quad (2.85)$$

where f is the surface which encloses the volume V, t denotes the time and $\boldsymbol{w}$ is the velocity of entropy transfer by molecular collisions through the surface f. We can see that $\boldsymbol{w}s$ is the entropy flux density. We assume that the thermodynamic equilibrium is preserved locally, so we have

$$\frac{ds}{dt} + div(\boldsymbol{w}s) = 0 \ , \quad (2.86)$$

or

$$\frac{\partial s}{\partial t} + \boldsymbol{v} grad s = -div(\boldsymbol{w}s) \ , \quad (2.87)$$

where $\boldsymbol{v}$ denotes the velocity. It is convenient to write this equation for heat $Q = \rho_0 T_0 s$ per unit volume, heat flux density $\boldsymbol{q} = \rho_0 \boldsymbol{w} T_0 s$ and introduce the temperature through $T_0 s = c_p(T - T_0)$, where c_p is the specific heat (per unit mass) at constant pressure. We note that we use variations with respect to equilibrium, *i.e.* at constant pressure. Equation (2.87) becomes

$$\rho_0 c_p \left(\frac{\partial T}{\partial t} + \boldsymbol{v} grad T \right) = -div \boldsymbol{q} \ . \quad (2.88)$$

The heat flux density is $\boldsymbol{q} = -\kappa grad T$, where κ the thermoconductivity, such that equation (2.88) becomes

$$\rho_0 c_p \left(\frac{\partial T}{\partial t} + \boldsymbol{v} grad T \right) = \kappa \Delta T \ . \quad (2.89)$$

We can see that the velocity $\boldsymbol{w}$ disappears from this equation. In our case equation (2.89) reduces to $\frac{d^2T}{dz^2} = 0$, which means that the exponential in equations (2.82) and (2.83) should be replaced by its linear approximation. Therefore, the temperature and density distributions become

$$\begin{aligned} T &= T_1 - \tfrac{g\rho_0 \Delta T}{K} z \ , \\ \rho &= \rho_0 \left(1 + \beta \Delta T - \tfrac{\beta g \rho_0 \Delta T}{K} z \right) \ . \end{aligned} \quad (2.90)$$

This approximation is valid for $g\rho_0 z/K \ll 1$, which is the condition for the relevant altitudes allowed by our approximation of small variations. Also, the condition of stable equilibrium (absence of a convective flow) is $-dT/dz < g\beta T/c_p$ (equation (2.68)), *i.e.* $\Delta T < \frac{\beta TK}{\rho_0 c_p}$. For air in normal conditions $\rho_0 = 1.2\times 10^{-3} g/cm^3$, $K = 10^6 dyn/cm^2$, $c_p = 7\times 10^{22}/g$ and $\beta = 10^{13}/erg$. For $T = 300K$ the allowed altitudes are up to $z \simeq 10km$ and the allowed temperature difference is up to $\Delta T \simeq 35K$. In this range of thermal gradients the hydrostatic equilibrium and the local thermodynamic equilibrium are stable for the atmospheric air. For water ($\rho_0 = 1g/cm^3$, $K = 2.2\times 10^{10} dyn/cm^2$, $c_p = 3\times 10^{23}/g$, $\beta = 10^{12}/erg$) at $T = 300K$ the condition of stability is $\Delta T < 20K$.

Similar calculations can be made for a layer with a uniform temperature $T_1 > T_0$ placed at altitude H, or for a fluid enclosed between two plane-parallel plates with different temperatures.

2.12 Circulation conservation

By using

$$\frac{1}{2} grad v^2 = \boldsymbol{v} \times curl \boldsymbol{v} + (\boldsymbol{v} grad)\boldsymbol{v} \ , \tag{2.91}$$

Euler's equation becomes

$$\frac{\partial \boldsymbol{v}}{\partial t} - \boldsymbol{v} \times curl \boldsymbol{v} = -grad\left(\frac{1}{2}v^2 + w\right) + \boldsymbol{F}/\rho \ , \tag{2.92}$$

where $\boldsymbol{F}$ is an external force per unit volume and w is the enthalpy per unit mass. The variation of the enthalpy per unit mass is $dw = Vdp + Tds$, where V is the volume of the unit mass, *i.e.* $V = 1/\rho$, where ρ is the density, and T is the temperature. For an ideal fluid the variation of the entropy per unit mass ds is zero. Therefore, $dw = dp/\rho$. The term $\rho\mathbf{v} \times curl\mathbf{v}$ is sometimes called Euler's force. Taking the *curl* of both sides we get

$$\frac{\partial}{\partial t} curl \boldsymbol{v} = curl(\boldsymbol{v} \times curl \boldsymbol{v}) + curl(\boldsymbol{F}/\rho) \ . \tag{2.93}$$

The quantity $\boldsymbol{\omega} = curl\mathbf{v}$ is called vorticity; for $\boldsymbol{\omega} = const$ we have $\boldsymbol{v} = \frac{1}{2}\boldsymbol{\omega} \times \boldsymbol{r}$. If the vorticity is vanishing, *i.e.* $curl\boldsymbol{v} = 0$, then

$\boldsymbol{v} = grad\Phi$, and the flow is called potential flow, or irrotational flow (it derives from a potential Φ). In that case the external force should also derive from a potential, $\boldsymbol{F} = -\rho grad\varphi$.

The change in time of the velocity circulation along a closed contour is

$$\frac{d}{dt}\oint \boldsymbol{v}d\boldsymbol{l} = \oint \frac{d\boldsymbol{v}}{dt}d\boldsymbol{l} + \oint \boldsymbol{v}d\boldsymbol{v} =$$

$$= \oint \frac{d\boldsymbol{v}}{dt}d\boldsymbol{l} + \oint d(v^2/2) = \oint \frac{d\boldsymbol{v}}{dt}d\boldsymbol{l} = \tag{2.94}$$

$$= \oint(-gradw + \boldsymbol{F}/\rho)d\boldsymbol{l} = \oint curl(\boldsymbol{F}/\rho)d\boldsymbol{f} \ .$$

We can see that the circulation of the velocity along closed paths is conserved for ideal flows, providing $\oint curl(\boldsymbol{F}/\rho)d\boldsymbol{f} = 0$.[37] The law can also be written as

$$\oint \boldsymbol{v}d\boldsymbol{l} = \oint curl\boldsymbol{v}d\boldsymbol{f} = \oint \boldsymbol{\omega}d\boldsymbol{f} = const \ . \tag{2.95}$$

The particles of the ideal fluid carry the vorticity with them along the flow, if the external force derives from a potential.

2.13 Potential flow

Let us consider an ideal fluid, which conserves the velocity circulation. We should note that the velocity circulation is conserved round contours which move together with the fluid particles. For a steady flow these contours encircle the streamlines, whose pattern is constant in time. Let us suppose that the vorticity is zero at some point on such a streamline, *i.e.* $curl\boldsymbol{v} \cdot \delta\boldsymbol{f} = 0$ at some point, where $\delta\boldsymbol{f}$ is a small area encircling the streamline at that point. Since the circulation is conserved, we have $curl\boldsymbol{v} = 0$ at any other point on that streamline. If the flow is not steady, the circulation is taken round any fluid particle in motion. A flow with $curl\boldsymbol{v} = 0$ in the whole space is potential flow;

[37] H. Helmholtz, "Ueber Integrale der hydrodynamischen Gleichungen, welche den Wirbelbewegungen entsprechen", J. reine und angew. Mathematik **55** 25 (1858); W. Thomson (Lord Kelvin), "On vortex motion", Trans. Roy. Soc. Edinburgh **25** 217 (1869); H. Poincare, *Theorie des Tourbillons*, Gabay, Paris (1893).

it is also called irrotational flow. The vorticity is zero at any time in this case, the flow is potential at any time.

It follows that we cannot put an ideal fluid in rotation. This is a serious drawback of the ideal fluids. The velocity $\boldsymbol{v}$ is derived from $grad\Phi$ and $curl\boldsymbol{A}$, where Φ is a scalar potential and $\boldsymbol{A}$ is a vector potential; if we use both, we need an additional condition $div\boldsymbol{A} = 0$. If $\boldsymbol{v} = grad\Phi$, the vorticity $curl\boldsymbol{v} = 0$; if $\boldsymbol{v} = curl\boldsymbol{A}$, there exists a non-vanishing vorticity. The vorticity can only be created through viscosity, such that for an ideal fluid we have only a potential flow. If the ideal fluid is incompressible, the condition $div\boldsymbol{v} = 0$ may require $\boldsymbol{v} = curl\boldsymbol{A}$, which, however, is not possible. In that case we have only $\Delta\Phi = 0$. Nevertheless, an ideal fluid can be put in rotation by the viscosity at its boundaries, the flow being potential or rotational at large distance from the boundary.

Also, it follows that a steady and uniform flow incident from infinity on a body is a potential flow. However, we cannot draw closed fluid contours round body's surface points, so the circulation conservation cannot be applied. In fact, the fluid particles adhere to body's surface, and, rigorously speaking, discontinuity surfaces appear, tangential to the streamlines. The adherence to body's surface generates a boundary layer, governed by viscosity. Usually, its thickness is small, such that it may be neglected; beyond the boundary layer the potential flow is a good approximation to the fluid flow.

Let us consider a body which performs small oscillations with velocity u and amplitude a in a fluid; the dimension l of the body is much larger than a, $l \gg a$. Since $\frac{\partial \boldsymbol{v}}{\partial t}$ is of the order of u^2/a and $(\boldsymbol{v}grad)\boldsymbol{v}$ is of the order of u^2/l in the neighbourhood of the body, we may neglect the advection term and the Euler equation can be written as $\partial\boldsymbol{v}/\partial t = -gradw$ for an ideal fluid. Hence, $curl\boldsymbol{v} = const$; however, in oscillatory motion this $const$ should be zero, so we get $curl\boldsymbol{v} = 0$. Therefore, the motion of a fluid which performs small oscillations is potential flow.

In potential flow $\boldsymbol{v} = grad\Phi$, where Φ is the velocity potential. From equation (2.92), where $\boldsymbol{F} = -\rho grad\varphi$, we get

$$\frac{\partial\Phi}{\partial t} + \frac{1}{2}v^2 + w + \varphi = f(t) \ , \tag{2.96}$$

where $f(t)$ is an arbitrary function of time. Since Φ is defined up to such a function, we can put $f = 0$. For steady flow

$$\frac{1}{2}v^2 + w + \varphi = const \ , \tag{2.97}$$

which is Bernoulli's equation, this time with the same constant for all the streamlines.

Another general observation relates to the equality $\oint \boldsymbol{v}d\boldsymbol{l} = \oint curl\boldsymbol{v} \cdot d\boldsymbol{f} = 0$ round a streamline. It tells that the streamlines in potential flow cannot be closed lines, because $\oint \boldsymbol{v}d\boldsymbol{l} \neq 0$ for a closed line. However, the Stokes theorem assumes a deformable contour up to an infinitesimal surface, *i.e.* a (simply) connected region; the result may not be applicable to multiply-connected regions.

Also, a flow rotating steadily in a cylinder, with velocity $\boldsymbol{v} = (-v\frac{y}{r}, v\frac{x}{r}, 0)$, has $curl\boldsymbol{v} = 0$ (for $\boldsymbol{r} \neq 0$) and $\oint \boldsymbol{v}d\boldsymbol{l} = 2\pi vr = const$, according to equation (2.94). Such a streamline is called vortex. From Bernoulli's equation (2.97) we get $const/r^2 - gz = 0$ in uniform gravitational field where z denotes the vertical coordinate. The fluid develops a free surface with equation $zr^2 = const$.

Finally, we note that potential flow of an incompressible fluid has a vanishing angular momentum. Indeed,

$$\begin{gathered} \int (\boldsymbol{r} \times grad\Phi)_i \, dV = \int \varepsilon_{ijk} x_j \partial_k \Phi dV = \\ = -\int \varepsilon_{ijk}\delta_{jk}\Phi dV = 0 \ . \end{gathered} \tag{2.98}$$

In general, the angular momentum is conserved for an incompressible ideal fluid.

2.14 Incompressible fluids

Let us write again the equations of a fluid without viscosity (equations (2.4), (2.10) and (2.12)):

$$\begin{gathered} \frac{\partial \rho}{\partial t} + div(\rho \boldsymbol{v}) = 0 \ , \\ \rho \left[\frac{\partial \boldsymbol{v}}{\partial t} + (\boldsymbol{v}grad)\boldsymbol{v}\right] = -gradp + \boldsymbol{F} \ , \\ \rho c_p \left(\frac{\partial T}{\partial t} + \boldsymbol{v}gradT\right) = \kappa \Delta T \ . \end{gathered} \tag{2.99}$$

Let us write $gradp$ according to equation (2.17), such that Euler's equation becomes

$$\rho\left[\frac{\partial \boldsymbol{v}}{\partial t}+(\boldsymbol{v}grad)\boldsymbol{v}\right]=-\frac{K}{\rho}grad\rho-K\beta gradT+\boldsymbol{F}\ . \tag{2.100}$$

For all common fluids the parameters K/ρ^2 and $K\beta/\rho$ are extremely large. For example, for the atmospheric air $\rho\simeq 10^{-3}g/cm^3$, $K\simeq 10^6 dyn/cm^2$, $\beta\simeq 10^{13}/erg$ (and $c_p\simeq 10^{23}/g$, $\kappa\simeq 10^{19}/cm\cdot s$). For water $\rho\simeq 1g/cm^3$, $K\simeq 10^{10}dyn/cm^2$, $\beta\simeq 10^{12}/erg$ (and $c_p\simeq 10^{23}/g$, $\kappa\simeq 10^{20}/cm\cdot s$). For any reasonable change of the velocity on the left in equation (2.100), *i.e.* for any reasonable flow, we need extremely small spatial variations of the density and the temperature on the right. We can see this also from the condition of absence of the convection (equation (2.67)), the Oberbeck-Boussinesq equations of free convection (equation (2.75)) and from the equilibrium in uniform gravitational field with thermal gradient (equation (2.90)). Therefore, we may take $\rho=const$, such that $gradp=(\partial p/\partial T)_V gradT$, where $\alpha=(\partial p/\partial T)_V$ is the thermal pressure coefficient, *i.e.* $\beta\rightarrow 0$, $K\rightarrow\infty$ and $\alpha=K\beta$ is finite. Equations (2.99) become

$$div\boldsymbol{v}=0\ ,$$

$$\rho\left[\tfrac{\partial \boldsymbol{v}}{\partial t}+(\boldsymbol{v}grad)\boldsymbol{v}\right]=-\alpha gradT+\boldsymbol{F}\ , \tag{2.101}$$

$$\rho c_p\left(\tfrac{\partial T}{\partial t}+\boldsymbol{v}gradT\right)=\kappa\Delta T\ .$$

These are the equations of an incompressible fluid (without viscosity). The variation of the temperature remains small. For an ideal fluid it is more convenient to write

$$\tfrac{\partial \boldsymbol{v}}{\partial t}+(\boldsymbol{v}grad)\boldsymbol{v}=-grad\left(\tfrac{p}{\rho}\right)+\boldsymbol{F}/\rho\ ,$$

$$\tfrac{ds}{dt}=0\ , \tag{2.102}$$

and the velocity is given in fact only by the external force. Also, the energy per unit mass is constant ($d\varepsilon=-pdV+Tds=0$) and we may take for the enthalpy per unit mass $w=p/\rho$. Multiplying the first equation (2.102) by $\boldsymbol{v}$ and using the second equation (2.102), we get the energy conservation

$$\frac{\partial}{\partial t}\left(\frac{1}{2}v^2\right)+div\left[\boldsymbol{v}\left(\frac{1}{2}v^2+\frac{p}{\rho}+\varphi\right)\right]=0 \ , \tag{2.103}$$

where φ is the external potential ($\boldsymbol{F}=-\rho grad\varphi$). For a steady flow we have Bernoulli's law

$$\frac{1}{2}v^2+\frac{p}{\rho}+\varphi=const \tag{2.104}$$

along streamlines. In uniform gravitational field this law becomes $\rho v^2/2+p+gz=const$ ($\varphi=gz$). If a fluid has velocity $\boldsymbol{v}$ and pressure p at infinity, at its boundary the velocity is zero (stagnation point), so the pressure there is $p_0=\frac{1}{2}\rho v^2+p$. We note that the pressure here is the external pressure.

We can use

$$grad\left(\frac{1}{2}v^2\right)=\boldsymbol{v}\times curl\boldsymbol{v}+(\boldsymbol{v}grad)\boldsymbol{v} \tag{2.105}$$

in equation (2.102), with $\boldsymbol{F}=-\rho grad\varphi$, and, by taking the $curl$, we get two equations only for velocity,

$$\begin{gathered}\frac{\partial}{\partial t}curl\boldsymbol{v}=curl(\boldsymbol{v}\times curl\boldsymbol{v}) \ , \\ div\boldsymbol{v}=0 \ .\end{gathered} \tag{2.106}$$

In potential flow $\boldsymbol{v}=grad\Phi$ and equations (2.106) reduce to

$$\Delta\Phi=0 \ . \tag{2.107}$$

This equation, known as Laplace equation, originates from Euler. It should be solved with boundary condition, which is zero normal velocity, $\boldsymbol{v}_n=\frac{\partial\Phi}{\partial\boldsymbol{n}}=0$.

Equation (2.96) for incompressible potential flow reads

$$\frac{\partial\Phi}{\partial t}+\frac{1}{2}v^2+\frac{p}{\rho}+\varphi=f(t) \ . \tag{2.108}$$

Since equation (2.107) does not contain the time, the latter enters the solution only through the boundary conditions, which include only the boundary velocity, so the velocity of the potential flow depends only on the velocity of the boundaries (*e.g.*, the fluid flow depends only on the velocity of the body moving in the fluid).

In a two-dimensional flow the second equation (2.106) is satisfied identically by

$$v_x = \frac{\partial \psi}{\partial y} \; , \; v_y = -\frac{\partial \psi}{\partial x} \; , \tag{2.109}$$

where the function ψ, called the stream function, is given by the first equation (2.106). The equation of the streamlines is $\frac{dx}{v_x} = \frac{dy}{v_y}$ (velocity tangent to the streamline), or

$$\frac{\partial \psi}{\partial x} dx + \frac{\partial \psi}{\partial y} dy = d\psi = 0 \; ; \tag{2.110}$$

the streamlines are given by $\psi(x, y) = const.$

In potential flow

$$v_x = \frac{\partial \Phi}{\partial x} = \frac{\partial \psi}{\partial y} \; , \; v_y = \frac{\partial \Phi}{\partial y} = -\frac{\partial \psi}{\partial x} \; , \tag{2.111}$$

which are the Riemann-Cauchy conditions of an analytic function $\Phi + i\psi$ (complex potential).

The conditions under which a fluid may be regarded as incompressible are obtained as follows: an adiabatic change in pressure gives a change in density $\Delta\rho = \left(\frac{\partial \rho}{\partial p}\right)_s \Delta p$; from Bernoulli's law, in steady flow, Δp is of the order of ρv^2; therefore, $\Delta\rho \simeq \rho \frac{v^2}{(\partial p/\partial \rho)_s}$. It follows that $(\partial p/\partial \rho)_s$ is the square of a velocity c. This is the sound velocity. Therefore, an incompressible fluid requires $v \ll c$. In general, when the flow is not steady, the flow velocity changes in time τ over a distance l, and we need to compare $\frac{l}{\tau}$ to c.

2.15 Rotating fluid

Let us consider an incompressible ideal fluid rotating with the angular velocity ω about the vertical z-axis in uniform gravitational field. The coordinates of a point are $x = r\cos\omega t$, $y = r\sin\omega t$ and z, where

$r = \sqrt{x^2 + y^2}$; the velocity components are $v_x = -\omega y$ and $v_y = \omega x$. The continuity equation is satisfied identically. Euler's equation reads

$$\rho\omega^2 x = \frac{\partial p}{\partial x} \ , \ \rho\omega^2 y = \frac{\partial p}{\partial y} \ ,$$
$$\frac{\partial p}{\partial z} + \rho g = 0 \ . \tag{2.112}$$

The solution is

$$p = \frac{1}{2}\rho\omega^2 r^2 - \rho g z + const \ . \tag{2.113}$$

For $p = 0$ we get the free surface, $z = \omega^2 r^2/2g$, which is a parabolloid, with the origin $z = 0$ at the lowest point of the surface. We note that $curl\boldsymbol{v} \neq 0$.

The forces acting upon a fluid element are the gravity force ρg and the centrifugal force $\rho\omega^2 r$. The equation of equilibrium is given by $dr \sim \rho g$ and $dz \sim \rho\omega^2 r$ (normal to the surface); hence, $\frac{dz}{dr} = \frac{\omega^2}{g} r$ and $z = \frac{\omega^2}{2g} r^2 + const$.

Let us consider a homogeneous sphere with density ρ, mass M and radius R. Its gravitational energy $E = \rho^2 G \int d\boldsymbol{R} d\boldsymbol{R}' \frac{1}{|\boldsymbol{R}-\boldsymbol{R}'|}$ gives $E = 2\pi\rho G M R^2$, where $G = 6.67 \times 10^{-8} cm^3/g \cdot s^2$ is the gravitational constant. Under a rotation with angular velocity ω the sphere may suffer a flattening, with a relative variation of the radius $\varepsilon = \delta R/R$. The change $\delta E = 4\pi\rho G M R^2 \cdot \varepsilon$ is compensated by the centrifugal energy $\frac{1}{2} M\omega^2 R^2$; hence, we get the change of the radius $\varepsilon = 10^{-5}$ for $\rho = 5g/cm^3$ and $\omega = 2\pi/24hours$. This is known as the figure of the Earth, a problem which comes from Newton. The actual figure is $\varepsilon = 3 \times 10^{-3}$ ($\pm 20km$ for Earth's mean radius $R = 6370km$), which is an appreciable discrepancy.

2.16 Flow past a sphere and a cylinder

Let us consider a sphere with radius R moving with velocity $\boldsymbol{u}$ in an incompressible ideal fluid. The fluid velocity vanishes at infinity. The flow may be viewed as a potential flow. The velocity derives from a potential, $\boldsymbol{v} = grad\Phi$, and the continuity equation reads $\Delta\Phi = 0$. The solutions vanishing at infinity are $1/r$ and the derivatives of $1/r$, where the radius r has the origin at the centre of the sphere. The symmetry of the problem requires only one vector in the solution, the linearity of the equations and the boundary conditions requires only a linear dependence on $\boldsymbol{u}$. Consequently, the potential has the form

$$\Phi = A\frac{\boldsymbol{ur}}{r^3} , \tag{2.114}$$

where the constant A is determined from the boundary conditions $v_n = \frac{\partial \Phi}{\partial n} = u_n$, *i.e.* the equality of the normal velocities at the surface of the sphere. These boundary conditions give $A = -R^3/2$. The velocity is

$$\boldsymbol{v} = \frac{R^3}{2r^3}\left[\frac{3(\boldsymbol{ur})\boldsymbol{r}}{2} - \boldsymbol{u}\right] , \; r > R . \tag{2.115}$$

If θ is the angle made by $\boldsymbol{u}$ with $\boldsymbol{r}$, the velocity at the two points $\theta = 0, \pi$ is the same, equal to u.

The pressure is given by equation (2.108),

$$p = p_0 - \frac{1}{2}\rho v^2 - \rho\frac{\partial \Phi}{\partial t} , \tag{2.116}$$

where p_0 is the pressure at infinity. The time derivative of the potential is computed from the time dependence of the parameters of the potential, which are $\boldsymbol{u}$ and the position of the origin $\boldsymbol{r}_0$ (centre of the sphere ($\Phi(\boldsymbol{u}, \boldsymbol{r} - \boldsymbol{r}_0)$; $\boldsymbol{r}_0$ moves with velocity $\boldsymbol{u}$. We get

$$\frac{\partial \Phi}{\partial t} = \frac{\partial \Phi}{\partial \boldsymbol{u}}\dot{\boldsymbol{u}} - \boldsymbol{uv} . \tag{2.117}$$

The pressure on the surface of the sphere is

$$p = p_0 + \frac{1}{8}\rho u^2(9\cos^2\theta - 5) + \frac{1}{2}\rho \boldsymbol{R}\dot{\boldsymbol{u}} . \tag{2.118}$$

For an infinite cylinder moving perpendicular to its axis the solutions of the Laplace equation are $\ln r$ and its derivatives. The potential is

$$\Phi = -R^2\frac{\boldsymbol{ur}}{r^2} , \tag{2.119}$$

the velocity is

$$\boldsymbol{v} = \frac{R^2}{r^2}\left[\frac{2(\boldsymbol{ur})\boldsymbol{r}}{r^2} - \boldsymbol{u}\right] \ , \ r > R \tag{2.120}$$

and the pressure on the surface is

$$p = p_0 + \frac{1}{2}\rho u^2(4\cos^2\theta - 3) + \rho \boldsymbol{R}\dot{\boldsymbol{u}} \ . \tag{2.121}$$

2.17 Stagnation point

A potential in the neighbourhood of a stagnation point has the expansion

$$\Phi = ax + by + cz + Ax^2 + By^2 + Cz^2 + Dxy + Eyz + Fzx \ . \tag{2.122}$$

The small region near the stagnation point can be approximated by a plane (x, y-coordinates). The potential should satisfy the Laplace equation, which gives $A + B + C = 0$. The boundary conditions are $v_z = \frac{\partial \Phi}{\partial z} = 0$ for $z = 0$ and all x, y and $\frac{\partial \Phi}{\partial x} = \frac{\partial \Phi}{\partial y} = 0$ for $x = y = z = 0$ (stagnation point). We get

$$\Phi = Ax^2 + By^2 - (A + B)z^2 \tag{2.123}$$

(the xy-term can be removed by a suitable choice of the x, y-axes). If the flow is axially symmetric about the z-axis, we have $A = B$. The velocities are

$$v_x = 2Ax \ , \ v_y = 2Ay \ , \ v_z = -4Az \ . \tag{2.124}$$

The streamlines are given by $dx/v_x = dy/v_y = dz/v_z$, which give $x^2 z = const$, $y^2 z = const$. If the flow is uniform along the y-direction, we have $B = 0$ and the streamlines $xy = const$.

2.18 Flow near an angle

Let us consider an angle α with polar coordinates r and θ. The Laplace equation for the potential in polar coordinates is

$$\Delta \Phi = \frac{1}{r}\frac{\partial}{\partial r}\left(r\frac{\partial \Phi}{\partial r}\right) + \frac{1}{r^2}\frac{\partial^2 \Phi}{\partial \theta^2} = 0 \ . \tag{2.125}$$

The solution is

$$\Phi = Ar^n \cos n\theta \ , \tag{2.126}$$

where n is positive; it corresponds to an angle with small dimensions (this is why the constant A remains undetermined). The normal velocity $v_\theta = \frac{\partial\Phi}{r\partial\theta}$ should be zero for $\theta = 0$ and $\theta = \alpha$; we get $n = \pi/\alpha$ and the velocity components

$$v_r = nAr^{n-1}\cos n\theta \ , \ v_\theta = -nAr^{n-1}\sin n\theta \ . \tag{2.127}$$

For $n > 1$ ($\alpha < \pi$) the velocity is vanishing at the origin, for $0 < n < 1$ ($\alpha > \pi$) the velocity is infinite at the origin. The stream function is given by $\frac{\partial\psi}{\partial r} r d\theta - \frac{\partial\psi}{r\partial\theta} dr = 0$, *i.e.* $\psi = Ar^n \sin n\theta$, and the streamlines are given by $r^n \sin n\theta = const.$

2.19 Hole in fluid

Let us assume a hole with radius a in an incompressible ideal fluid. The fluid flows towards its centre with radial velocity v.[38] The Euler equation is

$$\frac{\partial v}{\partial t} + v\frac{\partial v}{\partial r} = -\frac{1}{\rho}\frac{\partial p}{\partial r} \ . \tag{2.128}$$

The integration of the continuity equation $div\mathbf{v} = 0$ over a spherical surface of radius r gives $r^2 v = F(t)$, where $F(t)$ is a function of time. From equation (2.128) we get

$$\frac{\dot{F}}{r^2} + v\frac{\partial v}{\partial r} = -\frac{1}{\rho}\frac{\partial p}{\partial r} \ . \tag{2.129}$$

We integrate this equation from R to infinity, where the pressure is p_0. The pressure on the surface of the hole is zero. Equation (2.129) gives

$$-\frac{\dot{F}}{R} + \frac{1}{2}V^2 = \frac{p_0}{\rho} \ , \tag{2.130}$$

[38] W. H. Besant, *Hydrostatics and Hydrodynamics*, Deighton, Bell and Co., Cambridge (1859). Lord Rayleigh, "On the pressure developed in a liquid during the collapse of a spherical cavity", Phil. Mag. **34** 94 (1917).

where V is the velocity of the surface of the hole at R. We insert $\dot{F} = R^2\dot{V} + 2RV^2$ in this equation,

$$-\frac{3}{2}V^2 - \frac{1}{2}R\frac{dV^2}{dR} = \frac{p_0}{\rho} . \qquad (2.131)$$

The solution is $V^2 = const/R^3 - 2p_0/3\rho$. From the boundary condition $V = 0$ for $R = a$, we get

$$V = -\sqrt{\frac{2p_0}{3\rho}\left(\frac{a^3}{R^3} - 1\right)} . \qquad (2.132)$$

We note that $V \to \infty$ for $R \to 0$. The time needed to fill the hole is

$$\tau = \sqrt{\frac{3\rho}{2p_0}}a\int_1^\infty \frac{dx}{x^2\sqrt{x^3-1}} \simeq \sqrt{\frac{\rho}{p_0}}a . \qquad (2.133)$$

From equation (2.132) we can derive the function $R(t)$ (for zero pressure on the surface of the hole). If we know the function $R(t)$, equation (2.131) becomes

$$-\frac{3}{2}V^2 - \frac{1}{2}R\frac{dV^2}{dR} = \frac{p_0 - p}{\rho} . \qquad (2.134)$$

From this equation we get the pressure

$$p = p_0 + \frac{1}{2}\rho\left[\frac{d^2(R^2)}{dt^2} + \left(\frac{dR}{dt}\right)^2\right] . \qquad (2.135)$$

If we use the function $R(t)$ given by equation (2.132) (for zero pressure on the surface of the hole), we get a pressure of the order p_0 after the filling time τ. This is the phenomenon of cavity.

2.20 Drag force

Let us consider a body moving with velocity $\boldsymbol{u}$ in an incompressible ideal fluid. We are interested in the potential flow past the body, at large distances from the body. The velocity potential Φ is given by the Laplace equation $\Delta\Phi = 0$ with suitable boundary conditions on the surface of the body. The solution is

$$\Phi = \frac{a}{r} + \boldsymbol{A}grad\frac{1}{r} + ... \ , \tag{2.136}$$

where a and $\boldsymbol{A}$ are some constants, and only the leading terms are retained. The origin is taken somewhere inside the body. The laplacian of $\frac{1}{r}$ is $-4\pi\delta(\boldsymbol{r})$, which, integrated, gives a non-zero result; this would mean that the mass is not conserved. Therefore, $a = 0$. The laplacian of the second term is -$4\pi\boldsymbol{A}grad\delta(\boldsymbol{r})$, which, integrated, is zero. It follows the potential

$$\Phi = -\frac{\boldsymbol{An}}{r^2} \tag{2.137}$$

and the velocity

$$v = \frac{3(\boldsymbol{An})\boldsymbol{n} - \boldsymbol{A}}{r^3} \ , \tag{2.138}$$

where $\boldsymbol{n} = \boldsymbol{r}/r$ is the unit vector along the position vector $\boldsymbol{r}$.
The energy of the fluid is the integral of $\frac{1}{2}\rho v^2$ over the whole space occupied by the fluid, where ρ is the density. Since we do not know the velocity on the surface of the body, we write v^2 as $(\boldsymbol{v}+\boldsymbol{u})(\boldsymbol{v}-\boldsymbol{u}) + u^2$ and $\boldsymbol{v}+\boldsymbol{u} = grad(\Phi + \boldsymbol{ur})$, since $\boldsymbol{u}$ does not depend on the coordinates. The integral of the u^2-term gives $\frac{1}{2}\rho(V - V_0)u^2$, where V is the volume of a sphere with a large radius R and V_0 is the volume of the body. Moreover, since $div\boldsymbol{v} = 0$ and $div\boldsymbol{u} = 0$, we may write $grad(\Phi + \boldsymbol{ur}) \cdot (\boldsymbol{v} - \boldsymbol{u})$ as $div[(\Phi + \boldsymbol{ur})(\boldsymbol{v} - \boldsymbol{u})]$. We have the energy

$$\begin{aligned} E &= \tfrac{1}{2}\rho(V - V_0)u^2 + \textstyle\int dV div[(\Phi + \boldsymbol{ur})(\boldsymbol{v} - \boldsymbol{u})] = \\ &= \tfrac{1}{2}\rho(V - V_0)u^2 + \textstyle\oint d\boldsymbol{f}[(\Phi + \boldsymbol{ur})(\boldsymbol{v} - \boldsymbol{u})] \ , \end{aligned} \tag{2.139}$$

where the surface integral is performed both over the sphere with radius R and the surface of the body. On the surface of the body the normal velocities are equal ($v_n = u_n$), therefore we are left only with the integral over the surface at infinity. Keeping only the leading contributions we get

$$E = \frac{1}{2}\rho(V - V_0)u^2 + \frac{1}{2}\rho\int[3(\boldsymbol{An})(\boldsymbol{un}) - (\boldsymbol{un})^2R^3]do \ , \tag{2.140}$$

where we write the element of area as $d\boldsymbol{f} = R^2\boldsymbol{n}do$, with the element of the solid angle do. Effecting the integration, we get finally

$$E = \frac{1}{2}\rho(4\pi \boldsymbol{A}\boldsymbol{u} - V_0 u^2) \ , \tag{2.141}$$

where we put $V = \frac{4\pi}{3}R^3$.

The vector $\boldsymbol{A}$ is given by the boundary conditions. The boundary conditions are linear both in Φ and u. Therefore, the components of the vector $\boldsymbol{A}$ must be of the form $A_i = a_{ij}u_j$, where a_{ij} is some symmetric matrix. It follows that the energy can be written as

$$E = \frac{1}{2}m_{ij}u_i u_j \ , \ m_{ij} = \rho(4\pi a_{ij} - V_0\delta_{ij}) \ . \tag{2.142}$$

The matrix m_{ij} is called the induced-mass tensor. For a sphere with radius R_0 we get $a_{ij} = \frac{1}{2}R_0^3\delta_{ij}$ and $m_{ij} = \frac{2\pi}{3}\rho R_0^3$, *i.e.* half of the mass of the displaced fluid (equation (2.114)).

Let us compute now the momentum $\boldsymbol{P}$ of the fluid. It is the momentum given to the fluid by the moving body, *i.e.* $dE = \boldsymbol{u}d\boldsymbol{P}$, according to the conservation laws of energy and momentum; it is not the integral of $\rho\boldsymbol{v}$ over the whole space (which is divergent), because $\boldsymbol{v}d(\rho\boldsymbol{v}) = d(\frac{1}{2}\rho v^2)$ is an identity. We get $P_i = m_{ij}u_j$, or

$$\boldsymbol{P} = \rho(4\pi\boldsymbol{A} - V_0\boldsymbol{u}) \ . \tag{2.143}$$

The force acting upon the body is

$$\boldsymbol{F} = -\frac{d\boldsymbol{P}}{dt} \ . \tag{2.144}$$

The component of this force directed along the velocity $\boldsymbol{u}$ is called the drag force; the component transverse to $\boldsymbol{u}$ is called the lift force.

If the body is moving uniformly in the fluid ($\boldsymbol{u} = const$), the force acting upon it is zero. This counter-intuitive result is known as the paradox of d'Alembert. Indeed, a force is present only when momentum is lost by dissipation or at infinity, which is not the case in ideal fluids. However, this is true in infinite fluids. In finite fluids, energy is dissipated by surface waves (wave drag). Also, the friction forces acting at the surface of the body give a force. We can see again that the ideal fluid has serious drawbacks.

Let us assume that a large body executes small oscillations in a fluid under the action of an external force $\boldsymbol{f}$. The flow of the fluid may be viewed as potential, so the equation of motion of the body is

$$M\frac{d\boldsymbol{u}}{dt}+\frac{d\boldsymbol{P}}{dt}=\boldsymbol{f} \tag{2.145}$$

or

$$\frac{du_j}{dt}(M\delta_{ij}+m_{ij})=f_i \ , \tag{2.146}$$

where M is the mass of the body.

Let us now assume that the fluid executes an oscillatory motion. We consider that the velocity of the fluid varies slightly over the dimension of the body. Let $\boldsymbol{v}$ be the velocity of the fluid at the position of the body. If the body will be wholly carried by the fluid, then the force acting upon it would be $\rho V_0\frac{d\boldsymbol{v}}{dt}$, *i.e.* the force acting upon the mass of fluid displaced by the body (V_0 is the volume of the body, ρ is the density of the fluid). But the body has a relative motion with respect to the fluid, so the fluid will acquire an additional momentum with components $m_{ij}(u_j-v_j)$, where $\boldsymbol{u}$ is the velocity of the body. Therefore, the total force acting upon the body has the components

$$\rho V_0\frac{dv_i}{dt}-m_{ij}\frac{d}{dt}(u_j-v_j) \ , \tag{2.147}$$

and the equation of motion of the body is

$$M\frac{du_i}{dt}=\rho V_0\frac{dv_i}{dt}-m_{ij}\frac{d}{dt}(u_j-v_j) \ . \tag{2.148}$$

The integral of this equation gives

$$(m_{ij}+M\delta_{ij})u_j=(m_{ij}+\rho V_0\delta_{ij})v_j \ . \tag{2.149}$$

From this equation we can compute the velocity $\boldsymbol{u}$ of the body. We note that if the density of the body is equal to that of the fluid ($M=\rho V_0$), the body moves with the velocity of the fluid ($\boldsymbol{u}=\boldsymbol{v}$), as expected. For a sphere with density ρ_0, the velocity is

$$\boldsymbol{u}=\frac{3\rho}{\rho+2\rho_0}\boldsymbol{v} \ . \tag{2.150}$$

If $\rho < \rho_0$ the sphere lags behind the fluid, if $\rho > \rho_0$, the sphere goes ahead. The drag force acting upon the sphere is

$$\boldsymbol{F} = -\frac{2\pi}{3}\rho R_0^3 \frac{d\boldsymbol{u}}{dt} \tag{2.151}$$

(equation (2.144)).

Finally we note that a rotation with angle $\delta\boldsymbol{\theta}$ of the fluid, gives rise to a change of velocity $\delta\boldsymbol{u} = -\delta\boldsymbol{\theta} \times \boldsymbol{u}$ of the velocity of the body. Therefore, the change in energy becomes

$$\delta E = \delta\boldsymbol{u}\boldsymbol{P} = -\delta\boldsymbol{\theta}(\boldsymbol{u} \times \boldsymbol{P}) = \boldsymbol{M}\delta\boldsymbol{\theta} \ , \tag{2.152}$$

where

$$\boldsymbol{M} = -\boldsymbol{u} \times \boldsymbol{P} = 4\pi\rho\boldsymbol{A} \times \boldsymbol{u} \tag{2.153}$$

is the moment of forces acting upon the body (the torque).

2.21 Gravity waves

2.21.1 Surface waves

Under the action of the (uniform) gravitational field the free plane surface of a fluid can sustain waves (tidal waves).[39] The two terms in Euler's equation $\frac{\partial \boldsymbol{v}}{\partial t}$ and $(\boldsymbol{v}grad)\boldsymbol{v}$ are estimated by $\frac{v}{\tau}$ and $\frac{v}{l}v$, where τ is the time of variation of the velocity and l is the length of variation of the velocity. $a = v\tau$ is the length over which the fluid particles move in time τ. We can see that the ratio of the two terms is $\frac{a}{l}$. Obviously, a is of the order of the wave amplitude and l is of the order of the wavelength λ. If $a \ll \lambda$ we may neglect the term $(\boldsymbol{v}grad)\boldsymbol{v}$ in Euler's equation. In this case the flow is potential. Similarly, by dividing by τ the inequality $a \ll \lambda$, we get $v \ll \frac{\lambda}{\tau}$, which is the wave velocity. Consequently, we may neglect the squared velocity in comparison with velocity. Under these conditions, for an incompressible fluid, the integral of motion is given by equation (2.108)

$$\frac{\partial\Phi}{\partial t} + \frac{p}{\rho} + gz = 0 \ . \tag{2.154}$$

[39] H. Lamb, *Hydrodynamics*, Cambridge University Press, London (1879).

At the surface $z = 0$ the pressure is p_0 and the surface suffers a small vertical displacement ζ. The pressure p_0 can be absorbed in the time derivative of the potential. The above equation becomes

$$\left(\frac{\partial\Phi}{\partial t}\right)_\zeta + g\zeta = 0 \ . \tag{2.155}$$

On the other hand, the vertical velocity is

$$v_z = \frac{\partial\Phi}{\partial z} = \frac{\partial\zeta}{\partial t} = -\frac{1}{g}\left(\frac{\partial^2\Phi}{\partial t^2}\right)_\zeta \ , \tag{2.156}$$

or

$$\left(\frac{\partial\Phi}{\partial z}\right)_0 + \frac{1}{g}\left(\frac{\partial^2\Phi}{\partial t^2}\right)_0 = 0 \ , \tag{2.157}$$

an equality which can be viewed as a boundary condition for the Laplace equation $\Delta\Phi = 0$.

The equation $\Delta\Phi = 0$ with the boundary condition given by equation (2.157) is a vibration equation. The boundary condition arises from the Euler equation; it can be extended to any z and it may include an external force (see the Surface Waves chapter). We can impose causal conditions upon this equation. Since the constant gravitational field acts upon a limited depth, the solution is proportional to e^{kz}, for $k > 0$. Then, the equation $\Delta\Phi = 0$ gives plane waves $e^{i\mathbf{kr}}$, or Bessel functions, with the spatial dependence separated from the temporal dependence. Equation (2.157) becomes the equation of a harmonic oscillator, which may obey causal conditions. We follow the customary nomenclature and call the solutions waves. Since they are proportional to e^{kz}, they are surface waves. In addition, they are dispersive waves. Therefore, we adopt as an elementary solution

$$\Phi = Ae^{kz}\cos(kx - \omega t) \ , \tag{2.158}$$

where $k > 0$ and x is a coordinate along the surface; this is a surface wave. The boundary condition (2.157) gives

$$\omega^2 = gk \ . \tag{2.159}$$

The surface gravity waves are dispersive waves, which "propagate" with the group velocity $U = \frac{1}{2}\sqrt{g/k}$. From $v_x = \partial\Phi/\partial x$ and $v_z = \partial\Phi/\partial z$ we get the trajectory of a fluid particle in the gravity wave

$$\begin{aligned} x - x_0 &= -\tfrac{Ak}{\omega}e^{kz_0}\cos(kx_0 - \omega t) \ , \\ z - z_0 &= -\tfrac{Ak}{\omega}e^{kz_0}\sin(kx_0 - \omega t) \end{aligned} \tag{2.160}$$

in the neighbourhood of a point with coordinates x_0, z_0. The trajectories are circles with small radii, which decrease with increasing depth. The gravity waves are discussed also in the Surface Waves chapter.

2.21.2 Long gravity waves

The surface gravity waves have a much shorter wavelength than the depth of the fluid, which is supposed infinite. Now, let us assume a channel infinitely long in the x-direction, with a small cross-section S_0 in comparison with the wavelength. We consider waves which propagate along the x-direction. The component $v_x = v$ of the velocity of the fluid particles is much larger than the components $v_{y,z}$. The Euler equations are

$$\tfrac{\partial v}{\partial t} = -\tfrac{1}{\rho}\tfrac{\partial p}{\partial x} \ , \ \tfrac{\partial p}{\partial z} + \rho g = 0 \ . \tag{2.161}$$

The quadratic terms in velocity are omitted, because the amplitude of the waves is much shorter than their wavelengths. The second equation (2.161) gives

$$p = p_0 + \rho g(\zeta - z) \ , \tag{2.162}$$

where the free surface in equilibrium has the equation $z = \zeta$ and p_0 is the equilibrium pressure (incompressible fluids); of course, ζ is close to z. Inserting this pressure in the first equation (2.161) leads to

$$\frac{\partial v}{\partial t} = -g\frac{\partial \zeta}{\partial x} \ . \tag{2.163}$$

Let S be the cross-section of the fluid; the change in volume per unit time is

$$(Sv)_{x+dx} - (Sv)_x = \frac{\partial (Sv)}{\partial x}dx = -\frac{\partial S}{\partial t}dx \ , \tag{2.164}$$

hence the continuity equation

$$\frac{\partial S}{\partial t} + \frac{\partial (Sv)}{\partial x} = 0 \ . \tag{2.165}$$

The cross-section S can be written as $S = S_0 + b(\zeta - z)$, where b is the width of the channel. Equation (2.165) becomes

$$b\frac{\partial \zeta}{\partial t} + \frac{\partial (S_0 v)}{\partial x} = 0 \ , \tag{2.166}$$

or

$$\frac{\partial^2 \zeta}{\partial t^2} - \frac{g}{b}\frac{\partial}{\partial x}\left(S_0 \frac{\partial \zeta}{\partial x}\right) = 0 \ , \tag{2.167}$$

where equation (2.163) is used. For a uniform S_0 this is the wave equation for waves which propagate with velocity $U = \sqrt{gS_0/b}$.

We can consider waves propagating on an interface of two fluids. For an interface the pressure and the normal velocity should be continuous.

Let us assume now that we have a tank with the free surface in the x, y-plane and depth h_0. The component v_z of the velocity is small. The Euler equations read

$$\frac{\partial v_x}{\partial t} + g\frac{\partial \zeta}{\partial x} = 0 \ , \ \frac{\partial v_y}{\partial t} + g\frac{\partial \zeta}{\partial y} = 0 \tag{2.168}$$

and the equation of continuity is

$$\frac{\partial h}{\partial t} + \frac{\partial (hv_x)}{\partial x} + \frac{\partial (hv_y)}{\forall y} = 0 \ , \tag{2.169}$$

or, writing $h = h_0 + \zeta$,

$$\frac{\partial \zeta}{\partial t} + \frac{\partial (h_0 v_x)}{\partial x} + \frac{\partial (h_0 v_y)}{\forall y} = 0 \ . \tag{2.170}$$

By using equations (2.168) we get the wave equation

$$\frac{\partial^2 \zeta}{\partial t^2} - gh_0\left(\frac{\partial^2 \zeta}{\partial x^2} + \frac{\partial^2 \zeta}{\partial y^2}\right) = 0 \tag{2.171}$$

for waves which propagate with velocity $U = \sqrt{gh_0}$.

2.22 Internal gravity waves

In equilibrium the pressure and the density gradients are related by $grad p = (\partial p/\partial \rho)_s grad \rho$, where $\sqrt{(\partial p/\partial \rho)_s} = c$ is the sound velocity (s being the entropy). In uniform gravitational field $grad p$ is ρg, so the relative variation of the density over a height h is of the order $\delta\rho/\rho = gh/c^2$. For atmospheric air $h \simeq 10km$, for water $h \simeq 200km$. It follows that for reasonably small heights the density varies little, so we may view the fluid in equilibrium in uniform gravitational field as incompressible. On the other hand, the pressure and the entropy vary over these distances. Any fluid particle displaced in height carries with it its old entropy, which differs from the entropy at the new position; the mechanical equilibrium is destroyed and the particle may oscillate.[40] These oscillations may be termed free convection oscillations.

We assume that the wavelengths are small in comparison with distances over which the fluid in equilibrium may be viewed as incompressible. In motion, the pressure, the entropy and the density suffer small changes with respect to their equilibrium values. We write the entropy as $s = s_0 + s'$, where s_0 is the equilibrium entropy; as all the equilibrium quantities, it is a function of the vertical distance z. The equation of conservation of entropy is

$$\frac{\partial s'}{\partial t} + \boldsymbol{v} grad s_0 = 0 \ , \tag{2.172}$$

where the velocity is considered small in comparison with the wave velocity. Similarly, the quadratic term in Euler's equation is neglected. We write the pressure as $p = p_0 + p'$, the density as $\rho = \rho_0 + \rho'$, where all the quantities with suffix 0 are at equilibrium, and $grad p_0 = \rho_0 \boldsymbol{g}$. We note that $\boldsymbol{g}$ is directed downwards. Euler's equation reads

$$\begin{aligned} \frac{\partial \boldsymbol{v}}{\partial t} &= -\frac{1}{\rho} grad p + \boldsymbol{g} = \\ &= -\frac{1}{\rho_0} grad p_0 + \boldsymbol{g} - \frac{1}{\rho_0} grad p' + \frac{\rho'}{\rho_0^2} grad p_0 = \\ &= -\frac{1}{\rho_0} grad p' + \frac{\rho'}{\rho_0^2} grad p_0 \ . \end{aligned} \tag{2.173}$$

[40] L. Landau and E. Lifshitz, *Course of Theoretical Physics*, vol. 5 (*Fluid Mechanics*), Pergamon, Oxford (1987).

The change in density is due only to the change in entropy (buoyancy force), so we can write

$$\rho' = \left(\frac{\partial \rho_0}{\partial s_0}\right)_p s' \; ; \tag{2.174}$$

the Euler equation becomes

$$\frac{\partial \boldsymbol{v}}{\partial t} = \frac{\boldsymbol{g}}{\rho_0}\left(\frac{\partial \rho_0}{\partial s_0}\right)_p s' - grad\frac{p'}{\rho_0} \; . \tag{2.175}$$

The equation of continuity reads

$$div\boldsymbol{v} = 0 \; . \tag{2.176}$$

We have three equations (2.172), (2.175) and (2.176) with three unknowns s', p' and $\boldsymbol{v}$. These equations describe vibrations. As it is customary, we seek a solution of the form $s', p', \boldsymbol{v} \sim e^{i\boldsymbol{kr}-\omega t}$. Equation (2.176) gives $\boldsymbol{vk} = 0$, *i.e.* the velocity wave is transverse.

It is convenient to introduce the suffix $\|$ for vertical components and the suffix $\perp$ for horizontal components. By making use of the plane-wave solutions, we get

$$s' = -i\frac{v_\|}{\omega_0 \sin\theta}\frac{ds_0}{dz} \; , \; p' = \frac{gv_\|}{k\omega_0}\cot\theta\left(\frac{\partial \rho_0}{\partial s_0}\right)_p \frac{ds_0}{dz} \; , \qquad v_\perp = -v_\| \cot\theta \tag{2.177}$$

$(\theta \neq 0, \pi)$, where θ is the angle made by the wavevector $\boldsymbol{k}$ and the vertical axis; the frequency is given by the dispersion relation

$$\omega^2 = \omega_0^2 \sin^2\theta \; , \; \omega_0^2 = -\frac{g}{\rho_0}\left(\frac{\partial \rho_0}{\partial s_0}\right)_p \frac{ds_0}{dz} \; . \tag{2.178}$$

We show below that $\omega_0^2 > 0$. The velocity amplitude $v_\|$ remains undetermined. We can see that the frequency depends only on direction, not on wavelength. Also, waves with a vertical wavevector ($\theta = 0$) cannot exist. The group velocity can be calculated by using

$$\sin^2\theta = 1 - (\boldsymbol{\nu k}/k)^2 \; , \tag{2.179}$$

where $\boldsymbol{\nu}$ is the vertical unit vector. This relation gives $\delta\boldsymbol{\theta}$ as a function of $\delta\boldsymbol{k}$. We get

$$\boldsymbol{U} = -\frac{\omega_0^2}{\omega k}(\boldsymbol{\nu n})\left[\boldsymbol{\nu} - (\boldsymbol{\nu n})\boldsymbol{n}\right] \ , \tag{2.180}$$

where $\boldsymbol{n} = \boldsymbol{k}/k$. The group velocity is transverse, $\boldsymbol{Uk} = 0$, its vertical component is $U_{\parallel} = -\frac{\omega_0}{k}\sin\theta\cos\theta$ and its magnitude is $U = \frac{\omega_0}{k}\cos\theta$. Let us compute now the equilibrium thermodynamic quantities. *i.e.* those quantities which carry the suffix 0; we leave aside this suffix in our calculations below. First, it is easy to see that

$$\left(\frac{\partial\rho}{\partial s}\right)_p = \frac{T}{c_p}\left(\frac{\partial\rho}{\partial T}\right)_p \ , \tag{2.181}$$

where T is the temperature and c_p is the specific heat at constant pressure per unit mass. Then, the Gibbs free energy is given by $d\Phi = Vdp - SdT$. In equilibrium the temperature is constant, so the change in entropy is given only by the change in pressure,

$$\frac{ds}{dz} = \left(\frac{\partial s}{\partial p}\right)_T\frac{dp}{dz} = -\rho g\left(\frac{\partial s}{\partial p}\right)_T \ . \tag{2.182}$$

From the Gibbs free energy we get

$$\left(\frac{\partial s}{\partial p}\right)_T = -\left(\frac{\partial V}{\partial T}\right)_p = \frac{1}{\rho^2}\left(\frac{\partial\rho}{\partial T}\right)_p \ , \tag{2.183}$$

where V is the volume per unit mass. Therefore, we obtain

$$\frac{ds}{dz} = -\frac{g}{\rho}\left(\frac{\partial\rho}{\partial T}\right)_p \ . \tag{2.184}$$

By making use of equations (2.181) and (2.184), equation (2.178) gives

$$\omega_0^2 = \frac{g^2T}{\rho^2c_p}\left(\frac{\partial\rho}{\partial T}\right)_p^2 \ . \tag{2.185}$$

We can see that $\omega_0^2 > 0$. For a monoatomic ideal gas the equation of state is $p = \frac{\rho}{M}T$, where M is the atomic mass, such that $(\partial\rho/\partial T)_p = -\rho/T$ and $\omega_0^2 = g^2/c_pT$.

2.23 Inertial waves

Let us consider an incompressible ideal fluid rotating about the z-axis with a constant angular velocity Ω. It is convenient to use the rotating axes of coordinates, where the centrifugal term $\frac{1}{2}(\boldsymbol{\Omega}\times\boldsymbol{r})^2$ and the Coriolis force $2\boldsymbol{v}\times\boldsymbol{\Omega}$ occur (accelerations). We may include the centrifugal term in the pressure $P=p-\frac{1}{2}(\boldsymbol{\Omega}\times\boldsymbol{r})^2$ and write the Euler equation as

$$\frac{\partial\boldsymbol{v}}{\partial t}-2\boldsymbol{v}\times\boldsymbol{\Omega}=-\frac{1}{\rho}gradP \ , \tag{2.186}$$

where we neglect the quadratic term in velocity, since we consider waves with small amplitude in comparison with their wavelength (small oscillations). The equation of continuity is $div\boldsymbol{v}=0$. By taking the *curl* of this equation we get

$$\frac{\partial}{\partial t}curl\boldsymbol{v}=2(\boldsymbol{\Omega}grad)\boldsymbol{v} \ , \tag{2.187}$$

or

$$\frac{\partial}{\partial t}curl\boldsymbol{v}=2\Omega\frac{\partial\boldsymbol{v}}{\partial z} \ . \tag{2.188}$$

We seek a solution of this equation in the form $\boldsymbol{v}=\boldsymbol{A}e^{i(\boldsymbol{kr}-\omega t)}$, where $\boldsymbol{Ak}=0$. Equation (2.188) gives

$$\omega\boldsymbol{k}\times\boldsymbol{v}=2i\Omega k_z\boldsymbol{v} \ , \tag{2.189}$$

whence

$$-\omega k^2\boldsymbol{v}=2i\Omega k_z\boldsymbol{k}\times\boldsymbol{v} \tag{2.190}$$

and

$$\omega=2\Omega\cos\theta \ , \tag{2.191}$$

where θ is the angle made by the wavevector $\boldsymbol{k}$ and the z-axis ($\boldsymbol{\Omega}$). Therefore, equation (2.189) reads

$$\boldsymbol{n}\times\boldsymbol{v}=i\boldsymbol{v} \ , \tag{2.192}$$

where $\boldsymbol{n}=\boldsymbol{k}/k$. We write $\boldsymbol{A}=\boldsymbol{a}+i\boldsymbol{b}$, such that $\boldsymbol{n}\times\boldsymbol{b}=\boldsymbol{a}$. Therefore, the vectors $\boldsymbol{a}$ and $\boldsymbol{b}$ are equal in magnitude, perpendicular to each other and both are perpendicular to $\boldsymbol{k}$. We denote their axes by x and y, such that the velocity components are $v_x = a\cos(\boldsymbol{kr} - \omega t)$ and $v_y = -a\sin(\boldsymbol{kr} - \omega t)$. The velocity rotates in the plane perpendicular to $\boldsymbol{k}$, *i.e.* the velocity wave is circularly polarized. From $\cos\theta = \boldsymbol{\nu k}/k$, where $\boldsymbol{\nu}$ is the unit vector along the z-direction, we get the group velocity

$$\boldsymbol{U} = \frac{2\Omega}{k}\left[\boldsymbol{\nu} - (\boldsymbol{\nu n})\boldsymbol{n}\right] \; ; \tag{2.193}$$

it is perpendicular to the wavevector and its magnitude is $U = \frac{2\Omega}{k}\sin\theta$. These waves are called inertial waves.

One particular form of inertial waves are axially symmetric waves, which are in the form $e^{i(kz-\omega t)}$. Equation (2.186) gives

$$\begin{gathered} -i\omega v_r - 2\Omega v_\varphi = -\frac{1}{\rho}\frac{\partial P}{\partial r} \; , \\ -i\omega v_\varphi + 2\Omega v_r = 0 \; , \\ -i\omega v_z = -i\frac{k}{\rho}P \end{gathered} \tag{2.194}$$

for the polar components of the velocity. The continuity equation reads

$$\frac{1}{r}\frac{\partial}{\partial r}(rv_r) + ikv_z = 0 \; . \tag{2.195}$$

From these equations we can express v_φ, v_z and P as functions of v_r, which satisfies the equation

$$\frac{d^2 v_r}{dr^2} + \frac{1}{r}\frac{dv_r}{dr} + \left[k^2\left(\frac{4\Omega^2}{\omega^2} - 1\right) - \frac{1}{r^2}\right]v_r = 0 \; . \tag{2.196}$$

The solution of this equation is the Bessel function $J_1\left(kr(4\Omega^2/\omega^2 - 1)\right)$ (apart from $e^{i(kz-\omega t)}$); it is vanishing at $r = 0$. The waves are localized in concentric cylinders with the radii given by the zeros of the function J_1. The frequency is not determined, but it should be smaller than 2Ω. On a cylindrical boundary of radius a the velocity v_r should vanish for $r = a$, a condition which gives vibration frequencies.[41] The solutions are in fact vibrations.

[41] W. Thomson, "Vibrations of a columnar vortex", Proc. Roy. Soc. Edinburgh **10** 443 (1880).

2.24 Rapidly rotating fluid

In steady flow of a rotating fluid the term $(\boldsymbol{v}grad)\boldsymbol{v}$ is of the order u^2/l and the term $\boldsymbol{v}\times\boldsymbol{\Omega}$ in equation (2.186) is of the order $u\Omega$, where u is the velocity and l is the distance over which it varies. If $u/l\Omega \ll 1$ (rapidly rotating fluid), we may neglect the quadratic term in velocity, and Euler's equation becomes

$$2v_y\Omega = \frac{1}{\rho}\frac{\partial P}{\partial x} \ , \quad -2v_x\Omega = \frac{1}{\rho}\frac{\partial P}{\partial y} \ ,$$
$$\frac{\partial P}{\partial z} = 0 \ . \tag{2.197}$$

It follows that P does not depend on z and, consequently, $v_{x,y}$ do not depend on z. Moreover, since $\frac{\partial v_x}{\partial x}+\frac{\partial v_y}{\partial y}=0$ from the above equation, the continuity equation gives $\frac{\partial v_z}{\partial z}=0$. Therefore, the steady flow in rapid rotation consists of a transverse motion which does not depend on z and an independent uniform longitudinal flow.[42]

2.25 Entropic sound

According to the fundamental law of Thermodynamics $dE=-pdV+TdS$ the equilibrium variations of the energy E are governed by two independent variables: the volume V and the entropy S. The pressure $p=p(V,S)$ and the temperature $T=T(V,S)$ are functions of these two variables. A change δp in pressure is associated with a force, which produces a displacement of the particles, *i.e.* a change δV in volume. For a definite relationship between these two changes we get the motion which is called sound; the relationship between δp and δV gives the velocity of the sound. We can see that such a change is possible only for $\delta S=0$, *i.e.* an adiabatic sound, or $\delta T=0$, *i.e.* an isothermal sound. Since $\delta S\neq 0$ for $\delta T=0$, we prefer to call the latter an entropic sound.[43] In ideal continuous media the entropy is

[42] J. Proudman, "On the motion of solids in a liquid possessing vorticity", Proc. Roy. Soc. London **92** 408 (1916).

[43] B. F. Apostol, S. Stefan and M. Apostol, "Entropic sound in the atmosphere", arXiv: ao-sci/9606001v2 - 30 June 1996.

not changed, such that the sound in these media is adiabatic. In non-ideal media, other types of sound may appear, including the entropic sound.[44]

The sound is a succession of local compressions and rarefactions, which propagate. Let us suppose that a local compression occurs. It appears by an additional velocity given to the particles, so we expect an increase of the thermal agitation, *i.e.* an increase in temperature and a production of heat. It may happen that this increase of thermal agitation produces, entirely, a rarefaction (a dilatation), such that the compression is transmitted to the neighbouring particles, through pressure. The variations in volume, temperature and pressure are transmitted, *i.e.* the motion of the sound is transmitted integrally through the medium. The compensation of the volume change and the temperature change leaves the entropy constant, so we have an adiabatic sound. Indeed, the entropy is the logarithm of the number of available states, and this number is proportional to powers of the volume and the temperature. Also, it may happen that the thermal agitation is not transferred integrally upon the dilatation. For instance, it may be lost through thermoconduction, or, simply, the medium does not respond integrally to the motion of the particles, and then we get an uncompensated heat; this happens through viscosity. In this case the entropy changes. If the change is due entirely to the volume variations, then the temperature is constant, and the variations of the pressure transmit the motion. In this case we have the isothermal sound. If a particular relation exists between entropy and the temperature, as, for instance, arising from the interplay between thermoconductivity and the viscosity, we may have another sound, of a different nature. We can see that in ideal media, where the viscosity and the thermoconduction are absent, the entropy remains constant, and we have an adiabatic sound; while in non-ideal media, the origin of the sound is variable. As such, the entropic sound covers a larger area than the isothermal sound. However, the viscosity and the thermoconduction are irreversible processes and, in an isolated ensemble,

[44]M. Apostol and E. Preoteasa, "Density oscillations in a model of water and other similar liquids", Phys. Chem. Liquids **46** 653 (2008); M. Apostol, "Dynamics of collective modes in multi-component molecular mixtures", Phys. Chem. Liquids **47** 35 (2009); *Physical Kinetics*, Cambridge Scholars Publishing, Newcastle upon Tyne (2020).

they do not preserve the thermodynamic equilibrium, giving rise to absorption (damping) and retardation.

Let us suppose that in a continuous medium in equilibrium a field $\boldsymbol{u}(t,\boldsymbol{r})$ of local displacement occurs. It gives rise to a local change of volume $\delta V = V div\boldsymbol{u}$ and a local change of pressure

$$\delta p = -\frac{K}{V}\delta V = -K div\boldsymbol{u} \ , \tag{2.198}$$

where $K = -V\left(\frac{\partial p}{\partial V}\right)$ is the bulk modulus of compression. It follows that the equation of motion of the displacement $\boldsymbol{u}$ is

$$\frac{M}{V}\ddot{\boldsymbol{u}} = -grad\delta p = K grad\, div\boldsymbol{u} \ , \tag{2.199}$$

where M is the mass of a particle and V is the specific volume per particle. Since $\boldsymbol{u}$ derives from a gradient, $\boldsymbol{u} = grad\phi$, this equation is the wave equation for the potential ϕ, with the wave velocity

$$c_s^2 = \frac{KV}{M} \ . \tag{2.200}$$

Compressional waves (sound) propagating with this velocity occur for $\boldsymbol{u}$, δV and δp.

The change in pressure can also be written as

$$\begin{aligned}\delta p &= \left(\frac{\partial p}{\partial V}\right)_s \delta V + \left(\frac{\partial p}{\partial s}\right)_V \delta s = \\ &= -\frac{K_{ad}}{V}\delta V + \frac{T}{c_v}\alpha\delta s \ ,\end{aligned} \tag{2.201}$$

where s is the entropy per unit mass, K_{ad} is the adiabatic modulus of compression, c_v is the specific heat per unit mass at constant volume and $\alpha = \left(\frac{\partial p}{\partial T}\right)_V$ is the thermal pressure coefficient at constant volume. Making use of equations (2.198) and (2.200), we get

$$c^2 - c_s^2 = \frac{\alpha T V^2}{M c_v}\frac{\delta s}{\delta V} \ , \tag{2.202}$$

where c given by $c^2 = \frac{K_{ad}V}{M}$ is the velocity of the adiabatic sound. We can see from this equation that the change δs in entropy is a wave propagating with velocity c_s, like δV.

For $\delta s = 0$ we get from equation (2.202) the velocity of the adiabatic sound $c_s = c$. For the entropic sound we put $\frac{\delta s}{\delta V} = \left(\frac{\partial s}{\partial V}\right)_T$, where

$$\left(\frac{\partial s}{\partial V}\right)_T = \frac{1}{M}\left(\frac{\partial p}{\partial T}\right)_V = \frac{1}{M}\alpha \ . \tag{2.203}$$

Equation (2.202) becomes

$$c^2 - c_s^2 = \frac{\alpha^2 T V^2}{M^2 c_v} \ . \tag{2.204}$$

The relations $\frac{\alpha^2 TV}{MK_{th}} = c_p - c_v$ and $K_{th} = \frac{c_v}{c_p}K_{ad}$ are well known (see Appendix[45]), where K_{th} is the isothermal modulus of compression and c_p is the specific heat per unit mass at constant pressure. Therefore, we have

$$c^2 - c_s^2 = \left(\frac{c_p}{c_v} - 1\right)c_{th}^2 = \left(1 - \frac{c_v}{c_p}\right)c^2 \ , \tag{2.205}$$

where c_{th} given by $c_{th}^2 = \frac{K_{th}V}{M}$ is the velocity of the isothermal sound mentioned above. We can see that

$$c_s^2 = \frac{c_v}{c_p}c^2 \ , \tag{2.206}$$

i.e. the entropic sound is the isothermal sound, as long as the thermodynamic equilibrium relations ($\frac{\alpha^2 TV}{MK_{th}} = c_p - c_v$ and $K_{th} = \frac{c_v}{c_p}K_{ad}$) hold.

For the equation of state $(p + a/V^2)(V - b) = T$ of the van der Waals gas we use

$$p = \frac{T}{V}\left(1 + \frac{b}{V}\right) - \frac{a}{V^2} \ , \tag{2.207}$$

where the parameters a and b are small ($a \ll TV$, $b \ll V$).[46] The entropy of the van der Waals gas is given by

$$s = s_{id} + \frac{1}{M}\ln(1 - b/V) \simeq s_{id} - b/MV \ , \tag{2.208}$$

[45] L. Landau and E. Lifshitz, *Course of Theoretical Physics*, vol. 5 (*Statistical Physics*), Elsevier, Oxford (1980).

[46] M. Apostol, "On the van der Waals equation", J. Math. Theor. Phys. **1** 215 (2018); *Statistical Physics*, Cambridge Scholars Publishing, Newcastle upon Tyne (2021).

where $s_{id} = \frac{1}{M}\ln(VT^{3/2}) + const$ is the entropy per unit mass of the monoatomic ideal gas. By straightforward calculations we get

$$\begin{gathered}
\alpha = \left(\frac{\partial p}{\partial T}\right)_V = \frac{1}{V}\left(1+\frac{b}{V}\right) \ , \\
\beta = \frac{1}{V}\left(\frac{\partial V}{\partial T}\right)_p = \frac{1}{T}\left(1-\frac{b}{V}+\frac{2a}{TV}\right) \ , \\
c_v = \frac{3}{2M} \ , \ c_p = \frac{5}{2M}\left(1+\frac{4a}{5TV}\right)
\end{gathered} \tag{2.209}$$

and

$$\begin{gathered}
K_{ad} = -V\left(\frac{\partial p}{\partial V}\right)_s = \frac{5T}{3V}\left(1+\frac{2b}{V}-\frac{6a}{5TV}\right) \ , \\
K_{th} = -V\left(\frac{\partial p}{\partial V}\right)_T = \frac{T}{V}\left(1+\frac{2b}{V}-\frac{2a}{TV}\right) \ .
\end{gathered} \tag{2.210}$$

The velocity c_s of the entropic (isothermal) sound is

$$c_s^2 = \frac{T}{M}\left(1+\frac{2b}{V}-\frac{2a}{TV}\right) = \frac{3}{5}\left(1-\frac{4a}{5TV}\right)c^2 \ . \tag{2.211}$$

The thermodynamics of a simple liquid can be obtained in terms of the cohesion energy per particle, the spacing between the energy levels of the particle, a constraining volume for the motion of the particle and the mean inter-particle distance.[47] The velocities of the adiabatic sound and the entropic sound are given approximately by $c^2 \simeq 2T/M$ and $c_s^2 \simeq c^2/2$. Similar results can be obtained for other types of continuous media (condensed matter), like solids, plasmas, etc.

2.26 Vortices

2.26.1 Introduction

Around 1640 Descartes thought that swirling Planets move around the Sun like vortices in a universal fluid. His theory was soon superseded by Newton's universal attraction and theory of motion. In

[47] M. Apostol, "A model for the thermodynamics of simple liquids", Physica **B403** 3946 (2008); *Physical Kinetics*, Cambridge Scholars Publishing, Newcastle upon Tyne (2020).

the 18th century, among many other physical and mathematical topics, the fluid motion came into the attention of Bernoulli, Euler and Lagrange. Friction has been included in fluid motion through Navier-Stokes equations up to 1840. Around 1858 Helmholtz discovered a basic fact in fluid motion: the vortex. Fluid vortices are consistent swirling movements of fluids, which may possess their own dynamics. Possibly, Helmholtz had in mind the atmospheric cyclons. Soon, in 1865, another fluid was introduced by Maxwell, the electric and magnetic ethereal medium, and about the same year William Thomson (Lord Kelvin), based much on Helmholtz's work, revived the theory of vortices as ultimate matter constituents. Maxwell praised Kelvin's theory as an admirable advance coming from the interaction between mathematics and physics. Organized motion emerging from underlying motion caught much imagination. At the end of the 19th century Poincare synthesized the basic properties of vortices,[48] and a great deal of work in the applied mathematics of the fluids succeeded.[49]

Meanwhile, the 20th century witnessed the appearance of other *sui generis* fluids, like quantum-mechanical wavefunctions,[50] quantum fields, and even one more classical field, the gravitation, besides the electromagnetic one. In fact, fluids are (non-linear) fields of velocity. Emergent quantum or classical objects,[51] with their own dynamics, made out of an underlying (quantum or classical) swirling motion got a new interest, especially in connection with topological order in two dimensions, but, mainly, in relation to superconducting (and superfluid) vortices of Abrikosov.[52] Elementary particles made out of a non-linear dynamics of underlying fields caught the imagination of Einstein and Heisenberg in modern times, as well as today, when quarks, leptons or gauge bosons, this "fantastic reality",[53] are brought into play. Of course, the vortices should be closely related to the "turbulent" problem of the turbulence.[54]

---

[48]H. Poincare, *Theorie des Tourbillons*, Gabay, Paris (1893).

[49]See, for instance, H. Lamb, *Hydrodynamics*, Dover, NY (1945); G. K. Batchelor, *An Introduction to Fluid Dynamics*, Cambridge (1967) and References therein.

[50]M. Apostol, *Theory of Quanta*, Nova, NY (2019).

[51]P. W. Anderson, "More is different", Science **177** 393 (1972).

[52]A. A. Abrikosov, "On the magnetic properties of superconductors of the second group", Sov. Phys. JETP **5** 1174 (1957) (ZhETF **32** 1442 (1957)).

[53]F. Wilczek, *Fantastic Realities*, WorldSci (2006).

[54]M. Apostol, "Fluids, fluid vortices and the Theory of Electricity and Mag-

Lagrange and Euler equations. Fluid velocity is $\mathbf{v} = d\mathbf{r}/dt$, where t denotes the time and $\mathbf{r}$ is the position. The force acting upon the unit volume in a fluid is $-gradp$, where p is the pressure; adding an external force $\mathbf{f}$ per unit mass, Newton's equation reads $\rho d\mathbf{v}/dt = \rho\mathbf{f} - gradp$, where ρ is the fluid density, or

$$d\mathbf{v}/dt = \mathbf{f} - gradp/\rho \ . \tag{2.212}$$

This is Lagrange's equation of motion of the fluids.

The fluids flow continuously. This is why we prefer to view the position as a variable. The velocity becomes then a field $\mathbf{v}(t, \mathbf{r})$ of time and position. The continuity equation reads

$$\partial\rho/\partial t + div(\rho\mathbf{v}) = 0 \ . \tag{2.213}$$

Lagrange's equation (2.212) can then be written as

$$\partial\mathbf{v}/\partial t + (\mathbf{v}grad)\mathbf{v} = \mathbf{f} - gradp/\rho \ . \tag{2.214}$$

This is Euler's equation of motion of the fluids. It is a non-linear equation in velocity. It can also be written as

$$\partial\mathbf{v}/\partial t - \mathbf{v} \times curl\mathbf{v} = \mathbf{f} - grad(v^2/2) - gradp/\rho \ , \tag{2.215}$$

since $grad(v^2/2) = (\mathbf{v}grad)\mathbf{v} + \mathbf{v} \times curl\mathbf{v}$.[55] Euler's field of velocity is not the velocity of a particle, it is the velocity at point $\boldsymbol{r}$ and time t where fluid particles exist. In turbulence it may exhibit ruptures, discontinuities, and it may not even be defined, except as a statistical variable.

Static equilibrium. For static fluids $\mathbf{v} = 0$, the equilibrium is attained for $\mathbf{f} - gradp/\rho = 0$. This is the law of hydrostatics. It is the condition for mechanical equilibrium. It is worth noting that it may not have always a solution for pressure, since $\rho\mathbf{f}d\mathbf{r}$ is not a total differential in general. Very often, the external force $\mathbf{f}$ is conservative, *i.e.* it derives from a potential φ, through $\mathbf{f} = -grad\varphi$. If the

netism", J. Theor. Phys. **135** (2006).

[55]It can easily be checked by using the definition $(curlv)_i = \varepsilon_{ijk}\partial_j v_k$ of the curl and the definition $(a \times b)_i = \varepsilon_{ijk}a_j b_k$ of the vectorial product, where ε_{ijk} is the totally antisymmetric unit tensor of rank 3, $\varepsilon_{ijk}\varepsilon_{ilm} = \delta_{jl}\delta_{km} - \delta_{jm}\delta_{kl}$.

fluid is incompressible, *i.e.* if $\rho = const$ everywhere at any time, the equation of the mechanical equilibrium can be integrated for pressure p, and we get for p a certain function of position, according to the external potential φ. For instance, in the gravitational potential $\varphi = gz$, the pressure is given by $p = -\rho gz$ (up to a constant), where g is the gravitational acceleration and z denotes the altitude. This is the well-known Toricelli's barometric law for hydrostatic pressure. The fluids are usually in local thermal equilibrium, so we have a local equation of state which relates the pressure to density and temperature T, through some function f: $p = f(\rho, T)$. That means that the temperature changes with position, so there is no global equilibrium. Then, we may have convection, exchange of heat and entropy, and the mechanical equilibrium may be unstable, under certain conditions. In order to have a global thermal equilibrium the temperature must be constant, and, of course, the fluid must be incompressible. The Gibbs enthalpy Φ (per unit mass) can then be written as $d\Phi = -sdT + dp/\rho$, where s is the entropy per unit mass, so $d\Phi = dp/\rho$ is a total differential and mechanical equilibrium is reached for $\varphi + \Phi = const$. The equation of state gives then ρ as a function of p, and the pressure (and density) can be obtained. For an ideal gas, for instance, $p = \rho T/M$, where M is the molecular mass, and $p \sim \exp(-mgz/T)$ in gravitational potential. The density has a similar dependence on the altitude z, according to the equation of state. The entropy also varies with the altitude in this case; it is higher at higher altitudes; it is much more order at the bottom.[56]

Ideal fluids. Friction and thermal conductivity are supposed to be absent in Euler's equation. That means the flow is adiabatic. The fluid is said to be ideal in this case. The entropy is conserved locally, *i.e.* $ds = 0$, or $ds/dt = 0$, which makes

$$\partial s/\partial t + \mathbf{v} grads = 0 \ . \tag{2.216}$$

Together with the equation of continuity we get the conservation $\partial(\rho s)/\partial t + div(\rho s\mathbf{v}) = 0$ of the flow of entropy. Each infinitesimal bit of ideal fluid flows with its own constant entropy. The local equilibrium gives for the entropy s a function $s = g(\rho, T)$, which, together with the equation of state $p = f(\rho, T)$, can be written as $s = h(\rho, p)$,

[56]For an ideal classical gas $s \sim \ln(T^{3/2}/\rho)$.

where h is the function resulting from the combination of these two equations. The density and the pressure must be so as to satisfy the adiabatic equation (2.216).

The adiabatic flow $ds = 0$ of the ideal fluids allows the equation $dw = dp/\rho + Tds$ for the heat function w to be written as $dw = dp/\rho$, which makes $gradp/\rho = gradw$. Then, for conservative forces $\mathbf{f} = -grad\varphi$, Euler's equation (2.215) becomes

$$\partial\mathbf{v}/\partial t - \mathbf{v} \times curl\mathbf{v} = -grad(v^2/2 + \psi) \ , \tag{2.217}$$

where $\psi = w + \varphi$. $d\psi$ is a total differential. Equation (2.217) can also be written as $\partial(curl\mathbf{v})/\partial t - curl(\mathbf{v} \times curl\mathbf{v}) = 0$.

Now, we have three Euler's equations (2.217) for the three components of the velocity, one equation of continuity (2.213) and one equation (2.215) for the adiabatic flow; that makes five equations. And we have also three components of velocity, one density and one pressure (or entropy); that makes five unknowns. It follows that the flow of the ideal fluids is completely soluble; having determined ρ and p all the other thermodynamic quantities are determined, by making use of their definition and of the equation of state.

For incompressible fluids $\rho = const$, the continuity equation gives $div\mathbf{v} = 0$, the equation for the adiabatic flow becomes $\partial p/\partial t + \mathbf{v} gradp = 0$ (since p is given by s for $\rho = const$) and $w = p/\rho$ (up to a constant internal energy per unit mass $\varepsilon = const$; indeed, $d\varepsilon = -pdV + Tds = (p/\rho^2)d\rho + Tds = 0$ for incompressible ideal fluids).

In general, the adiabatic flow of the ideal fluids implies an entropy which depends on time and position. Again, each bit of fluid flows with its own constant entropy. The temperature may also depend on time and position, and we have not a global thermal equilibrium, as expected. However, usually the entropy is constant in the whole ideal fluid (the flow is isentropic).

We can recognize the energy per unit mass in the *rhs* of equation (2.217). The first term in equation (2.217) is the (local) acceleration, the second term can be viewed as a force (it will shortly be termed the Euler force; also, it is known as Lamb's vector), so this equation tells that the variation of the energy in its *rhs* equals the (local) kinetic energy and the mechanical work. It is Newton's equation, as expected, for velocity fields.

It can easily be shown that the energy flows continuously, according to

$$\frac{\partial}{\partial t}(\rho v^2/2+\rho\varepsilon)+div[\rho\mathbf{v}(v^2/2+w)]=0 \ . \tag{2.218}$$

where ε is the energy of the fluid per unit mass and $\rho\mathbf{v}(v^2/2+w)$ is the energy flow (for an ideal fluid). Similarly, the momentum flows continuously, according to

$$\partial(\rho v_i)/\partial t+\partial\Pi_{ik}/\partial x_k=0 \ , \tag{2.219}$$

where $\Pi_{ik}=p\delta_{ik}+\rho v_i v_k$ is the momentum flow. The fluid carries momentum flow $p+\rho v^2$ along its velocity.

Streamlines. Bernoulli's law. Since $\mathbf{v}=d\mathbf{r}/dt$, the curve

$$dx/v_x=dy/v_y=dz/v_z=dt \tag{2.220}$$

is tangent to velocities. It is called streamline. For steady flow it does not change in time, and coincides with the path of the fluid particles. For non-steady flow the path of the fluid particles differs from the streamline. The action $\int dt\cdot(v^2/2-\psi)$ of the lagrangian along a path is an extremum, in view of Lagrange's equation (2.212). For steady flow the energy per unit mass $v^2/2+\psi$ is constant along a streamline, in view of the same equation of motion. The streamline does not move in this case. The constant, in general, differs from a streamline to another. The constancy of the energy along the streamlines in steady flow is Bernoulli's law. It comes also from equation (2.217) by taking its projection along a steady-flow streamline.

Potential flow. As any vector, the velocity can be represented as a sum $\mathbf{v}=\mathbf{v}_1+\mathbf{v}_2$ of two vectors $\mathbf{v}_{1,2}$ enjoying the properties: $div\mathbf{v}_1=0$ and $curl\mathbf{v}_2=0$. That means that $\mathbf{v}_1$ comes from a *curl*, $\mathbf{v}_1=curl\mathbf{A}$ (since $div\, curl\mathbf{A}=0$) and $\boldsymbol{v}_2$ comes from a gradient, $\boldsymbol{v}_2=grad\Phi$ (since $curl\, grad\Phi=0$); $\mathbf{A}$ and Φ are some vector and scalar potentials.

Let us assume that $\mathbf{v}_1=0$, *i.e.* $\mathbf{v}=grad\Phi$ (and $curl\mathbf{v}=0$). This is called the irrotational motion or the potential flow. Euler's equation (2.217) leads then to $grad(\partial\Phi/\partial t+v^2/2+\psi)=0$, *i.e.*

$$\partial\Phi/\partial t+v^2/2+\psi=0 \tag{2.221}$$

everywhere (up to a function of time). This is the generalized Bernoulli law, and a first integral of motion. For steady flow it gets simplified to $v^2/2 + \psi = const$ everywhere.

A further simplification is achieved for an incompressible fluid, *i.e.* a fluid for which $\rho = const$ everywhere. The equation of continuity reads then $div\mathbf{v} = div\, grad\Phi = \Delta\Phi = 0$. With suitable boundary conditions this equation can be solved for Φ, *i.e.* for $\mathbf{v}$, and from equation (2.221) we get the pressure p. The problem is then solved completely, providing the equation of adiabatic flow is satisfied.

It can be shown easily that if the velocity of the fluid is much smaller than the sound velocity in that fluid, $v \ll c$, then the flow may be viewed as being incompressible. If velocity changes appreciably in time τ over distances l, then the flow may be viewed as steady for $\tau \ll l/c$, *i.e.* for interaction propagating "instantly".

An example of potential flow is the small oscillations of a large body. Then, $\partial\mathbf{v}/\partial t \sim \omega v \sim v^2/a$, where ω is the frequency of oscillations and a is the amplitude; on the other hand, $(\mathbf{v}grad)\mathbf{v} \sim v^2/l$, where l is the size of the body; since $l \gg a$ we may neglect $(\mathbf{v}grad)\mathbf{v}$ in Euler's equation, which reads $\partial\mathbf{v}/\partial t = -grad(v^2/2 + \psi)$, or $\partial(curl\mathbf{v})/\partial t = 0$. However, the mean velocity in oscillatory motion vanishes, such that $curl\mathbf{v} = 0$, *i.e.* the motion is potential.

Helmholtz's law of circulation. Multiplying equation (2.212) by $d\mathbf{r}$ we get $(d\mathbf{v}/dt)d\mathbf{r} = -d\psi$. Integrating this equation along a closed curve we get

$$\oint (d\mathbf{v}/dt)d\mathbf{r} = -\oint d\psi = 0 \ . \qquad (2.222)$$

The curve moves with the fluid, so we can write

$$\oint (d\mathbf{v}/dt)d\mathbf{r} = \frac{d}{dt}\oint \mathbf{v}d\mathbf{r} - \oint \mathbf{v}d\mathbf{v} = \frac{d}{dt}\oint \mathbf{v}d\mathbf{r} - \oint d(v^2/2) \ , \qquad (2.223)$$

and, together with equation (2.222),

$$\frac{d}{dt}\oint \mathbf{v}d\mathbf{r} = \oint d(v^2/2 - \psi) = 0 \ , \qquad (2.224)$$

or

$$\oint \mathbf{v}d\mathbf{r} = l = const \ . \qquad (2.225)$$

This is the Helmholtz theorem: the circulation (l) of the velocity along any closed fluid curve is constant. We emphasize that it holds only for ideal fluids.

Making use of the Stokes theorem, we get also

$$\oint \mathbf{v}d\mathbf{r} = \int d\mathbf{f}curl\mathbf{v} \ , \tag{2.226}$$

for any surface limited by the closed curve.

Suppose $curl\mathbf{v} = 0$ at some instant on a streamline. It is preserved in time along that streamline, so the potential flow ($curl\mathbf{v} = 0$) is preserved in time. Except for the streamlines ending up on a solid surface, like the surface of a body immersed in the flow. There, the velocity which is tangential to the surface is discontinuous, and we are not able to draw anymore a closed curve in the fluid encircling the streamline. An entire family of tangential discontinuities develops behind that body (in the wake of the flow), which makes the solution of the fluid flow non-uniquely determined. Actually, this holds only for ideal fluids. In real fluids, the friction makes the velocity continuous, and the boundary conditions determine the solution uniquely, together with the equations of motion.

In a potential flow the streamlines are open curves. Indeed, if they would be closed, the circulation of the velocity along them would not be zero, *i.e.* the flow would have a non-vanishing *curl*, *i.e.* it would not be a potential flow. This is however true for a simply-connected regions. If the region is multiply-connected, the streamlines may be closed curves; the circulation of the velocity along such a streamline is non-vanishing, but it may be compensated by another closed streamline which together with the former may define a surface where the *curl* is vanishing.

2.26.2 Vortices

Vortex. Suppose a purely rotational motion, *i.e.* $\mathbf{v} = curl\mathbf{A}$, where $\mathbf{A}$ is a vector potential, such that $div\mathbf{v} = 0$ (incompressible fluids). We denote $curl\mathbf{v} = 2\boldsymbol{\omega}$, and call $\boldsymbol{\omega}$ (or $curl\mathbf{v}$) the vorticity. The analogy with the static magnetic field is obvious: the velocity $\mathbf{v}$ is the magnetic field $\mathbf{H}$ and the vorticity $\boldsymbol{\omega}$ is $2\pi\mathbf{j}/c$, where $\mathbf{j}$ is the density of electric current and c is the velocity of light. In contrast to the magnetic field, it is not necessary to have $div\boldsymbol{A} = 0$.

The curves

$$\frac{dx}{\omega_x} = \frac{dy}{\omega_y} = \frac{dz}{\omega_z} \tag{2.227}$$

are lines of vortex, and the surface $\Phi = const$ with $\boldsymbol{\omega} grad\Phi = 0$ is the surface of the vortex tube.

Let us assume $div\boldsymbol{A} = 0$. Then, since $curl\mathbf{v} = curl\, curl\mathbf{A} = -\Delta\mathbf{A} + grad\, div\mathbf{A} = -\Delta\mathbf{A} = 2\boldsymbol{\omega}$, we get Poisson's equation $\Delta\mathbf{A} = -2\boldsymbol{\omega}$, whose solution vanishing at infinity is given by

$$\mathbf{A}(\mathbf{r}) = \frac{1}{2\pi}\int d\mathbf{r}' \cdot \frac{\boldsymbol{\omega}(\mathbf{r}')}{R} \ , \tag{2.228}$$

where $\mathbf{R} = \mathbf{r} - \mathbf{r}'$. This is the law of Biot and Savart. The velocity $\mathbf{v} = curl\mathbf{A}$ is then given by

$$\mathbf{v}(\mathbf{r}) = \frac{1}{2\pi}\int d\mathbf{r}' \cdot \frac{\boldsymbol{\omega}(\mathbf{r}') \times \mathbf{R}}{R^3} \ . \tag{2.229}$$

The streamlines are closed, solenoidal curves, like the magnetic field, perpendicular to the vorticity $\boldsymbol{\omega}$ and to the position vector $\mathbf{R}$. Their intensity decreases to infinity. This is a vortex.

In general, having determined the velocity, we use the equation of continuity (2.213) and Euler's equation (2.217) to get the density ρ and the heat function w. ρ and p must satisfy in addition equation (2.216) of adiabatic flow. For an incompressible fluid the equation of continuity is already satisfied ($div\mathbf{v} = div\, curl\mathbf{A}) = 0$), and we get directly the pressure from Euler's equation.

The vortices are consistent objects in an ideal fluid, in the sense that the circulation of the velocity along any closed curve is constant in time. They are open tubes, or closed tubes like torri, they may have their own motion, in particular the motion associated with their centre-of-vorticity.

Special vortices. Let the fluid particles rotate with constant angular velocity $\boldsymbol{\omega} = (0, 0, \omega)$ about an axis passing through some point. The velocity is given by $\mathbf{v} = \boldsymbol{\omega} \times \mathbf{r} = (-\omega y, \omega x, 0)$, where $\mathbf{r}$ is the position with respect to the rotation axis. The circulation of the velocity along the circle of radius r is $2\pi\omega r^2$. The *curl* of velocity is $curl\mathbf{v} = 2\boldsymbol{\omega} = (0, 0, 2\omega)$, and its integral over the surface limited by the circle is $2\omega\pi r^2$, *i.e.* the circulation of the velocity, as expected. This is a special vortex, whose vector potential is $\boldsymbol{A} = (0, 0, -\omega r^2/2)$. It satisfies the Poisson equation $curl\mathbf{v} = 2\boldsymbol{\omega}$, where $\boldsymbol{\omega}$ is constant everywhere. The vector potential increases indefinitely with increasing radius r. The velocity also has a similar indefinite increase. The flow is steady, and Euler's equation $(\mathbf{v}grad)\mathbf{v} = -grad\psi$ gives $\psi = \omega^2 r^2/2$. For an incompressible fluid $div\mathbf{v} = 0$ is satisfied, and the pressure is given by $p = \rho\omega^2 r^2/2$. This is a fairly unrealistic pressure. It increases toward the circumference of the rotating fluid, according to the centrifugal force. We can check the equation of adiabatic flow $\boldsymbol{v}gradp = 0$. It is worth noting that the temperature varies with position. In general, the motion of a rotating fluid is described more conveniently in the rotating frame, where the centrifugal force must be added to Euler's equation. If the fluid has also a velocity relative to the moving coordinates, then the Coriolis force must also be added. This inertial force determines an inhomogeneity in the pressure, so that there exist inertial waves in a rotating fluid. Similar internal waves are due to gravity, besides surface waves, which are also caused by gravity.

A more complex vortex is given by velocities $\mathbf{v} = \boldsymbol{\omega} \times \mathbf{r} \cdot f(r)$, where $\boldsymbol{\omega} = (0, 0, \omega)$ is a constant vector and $f(r)$ is an arbitrary function of $r = (x^2+y^2)^{1/2}$. The velocity reads $\mathbf{v} = (-\omega yf, \omega xf, 0)$, and a circle of radius d must be used for the integration contour, besides the current circle of radius r, if function f is singular at the origin; in general, we must use multiply-connected regions where singularities are present, instead of simply-connected ones for finite velocities. The circulation of the velocity is therefore $2\pi\omega[r^2 f(r) - d^2 f(d)]$. The vorticity is given by $curl\mathbf{v} = (0, 0, 2\omega f + \omega r f')$, and one can check that the circulation is given by Stokes' theorem. The special case $f = 1/r^2$ gives a vanishing circulation. It is worth noting that the vorticity depends in general on radius r, *i.e.* it is not constant in the flowing fluid. Such a complex vortex looks like a superposition of elementary vortices. The forces are given by $-\partial\psi/\partial x = (\mathbf{v}grad)v_x = -2\omega^2 f^2 x$, etc, and, like the velocity, they may be singular at the origin. We get $\psi = 2\omega^2 \int rf^2 dr$, and the pressure $p = 2\rho\omega^2 \int rf^2 dr$ for an incompressible fluid. The continuity equation $div\mathbf{v} = 0$ is satisfied. We can check $\boldsymbol{v}grad p = 0$.

The cyclon. Filament. Suppose a thin, rectilinear distribution of vorticity given by $\boldsymbol{\omega}(\mathbf{r}) = \boldsymbol{\omega}a^2\delta(\mathbf{r})$ and directed along the z-axis, where a is a characteristic thickness of this distribution and $\mathbf{r}$ is the position with respect to the z-axis. Such a distribution of vorticity is called a filament; it resembles very much a rectilinear, infinitely thin, electric current. The vector potential given by equation (2.228) can easily be calculated; it is given by

$$\begin{aligned} \boldsymbol{A}(r) &= (\boldsymbol{\omega}a^2/2\pi)\ln(z/r + \sqrt{z^2/r^2+1}\,\Big|_{-L}^{+L} \simeq \\ &\simeq -(\boldsymbol{\omega}a^2/\pi)\ln(r/2L)\ , \end{aligned} \tag{2.230}$$

where $2L$ is the length of the filament and r is the distance to the filament. Indeed, $\Delta \ln r = 2\pi\delta(\mathbf{r})$. The velocities are given by $\mathbf{v} = (-\omega a^2 y/\pi r^2, \omega a^2 x/\pi r^2, 0) = (a^2/\pi)\boldsymbol{\omega} \times grad(\ln r)$. One can check easily that $curl(\boldsymbol{\omega} \times \mathbf{f}) = \boldsymbol{\omega} div\mathbf{f} - (\boldsymbol{\omega} grad)\mathbf{f}$, where $\mathbf{f}$ is some vector, so $curl\mathbf{v} = (a^2/\pi)\boldsymbol{\omega}\Delta(\ln r) = 2a^2\boldsymbol{\omega}\delta(\mathbf{r})$.

One can also check easily that $div\mathbf{v} = 0$, and, for an incompressible fluid, $p = -\rho\omega^2 a^4/2\pi^2 r^2$ and $\mathbf{v}grad p = 0$ (adiabatic flow), for a steady flow. This is quite suggestive of an atmospheric cyclon.

The force $-\mathbf{v} \times curl\mathbf{v} = -(2a^4/\pi)\omega^2\delta(\mathbf{r})grad(\ln r)$ brought by Euler's equation (per unit mass) vanishes everywhere, except at the origin, where it is divergent. It follows that $v^2/2 + \psi = const$, *i.e.* the pressure changes such as to accommodate the kinetic energy $v^2/2 = \omega^2 a^4/2\pi^2 r^2$, as found above. We note that the energy $\int d\mathbf{r} \cdot v^2/2$ is infinite, due to the vortex singularity. This is the vortex self-energy. Similarly, the self-force $-\mathbf{v} \times curl\mathbf{v} = -(2a^4/\pi)\omega^2\delta(\mathbf{r})grad(\ln r)$ found above is singular at the origin. This is not as strange as it may look. If we want to create a finite vorticity along a line of vanishing thickness it is natural to give an infinite energy to the rotating fluid of vanishing mass localized on that line, and an infinite force to enable such a motion. Of course, the δ-function is not singular in fact, but it is some finite function localized on a small range. Therefore, we should not worry too much about the singularities displayed by localized vortices.

Apart from the free motion of the line of vorticity of a filament, we can consider the small deviations of this vorticity from a straight line. Then, we may have oscillations and waves of vorticity propagating along the filament.

Parallel filaments. Let us assume an ensemble of parallel filaments described by vorticity

$$\boldsymbol{\omega} = a^2 \sum_i \boldsymbol{\omega}_i \delta(\mathbf{r} - \mathbf{r_i}) \ , \tag{2.231}$$

where $\mathbf{r}_i$ are positions of these filaments in a plane perpendicular to their direction. We get the vector potential

$$\mathbf{A} = -(a^2/\pi) \sum_i \boldsymbol{\omega}_i \ln |\mathbf{r} - \mathbf{r}_i| \tag{2.232}$$

and velocity

$$\begin{aligned} \mathbf{v} = curl\mathbf{A} = (a^2/\pi) \textstyle\sum_i \boldsymbol{\omega}_i \times grad(\ln |\mathbf{r} - \mathbf{r}_i|) = \\ = (a^2/\pi) \textstyle\sum_i \frac{\boldsymbol{\omega}_i \times \mathbf{R}_i}{R_i^2} \ , \end{aligned} \tag{2.233}$$

where $\mathbf{R}_i = \mathbf{r} - \mathbf{r}_i$.

Euler's equation brings a force (per unit mass)

$$\begin{aligned} -\mathbf{v} \times curl\mathbf{v} = 2\boldsymbol{\omega} \times \mathbf{v} = \\ = -(2a^4/\pi) \textstyle\sum_{ij} \delta(\mathbf{r} - \mathbf{r}_i) \omega_i \omega_j grad(\ln |\mathbf{r} - \mathbf{r}_j|) = \\ = -(2a^4/\pi) \textstyle\sum_{ij} \delta(\mathbf{R}_i) \omega_i \omega_j \mathbf{R}_j / R_j^2 \end{aligned} \tag{2.234}$$

which vanishes everywhere, except at the filaments. The $v^2/2$-contribution to the *rhs* of Euler's equation reads

$$v^2/2 = \frac{1}{2}(a^2/\pi)^2 \sum_{ij} \omega_i \omega_j \boldsymbol{R}_i \mathbf{R}_j / R_i^2 R_j^2 \ , \tag{2.235}$$

and its gradient is given by

$$grad(v^2/2) = (a^2/\pi)^2 \sum_{ij} \omega_i \omega_j \frac{R_i^2 - 2(\mathbf{R}_i \mathbf{R}_j)}{R_i^4 R_j^2} \mathbf{R}_i \ . \tag{2.236}$$

According to equation (2.233), we may look for a steady-flow solution, as we assume ω_i to be constant; then, since the force given by equation (2.234) vanishes outside any $\mathbf{r}_i$, the *rhs* of Euler's equation vanishes for any $\mathbf{r} \neq \mathbf{r}_i$, so that the pressure is so as to accommodate the $v^2/2$, *i.e.* $p/\rho = -v^2/2$, for instance. In general, such a solution of Euler's equation may exist, though for an incompressible fluid it does not: the adiabatic equation $\mathbf{v}grad p = 0$ is not satisfied in that case, as it can be easily checked directly by making use of equations (2.233) and (2.236). We may neglect the adiabatic equation, since it brings very small contributions, but the corresponding pressure is fairly unrealistic. In general, such a solution does not exist for the points $\mathbf{r} = \mathbf{r_i}$, where the force $-\mathbf{v} \times curl\mathbf{v}$ is singular.

The singular points are dealt with by encircling any point $\mathbf{r}_i$ by a small circle of radius ε, integrating over its surface and letting then $\varepsilon \rightarrow a$. By doing so in equation (2.234), we get a force

$$\mathbf{f}_i = -(2a^2/\pi) \sum_j{}' \omega_i \omega_j \mathbf{R}_{ij}/R_{ij}^2 \ , \tag{2.237}$$

acting upon the i-th filament, as resulting from $-\mathbf{v} \times curl\mathbf{v}$, where $\mathbf{R}_{ij} = \mathbf{r}_i - \mathbf{r}_j$ and the prime over the sum means $j \neq i$. We note that the singular self-force is not included. The force the j-th filament acts upon the i-th filament is therefore

$$f_{ij} = -(2a^2/\pi) \omega_i \omega_j \mathbf{R}_{ij}/R_{ij}^2 \ , \tag{2.238}$$

while the force the i-th filament acts upon the j-th filament is

$$f_{ji} = -(2a^2/\pi) \omega_i \omega_j \mathbf{R}_{ji}/R_{ij}^2 = -f_{ij} \ . \tag{2.239}$$

Obviously, it is the analog of Ampere's electrodynamic force acting between two parallel electric currents, as $-\mathbf{v} \times curl\mathbf{v} \rightarrow \mathbf{j} \times \mathbf{H}$ is analogous to Lorentz's force. We call $-\mathbf{v} \times curl\mathbf{v} = 2\omega \times \mathbf{v}$ the Euler force. Formally it is analogous to the Lorentz force $\sim \mathbf{j} \times \mathbf{H}$ acting upon an electric current $\mathbf{j}$ placed in a magnetic field $\mathbf{H}$.[57]

[57] A. Sommerfeld, *Vorlesungen ueber Theoretische Physik*, Band II (*Mechanik der Deformierbaren Medien*), Akad. Verlagsgesellschaft, Leipzig (1970).

It is easy to see that the force given by equation (2.239) derives from a potential

$$U_{ij} = (2a^2/\pi)\omega_i\omega_j \ln R_{ij} \ , \tag{2.240}$$

so the energy of the interacting filaments (per unit length) reads

$$U = (a^2/\pi)\sum_{i\neq j}\omega_i\omega_j \ln R_{ij} \ . \tag{2.241}$$

It is worth noting that the total force $\sum_{i\neq j} f_{ij}$ on the ensemble vanishes. We apply the same procedure to all the other contributions to Euler's equation. We find easily, by direct integration, that $(\mathbf{v})_i \to 0$, $(\partial\mathbf{v}/\partial t)_i \to 0$, and $(grad(v^2/2))_i \to 0$, where the subscript i denotes the result of integration over a circle of vanishing radius ε drawn around the i-th filament. It follows that Euler's equation requires the equilibrium condition $\mathbf{f}_i = -(grad\psi)_i$, and we may take any pressure that satisfies the equation of adiabatic flow; then external forces are needed to satisfy the equilibrium condition. The ensemble of filaments is not stable.

The fact that Euler's equation has not solution for the given distribution of vorticities (equation (2.231)) is not surprising: not any velocity makes a solution. Our distribution depends on the parameters $\mathbf{r}_i$, and in order to look for a solution, we have, at least, to give a dynamics to these parameters, *i.e.* to let the filaments move around. Indeed, the force $\mathbf{f}_i$ acting upon any i-th filament is not vanishing, so there is a motion of the filaments.[58] We emphasize that a velocity $\mathbf{u}_i = \dot{\mathbf{r}}_i$ must be attributed to any i-th filament, as distinct from the fluid velocity $(\mathbf{v})_i$, and, similarly $(grad\psi)_i = 0$, as for line-like filaments. We get then a distinct dynamics of the filaments, as emerging from Euler's fluid equation, and coupled in a certain way to the fluid motion, as it will be seen shortly. First, we note the equation of motion

$$d\mathbf{u}_i/dt = \mathbf{f}_i = -(2a^2/\pi)\sum_j{}' \omega_i\omega_j\mathbf{R}_{ij}/R_{ij}^2 \tag{2.242}$$

for filaments. It is conceivable that the evolution of the positions $\mathbf{r}_i$ of the filaments might be so as to attain a minimum of the "mechanical",

[58] This is another example of emergent dynamics (P. W. Anderson, "More is different", Science **177** 393 (1972)).

potential energy U given by equation (2.241), *i.e.* for vanishing forces on any filament,

$$\sum_j{}' \omega_j \mathbf{R}_{ij}/R_{ij}^2 = 0 \ . \qquad (2.243)$$

However, such a solid state of filaments is not stable. This can be checked by direct calculation, though it would imply a long-range order which is not attainable, in general, in two dimensions. On the other hand, the motion towards such a state may not preserve the ensemble of vortices. The $v^2/2$ gives an energy per unit length

$$\begin{aligned} E &= \int d\mathbf{r} \cdot v^2/2 = \int d\mathbf{r} \cdot \mathbf{v} curl \mathbf{A}/2 = \int d\mathbf{r} \cdot \mathbf{A} curl \mathbf{v}/2 = \\ &= \int d\mathbf{r} \cdot \mathbf{A}\omega = -(a^4/\pi) \textstyle\sum_{i \neq j} \omega_i \omega_j \ln R_{ij} = -a^2 U \ . \end{aligned} \qquad (2.244)$$

One can see that the "mechanical" energy U due to the forces acting between the filaments changes at the cost of the "field" energy E stored by the velocities, such as $E/a^2 + U = const$, as for magnetic fields. The analogy of $\mathbf{A}\boldsymbol{\omega}$ with the interaction energy $\mathbf{jA}$ of a current in a magnetic field is obvious. This is the way the motion of the filaments is coupled to the motion of the velocity field. A total angular momentum and a total torque can be computed.

During this evolution, Euler's equation may be satisfied for a non-steady flow of velocities, though it is not necessary for a given ensemble of vorticities. The problem of the continuity of the solution does not appear anymore, since the line-like (point-like) filaments are now removed from the fluid. This is the basic decomposition into particles and fields, as allowed by the vortex concept. The fields are produced by particles, the particles are localized fields, singularities in the fields, both the fields and the particles may have their own dynamics, and the fields interact with particles. This interaction relates the motion of the fields to the motion of the particles. We note that the kinetic energy E_{kin} of the particles, associated with velocities $\mathbf{u}_i$, is conserved against the potential energy U, so it is a measure of the "field" energy E. This is distinct from electromagnetism. The particles are singularities of the fields, and as long as the self-interaction is avoided by the point-like (line-like) nature of the particles, there is no problem of infinities.

The problem is however unavoidable in electromagnetism, where we start with two distinct things, pointlike charges and fields, and end up with one, which is the union of the two.

For an (unstable) equilibrium of the ensemble of the filaments, according to equation (2.243), which may look like an ordered, "decorated", two dimensional lattice of filaments, *i.e.* a lattice of filaments with opposite nearest-neighbours vorticities, the *rhs* of Euler's equation vanishes, and the pressure follows the $v^2/2$. Let the origin be on some i-th filament, and let $\boldsymbol{\rho}$ be the position vector with respect to this origin. For ρ close to the origin we have

$$\mathbf{v} = (a^2/\pi)\frac{\boldsymbol{\omega}_i \times \boldsymbol{\rho}}{\rho^2} + ..., \tag{2.245}$$

from equation (2.233), and

$$p = -v^2/2 = -\frac{1}{2}(a^2/\pi)^2\omega_i^2/\rho^2 + ..., \tag{2.246}$$

for the dominant contributions, providing the equilibrium condition (2.243) is fulfilled. The filaments behave then mainly as independent, "one-particle", objects. It is easy then to see that the equation of adiabatic flow $\mathbf{v}grad p = 0$ is satisfied. It is not obvious that it is also satisfied for finite $\boldsymbol{\rho}$ (and in fact it is not).

The ensemble of interacting vortex filaments introduced here can be viewed as a liquid of filaments, or a plasma. We may call it a vorticial liquid. Obviously, the length a plays the role of a short-range cutoff, as length L plays the role of a long-range cutoff. Two filaments with opposite vorticities repel each other, so, according to equation (2.241), we may estimate the increase in energy associated with introducing such an anti-filament in an environment of nearest-neighbouring parallel filaments as $\delta U \sim -(a^2/\pi)\omega^2 \ln(a/L)$, for vorticities equal in absolute value. At the same time, the gain in entropy is of the order $\delta S \sim \ln(L/a)$. A phase transition is therefore possible at temperature $T \sim a^2\omega^2/\pi$. This is the well-known Berezinsky-Kosterlitz-Thouless transition in two dimensions from a high-temperature disordered liquid (plasma) to a low-temperature "condensed" phase, where filament-anti-filament pairs are present.[59] As it is well-known, it is a topologically ordered phase, not a long-range ordered one.

[59] V. I. Berezinsky, "Destruction of long-range order in one-dimensional and two-

We may consider an ensemble of filaments with various orientations, but the incompressibility condition is not fulfilled. Suppose that the integral in equation (2.230) is carried along the filament between $z-\varepsilon$ and $z+\varepsilon$, as for a small filament length 2ε. Then, we get $\mathbf{A}=(\omega a^2\varepsilon/\pi)\cdot 1/R$, where R is the distance to that small portion of filaments. Obviously, $\Delta\mathbf{A}=-4\boldsymbol{\omega} a^2\varepsilon\delta(\mathbf{R})$, as if the vorticity is $\boldsymbol{\omega}(\mathbf{R})=2\boldsymbol{\omega} a^2\varepsilon\delta(\mathbf{R})$. We get therefore localized particles of vortices. However, $div\boldsymbol{A}\neq 0$ in this case.

Coulombian vorticial liquid. Let us consider the velocity

$$\mathbf{v}=-(a^3/2\pi)\cdot\boldsymbol{\omega}\times grad(1/r)\ , \tag{2.247}$$

where $\boldsymbol{\omega}$ is a constant vector and a is a small length. We can check easily that $div\boldsymbol{v}=0$ and

$$\boldsymbol{v}=(a^3/2\pi)curl(\boldsymbol{\omega}/r)\ . \tag{2.248}$$

Therefore, the vector potential is

$$\mathbf{A}(\mathbf{r})=(a^3/2\pi)\cdot\frac{\boldsymbol{\omega}}{r}\ ; \tag{2.249}$$

it satisfies the equation $\Delta\boldsymbol{A}=-2a^3\boldsymbol{\omega}\delta(\boldsymbol{r})$, though $divA\neq 0$ and

$$curl\boldsymbol{v}=2a^3\boldsymbol{\omega}\delta(\boldsymbol{r})+(a^3/2\pi)grad\left[\boldsymbol{\omega} grad(1/r)\right]\ . \tag{2.250}$$

The integral of the *grad*-term in equation (2.250) over a sphere of small radius centered at $\boldsymbol{r}=0$ is zero; on integrating the contributions of this term over such spheres when calculating the Euler force and the energy for an ensemble of vortices, we get either zero, or infinities, which are discarded. Therefore we may left aside this term and use a vorticity

$$\boldsymbol{\omega}(\boldsymbol{r})=a^3\boldsymbol{\omega}\delta(\boldsymbol{r})\ . \tag{2.251}$$

The Euler force $\mathbf{F}=-\mathbf{v}\times curl\mathbf{v}$ does not vanish, and Euler's equation is not satisfied.

---

dimensional systems having a continuous symmetry group. I. Classical systems", Sov. Phys.-JEPT **32** 493 (1971); *ibid*, **34** 610 (1972) (ZhETF **59** 907 (1971); **61** 1144 (1972)); J. M. Kosterlitz and D. J. Thouless, "Ordering, metastability and phase transitions in two-dimensional systems", J. Phys. **C6** 1181 (1973).

Let us assume an ensemble of vortex particles, with the vector potential

$$\mathbf{A}(\mathbf{r}) = (a^3/2\pi)\sum_i \frac{\boldsymbol{\omega}_i}{|\mathbf{r}-\mathbf{r}_i|} \tag{2.252}$$

and the velocity

$$\mathbf{v} = -(a^3/2\pi)\cdot\sum_i \boldsymbol{\omega}_i \times grad(1/|\mathbf{r}-\mathbf{r}_i|) \ . \tag{2.253}$$

The Euler force (per unit mass) $-\boldsymbol{v} \times curl\boldsymbol{v}$ is singular. We can attempt to extract finite contributions by integrating over the whole space and leaving aside the divergent terms. This is a "renormalization" procedure; it is applied already to equation (2.237). In general, we leave aside the result of integration over small regions around each i-th vortex, since it is a self-force or a self-energy. By doing so, we get

$$\begin{aligned}\mathbf{F} &= -\int d\boldsymbol{r}\mathbf{v} \times curl\mathbf{v} = \\ &= (a^6/\pi)\textstyle\sum_{i\neq j} grad_i \frac{\boldsymbol{\omega}_i\boldsymbol{\omega}_j}{|\mathbf{r}_i-\mathbf{r}_j|} \ .\end{aligned} \tag{2.254}$$

We can see that the j-th particle acts upon the i-th particle with a force

$$\mathbf{f}_{ij} = -(a^6/\pi)(\boldsymbol{\omega}_i\boldsymbol{\omega}_j)\mathbf{R}_{ij}]/R_{ij}^3 \ , \tag{2.255}$$

where $\mathbf{R}_{ij} = \mathbf{r}_i - \mathbf{r}_\mathbf{j}$, while the force on the j-th particle on behalf of the the i-th particle is

$$\mathbf{f}_{ji} = (a^6/\pi)(\boldsymbol{\omega}_i\boldsymbol{\omega}_j)\mathbf{R}_{ij}]/R_{ij}^3 \ . \tag{2.256}$$

We note that the force given by equation (2.255), although of Coulomb-law type, implies vorticities but not scalar electric charges as in the theory of electricity and magnetism.

In general, such an ensemble of vorticial particles is not in equilibrium; it exhibits internal movements. An ordered combination of parallel and antiparallel vortices, *i.e.* an ordered combination of vortices and anti-vortices, may get equilibrium, but such an equilibrium is unstable. Indeed, one can check by direct calculation that the force acting upon each particle may vanish, but the matrix of the force derivatives (the force matrix) is not positive definite. Such a stable state would correspond to a latticial screening, which is not attainable for a purely-Coulomb interacting ensemble of particles. We may call such an ensemble of vortices a coulombian vorticial liquid; the ensemble of filaments discussed above is also a (filamentary) vorticial liquid.

The force given by equation (2.255) arises from a potential $U_{ij} = -(a^6/\pi)(\boldsymbol{\omega}_i\boldsymbol{\omega}_j)/R_{ij}$. The total kinetic energy is

$$E = \frac{1}{2}\int drv^2 = \frac{a^6}{2\pi}\sum_{i\neq j}\frac{\boldsymbol{\omega}_i\boldsymbol{\omega}_j}{R_{ij}} \ , \qquad (2.257)$$

such that the total energy $U + E = 0$ (or *const*) is conserved, where $U = (1/2)\sum_{i\neq j} U_{ij}$. In computing both $\boldsymbol{F}$ and E we use $\boldsymbol{v} = curl\,\boldsymbol{A}$, the properties of $\boldsymbol{A}$ and the "renormalization" procedure (not only the singular term of vorticity in equation (2.250)).

Dipolar vorticial liquid. Let us consider a constant vector $\boldsymbol{m}$ and a distribution of vorticities

$$\boldsymbol{\omega}(\boldsymbol{r}) = -2\pi\boldsymbol{m}\times grad\delta(\boldsymbol{r}) \ . \qquad (2.258)$$

The equation $\Delta\boldsymbol{A} = -2\omega(\boldsymbol{r})$ becomes

$$\Delta\boldsymbol{A} = -2\omega(\boldsymbol{r}) = 4\pi\boldsymbol{m}\times grad\delta(\boldsymbol{r}) = \\ = -\boldsymbol{m}\times grad\Delta(1/r) \ , \qquad (2.259)$$

which gives

$$\mathbf{A} = \frac{\mathbf{m}\times\mathbf{r}}{r^3} \qquad (2.260)$$

($div\boldsymbol{A} = 0$). We can imagine the function $\delta(\boldsymbol{r})$ as a function localized in a small, spherical volume; then $\boldsymbol{\omega}(\boldsymbol{r})$ is a vector parallel to the latitude lines. The analogy with the magnetostatics is obvious: $\boldsymbol{m}$ may be viewed as a magnetic moment, and the potential $\boldsymbol{A}$ given by equation (2.260) is the vector potential of a magnetic dipole. Also, $\boldsymbol{m}$ may be viewed as proportional to an intrinsic angular momentum similar with the spin of a particle, which is related to its magnetic momentum. Therefore, we may call such a vortex a dipolar vortex.

The velocity $\mathbf{v} = curl\mathbf{A}$ is given by

$$\mathbf{v} = -\mathbf{m}/r^3 + 3\mathbf{r}(\mathbf{mr})/r^5 \ . \qquad (2.261)$$

It is easy to recognize here the dipole field, *i.e.* the magnetic or electric field produced by a magnetic or an electric dipole.

We consider a distribution

$$\boldsymbol{\omega}(\mathbf{r}) = -2\pi \sum_i \mathbf{m}_i \times grad\delta(\boldsymbol{r} - \boldsymbol{r}_i) \tag{2.262}$$

of dipolar vorticities with centres at $\mathbf{r}_i$. It gives a vector potential

$$\mathbf{A}(\mathbf{r}) = \sum_i \frac{\mathbf{m}_i \times \mathbf{R}_i}{R_i^3} \tag{2.263}$$

and a velocity

$$\mathbf{v}(\mathbf{r}) = \sum_i [-\mathbf{m}_i/R_i^3 + 3\mathbf{R}_i(\mathbf{m}_i\mathbf{R}_i)/R_i^5] \ , \tag{2.264}$$

where $\boldsymbol{R}_i = \boldsymbol{r} - \boldsymbol{r}_i$. We can now compute the Euler force $-\mathbf{v} \times curl\mathbf{v} = 2\boldsymbol{\omega}(\mathbf{r}) \times \mathbf{v}(\mathbf{r})$. Leaving aside the singularities occurring at the positions of the vorticities, we get

$$-\int d\boldsymbol{r}\boldsymbol{v} \times curl\boldsymbol{v} \Longrightarrow -\sum_{i \neq j} grad_i U_{ij} \ , \tag{2.265}$$

where

$$U_{ij} = -4\pi \boldsymbol{m}_i grad_i \left[\boldsymbol{m}_j grad(1/R_{ij})\right] \ . \tag{2.266}$$

We can see that there exist internal forces. Such an ensemble of particles is another example of a vorticial liquid. This vorticial liquid is also unstable. The dipolar interaction between dipoles in fixed positions, may lead in certain circumstances to ferroelectricity.[60]

It is easy to check by direct calculation that the "field" energy $\int d\mathbf{r} \cdot v^2/2$ of the velocity field counter-balances the "mechanical" potential energy of the torri distribution, *i.e.*

$$\int d\boldsymbol{r}(v^2/2) \Longrightarrow -\frac{1}{2}\sum_{i \neq j} U_{ij} \ ; \tag{2.267}$$

the integral $\int d\mathbf{r} \cdot v^2/2$ is computed by $\int d\mathbf{r} \cdot \boldsymbol{\omega}\boldsymbol{A}$.

---

[60] M. Apostol, *Magnetic and Electric Resonance*, Cambridge Scholars Publishing, Necastle upon Tyne (2018).

2.26.3 Electromagnetism

Electric charge. It is tempting to construct a theory of electricity and magnetism based on fluid vortices. However, such a construction is untenable.

For instance, one may view the magnetic field $\mathbf{H}$ as a field of velocities whose *curl* is proportional to the density $\mathbf{j}$ of electric currents. We get then the correct laws of magnetostatics. Indeed, if we put $\mathbf{v} \rightarrow \mathbf{H}$ and $curl\mathbf{H} = 4\pi\mathbf{j}/c$, we have already Ampere's law. With $\mathbf{H} = curl\mathbf{A}$ and $div\mathbf{A} = 0$, we get Green's law $div\mathbf{H} = 0$, and the Biot and Savart's solution $\mathbf{A} = \int \mathbf{j}/R$, $\mathbf{H} = -(1/c)\int \mathbf{j} \times grad(1/R)$ for Poisson equation $\Delta\mathbf{A} = -4\pi\mathbf{j}/c$. Moreover, Euler's force $\mathbf{F} = -\mathbf{v} \times curl\mathbf{v}$ in Euler's equation becomes Lorentz's $\mathbf{j} \times \mathbf{H}$ force, and its mechanical work may compensate the variation of the field energy $\mathbf{H}^2/2$. However, it is impossible to get stable localized vortices obeying the laws of electricity and magnetism, as shown by the vorticial liquids described above. In addition, the time dependence of the magnetic and the electric fields is not supported by the analogy with vortices and fluid dynamics. The basic difference between fluid dynamics and electricity and magnetism arises from the existence of localized electric charge.

Maxwell equations. Newton's laws of motion arise from a basic perception of motion in space and time which is the inertial motion. A similar perception would be satisfactory for electricity and magnetism.

The basic fact known about electricity and magnetism is that some force fields do occur, are produced, by an electric charge with density, say, $\rho(\mathbf{r})$. To produce something, we have two means: use a gradient $grad\varphi$ of some scalar potential φ, and use a time derivative $\partial\mathbf{A}/\partial t$ of some vector potential $\mathbf{A}$. Indeed, something is produced by changes of something. It is worth noting that such a production is not a flow $\partial/\partial t + (\mathbf{v}grad)$ with velocity $\mathbf{v}$ of something. The two changes $grad\varphi$ and $\partial\mathbf{A}/\partial t$ together require, obviously, a universal velocity c; it is the light velocity. Then, we may define an electric field through

$$\mathbf{E} = -grad\varphi - \frac{1}{c}\frac{\partial\mathbf{A}}{\partial t} \; . \tag{2.268}$$

This is the basic law of the theory of electricity and magnetism. Obviously, it is a consistent law for inertial frames, if we view the passing from a frame to another moving with a constant velocity $\mathbf{v}$ as a rotation in the four-dimensional space defined by $(\mathbf{r}, ict)$. Then, we have at once the principle of relativity, the Lorentz transformations and the theory of relativity.[61]

Equation (2.268) requires the existence of another field $\mathbf{H}$, defined by $\mathbf{H} = curl\mathbf{A}$. Indeed, if we take the $curl$ in both sides of equation (2.268) we get $curl\mathbf{E} = (1/c)[\partial(curl\mathbf{A})/\partial t]$, since $curl\, grad = 0$. We have immediately the Maxwell equation

$$curl\mathbf{E} = -\frac{1}{c}\frac{\partial \mathbf{H}}{\partial t} \tag{2.269}$$

for the Faraday and Lenz law of induction. In addition, since $\mathbf{H} = curl\mathbf{A}$, we get the Maxwell equation

$$div\mathbf{H} = 0 \tag{2.270}$$

for Green's law of the rotational magnetic field.

We take now the divergence of equation (2.268): $div\mathbf{E} = -\Delta\varphi - (1/c)[\partial(div\mathbf{A})/\partial t]$; it is easy to see that we can arrive at once to a wave equation with sources and a law of producing electric fields from such sources by imposing the Lorentz gauge

$$div\mathbf{A} + \frac{1}{c}\frac{\partial\varphi}{\partial t} = 0\ ; \tag{2.271}$$

Then we get the wave equation

$$\Delta\varphi - \frac{1}{c^2}\frac{\partial^2\varphi}{\partial t^2} = -4\pi\rho \tag{2.272}$$

for the scalar potential generated by electric charges, and the Maxwell equation

$$div\mathbf{E} = 4\pi\rho \tag{2.273}$$

for the Gauss law of the electric fields.

It remains to take the time derivative of equation (2.268), use the Lorentz gauge: $\partial\mathbf{E}/\partial t = cgrad\, div\mathbf{A} - (1/c)(\partial^2\mathbf{A}/\partial t^2)$, and use in addition $curl\mathbf{H} = curl\, curl\mathbf{A} = -\Delta\mathbf{A} + grad\, div\mathbf{A}$; we get $(1/c)(\partial\mathbf{E}/\partial t) -$

[61] The relativistic invariance requires $(\mathbf{p}, iE/c)$ to be a negative invariant for a particle; it gives Einstein's relativistic energy $E = \sqrt{m^2c^4 + c^2p^2}$, where E is the energy of the particle, $\mathbf{p}$ is its momentum and m is its mass.

$curl\mathbf{H} = \Delta\mathbf{A} - (1/c^2)(\partial^2\mathbf{A}/\partial t^2)$. In order to get a new wave equation for the vector potential produced by electric currents $\mathbf{j}$, and a production of magnetic field by electric currents, we put

$$\Delta\mathbf{A} - \frac{1}{c^2}\frac{\partial^2\mathbf{A}}{\partial t^2} = -\frac{4\pi}{c}\mathbf{j} \qquad (2.274)$$

and get the Maxwell equation

$$curl\mathbf{H} = \frac{1}{c}\frac{\partial\mathbf{E}}{\partial t} + \frac{4\pi}{c}\mathbf{j} \qquad (2.275)$$

for Ampere's law of producing magnetic fields through electric currents. $(1/4\pi)\partial\mathbf{E}/\partial t$ is Maxwell's displacement current. In addition, taking the time derivative of equation (2.273) and using equation (2.275) we get the continuity equation

$$\frac{\partial\rho}{\partial t} + div\mathbf{j} = 0 \ , \qquad (2.276)$$

which shows that electric curents are moving electric charges,

$$\mathbf{j} = \rho\mathbf{v} \ . \qquad (2.277)$$

With this, the derivation of the Maxwell equations of electricity and magnetism is completed. By adequate notations, it is easy to show that they are relativistically invariant. We note that they are linear equations in fields. They obey the principle of superposition. In addition, in the non-relativistic limit $v/c \to 0$ the vector potential vanishes, so the magnetic field is a purely relativistic effect. Moreover, they contain singularities for pointlike charges and currents, like Euler's equation of fluids for point-like vortices.

However, there is a profound difference between Euler's equation for fluids and the Maxwell equations. Euler's equation has only one object, the field of velocities, and particular forms of this field, localized vortices, can be consistently separated from the velocity field, providing thereby an image for particles and fields. The two objects thus separated have their own dynamics, even possibly inconsistent in general, and possibly inter-related in some particular cases. At least in simple cases like one vortex such a separation is fully consistent.

This is not so for the Maxwell equations. The theory of electricity and magnetism recognizes two separate, distinct entities, fields and pointlike charges, but inter-relates them through Maxwell equations, which amounts to identify them as one and the same object. This looks like a logic fault, albeit supported by a natural empirical situation. It has troubling consequences, as expected, in the well-known "self-interacting infinities". The unavoidable logical consequence of such a self-contradictory position, is that such a situation, as described by the theory of electricity and magnetism, does not exist, under certain conditions. This is implied by the quantum-mechanical delocalization of the pointlike charges during their relativistic existence.

The wave equations (2.272) and (2.274) suggest that fields are coupled to matter through an action reading

$$S_{fm} = \int dt \cdot (-q\varphi + \mathbf{i}\mathbf{A}/c) \ , \tag{2.278}$$

where q is a pointlike charge and $\mathbf{i} = q\mathbf{v}$ is the corresponding electric current. We get

$$-qgrad\varphi - (q/c)(\mathbf{A}grad)\mathbf{v} \tag{2.279}$$

for the variation of this action with respect to particle coordinate $\mathbf{r}$. Now, we use $grad(\mathbf{ab}) = (\mathbf{a}grad)\mathbf{b} + (\mathbf{b}grad)\mathbf{a} + \mathbf{a} \times curl\mathbf{b} + \mathbf{b} \times curl\mathbf{a}$ for $\mathbf{v}$ and $\mathbf{A}$ (which can be checked by using again the tensor ε_{ijk} for the definition of the *curl* and of the vectorial product) and get

$$\begin{gathered} -qgrad\varphi + (q/c)(\mathbf{v}grad)\mathbf{A} + (q/c)\mathbf{v} \times curl\mathbf{A} = \\ = -qgrad\varphi - (q/c)\partial\mathbf{A}/\partial t + (q/c)d\mathbf{A}/dt + (1/c)\mathbf{i} \times \mathbf{H} = \\ = q\mathbf{E} + (1/c)\mathbf{i} \times \mathbf{H} + (q/c)d\mathbf{A}/dt \end{gathered} \tag{2.280}$$

for this variation. On the other hand, the variation of the free-particle action is $-d\mathbf{p}/dt$, and from equation (2.278) we get an additional term $-(q/c)d\mathbf{A}/dt$, so that $\mathbf{P} = \mathbf{p} + q\mathbf{A}/c$ is the canonical momentum of a particle in the field (and its hamiltonian reads $H = p^2/2m + q\varphi = (\mathbf{P} - q\mathbf{A}/c)^2/2m + q\varphi$). Then, by equating $d(\mathbf{p} + q\mathbf{A}/c)$ with equation (2.280) we get the Lorentz force

$$d\mathbf{p}/dt = q\mathbf{E} + (1/c)\mathbf{i} \times \mathbf{H} \tag{2.281}$$

acting upon a charged particle moving in an electromagnetic field. The magnetic field brings no contribution to the rate $\mathbf{v}d\mathbf{p}/dt$ of change of the kinetic energy E_{kin} of a particle.

The electromagnetic waves propagate with the velocity of light c, as described by equations (2.272) and (2.274). It is easy to show[62] that the field energy $E_f = (1/4\pi)(E^2/2 + H^2/2)$ propagates according to $\partial E_f/\partial t = -\mathbf{jE} - div\mathbf{S}$, where $\mathbf{S} = (c/4\pi)\mathbf{E}\times\mathbf{H}$ is the Poynting vector. By integrating over space, $\mathbf{jE}$ leads to the rate $\mathbf{v}d\mathbf{p}/dt$ of change of the kinetic energy E_{kin} of the charges, so that $\partial(E_f + E_{kin})/\partial t + div\mathbf{S} = 0$, *i.e.* the rate of change of the field energy plus the kinetic energy of the charges is equal to the flow of the field energy. This is quite different from the field of velocities and the motion of fluid vortices.

The action of the field is $\int dtd\mathbf{r} \cdot (E^2 - H^2)/8\pi$. Its variation (including the field-matter part S_{fm}) leads to Maxwell equations. Its flow lead to the momentum conservation. The Poynting vector carries a momentum density $\mathbf{S}/c$, while the field contains the Maxwell stress $(E_iE_j + H_iH_j)/4\pi - \delta_{ij}E_f$. The momentum conservation looks similar to the motion of an elastic, rigid, ethereal medium, rather than a fluid.

Particles and fields. Infinities. Electric charges are pointlike. Consequently, their electromagnetic energy is infinite. The momentum carried by the electromagnetic field produced by a moving pointlike charge gives an "electromagnetic mass" of the charge, which again is infinite, due to the vanishing size of the particle. The rate of change in time of such an electromagetic momentum gives a force. It acts upon the particle itself, and it is Lorentz's damping force. It has infinite effects, too, because of the vanishing size of the particle. The charge interacts with its own field, since the electromagnetic field has its own wave-like motion with the velocity of light, which is different from the velocity of the moving charge. In the non-relativistic limit $v/c \to 0$, there is no such a damping force. The retarded electromagnetic interaction looks like the field produced by the particle now, propagates, and the particle meets it again, in its motion, after a while.

These infinities are the origin of the well-known ultraviolet divergences in quantum field theory. A cutoff should be used in such computa-

[62] Since $div(\mathbf{a}\times\mathbf{b}) = \mathbf{b}curl\mathbf{b} - \mathbf{a}curl\mathbf{b}$.

tions, arising from a finite size of particle. It is the distance over which the light propagates in the time uncertainty corresponding to the rest energy of the particle, *i.e.* the Compton length $c \cdot \hbar/mc^2 = \hbar/mc$, where $\hbar$ is Planck's constant.[63] It means that we should do perturbation calculations as long as the particles do exist.

Summary. Let us summarize. The fluids may support vortices, although, in general, they are unstable. This was shown by Helmholtz. Such vortices may be localized, either, for instance, as filaments or as vortex particles, or as particles endowed with dipolar vorticities. Such a localization may be viewed as singular, giving the idea of particles made up of fields, in particular fields of velocity. Then a problem appears: the singularities in Euler's equation. We may get rid of them by assuming that particles have their own dynamics, related to the dynamics of fields by the conservation of energy in some cases. This is satisfactory. On the other hand, such a picture is quite similar with static magnetic fields, though Euler's force may bring appreciable differences. The differences arise from the non-linear nature of the fluid dynamics arising from one single object - the velocity field -, in contrast with the linear Maxwell equations. It is also related to the definition of the electric currents as being proportional to the velocity of the moving charges, in contrast to the fluid vorticity, which arises precisely from the *curl* of its field, *i.e.* the *curl* of the velocity. Nevertheless, it is tempting to liken the currents with vortices, and magnetic fields to fields of velocity in some ethereal fluid. Unfortunately, such a picture is untenable for electric fields and charges, and the time behaviour of the electric and magnetic fields has no connection to any fluid motion. Moreover, the electromagnetic field seems to require an ethereal rigid solid rather than a fluid. The electromagnetism is fundamentally different from the fluid dynamics, though it admits the distinction between particles and fields. However, this admission is not consistently kept in Maxwell equations, which, actually, end up with removing it. This logical inconsistency of the theory of electricity and magnetism leads to well-known difficulties with the infinities, which the relativity and the quantum motion may deal with, to some extent.

The fluid motion is by far richer than the motion of the electricity and

[63] M. Apostol, *Radiation and Matter*, Nova, NY (2022).

magnetism, as a consequence of the non-linear Euler's force. It allows the contemplation of new types of matter, like the vorticial liquids. The theory of electricity and magnetism, with its logical inconsistencies, makes unavoidable conclusions like quantum and relativistic behaviour, and, especially, the limited existence of the substantial matter.

There is another limited, very interesting, analogy/difference between fluids and electromagnetism. Let us suppose that we rotate a long, thin bar in a fluid, or we set in rotation a fluid layer placed at some distance r from the origin. The adjacent fluid layers are set in rotation, due to the friction (viscosity) force. Let us assume that the viscosity force between adjacent layers is vanishing. This leads to the equation $(\Delta \boldsymbol{v})_\varphi = 0$, where the velocity $\boldsymbol{v}$ has only one component $v = v_\varphi$. This equation has the solution $v = ar + b/r$, as in the Couette flow, where a and b are two constants. For an infinite fluid $a = 0$, and we are left with $v \sim 1/r$. This is precisely the filamentary vortex, singular at the origin, with the vector potential $\boldsymbol{A} = -\boldsymbol{\omega} \ln r$, where $\boldsymbol{\omega}$ is a constant vector perpendicular to $\boldsymbol{r}$, and the velocity is given by $\boldsymbol{v} = curl \boldsymbol{A}$. We get $\Delta \boldsymbol{A} = -\boldsymbol{\omega} \Delta \ln r = -2\pi \boldsymbol{\omega} \delta(\boldsymbol{r})$. This equation is the same as the equation of the vector potential generated by an electric current. The difference is that in electromagnetism we need only the smoothness assumption for the potentials, while for fluids we need in addition the viscosity. The fluids are material media, while the electromagnetic field is an "ethereal" medium. In addition, the filamentary vortex produces an infinite depression $p = -\rho\omega^2/2r^2$ at the origin, and it may lead to turbulence.

3 Viscous Fluids

3.1 Viscosity

The pressure force $\boldsymbol{F} = -grad p$ occurring in Euler's equation

$$\rho \frac{d\boldsymbol{v}}{dt} = -grad p \tag{3.1}$$

can be viewed as arising from a stress tensor $\sigma_{ij} = -p\delta_{ij}$, according to $F_i = \partial_j \sigma_{ij}$, where $F_i = -\partial_i p$ and $\boldsymbol{F} = -grad p$. The expression $F_i = \partial_j \sigma_{ij}$ accounts for the surface character of these forces. Indeed,

$$\int dV F_i = \int dV \partial_j \sigma_{ij} = \oint df_j \sigma_{ij} \ ; \tag{3.2}$$

since the torque of this force must be vanishing inside any domain (its only contribution arises from surface), the stress tensor should be symmetric.

The viscosity generates a force which arises from the spatial variations of the velocity; therefore, we can write the corresponding viscosity tensor as

$$\sigma'_{ij} = \eta \left(\partial_i v_j + \partial_j v_i - \frac{2}{3} \partial_k v_k \delta_{ij} \right) + \zeta \partial_k v_k \delta_{ij} \ , \tag{3.3}$$

where η and ζ are viscosity coefficients; they may be taken constant and uniform in fluid, and have positive values. The equation of motion is the Navier-Stokes equation[1]

$$\rho\frac{dv_i}{dt}=\rho\frac{\partial v_i}{\partial t}+\rho(\boldsymbol{v}grad)v_i= \\ =-\partial_i p+\partial_j\sigma'_{ij}\ , \tag{3.4}$$

or

$$\rho\frac{d\boldsymbol{v}}{dt}=\rho\frac{\partial\boldsymbol{v}}{\partial t}+\rho(\boldsymbol{v}grad)\boldsymbol{v}= \\ =-gradp+\left(\frac{1}{3}\eta+\zeta\right)grad\,div\boldsymbol{v}+\eta\Delta\boldsymbol{v}\ . \tag{3.5}$$

The momentum conservation reads

$$\frac{\partial}{\partial t}(\rho v_i)+\partial_j(p\delta_{ij}+\rho v_i v_j-\sigma'_{ij})=0\ , \tag{3.6}$$

where

$$\Pi_{ij}=p\delta_{ij}+\rho v_i v_j-\sigma'_{ij} \tag{3.7}$$

is density of momentum flux. We can see that the viscosity tensor is a momentum flux density tensor: part of the momentum is lost by viscosity (collisions).

Once the viscosity is taken into account, water is no longer dry: now it is wet, and it wets the walls of the container. Therefore, the boundary condition for the Navier-Stokes equation is a vanishing velocity.

A viscous fluid acts by a force on the surface bounding the fluid. It is given by the momentum flux density multiplied by the surface element,

$$dP_i=(p\delta_{ij}+\rho v_i v_j-\sigma'_{ij})df_j=pdf_i-\sigma'_{ij}df_j \tag{3.8}$$

(since $v_j df_j=0$). The first term in equation (3.8) is the pressure, while the second term is due to friction (viscosity). For a surface of separation between two immiscible fluids the velocities should be equal and the forces must be in equilibrium. At a free surface $P_i=0$. In

[1]C. L. Navier, "Memoire sur les lois du mouvement des fluides" (read in Paris Academy in 1822), Mem. de l'"Academie royale des sciences de l'Institut de France **6** 389 (1823); G. G. Stokes, "On the theories of the internal friction of fluids in motion, and of the equilibrium and motion of elastic solids", Trans. Cambridge Phil. Soc. **8** 287 (1845) (in *Mathematical and Physical Papers by G. G. Stokes*, vol. 1, 75, Johnson Reprint Corp., NY (1966)).

equation (3.8) we can introduce the unit vector normal to the surface $\boldsymbol{n}$, through $df_i = n_i df$, and get the force per unit area of the surface.

It is useful to have the viscosity tensor in a curvilinear system of coordinates (see Appendix[2]).

The viscosity coefficients η and ζ are of the same order of magnitude. For air $\eta \simeq 2 \times 10^{-4} g/cm \cdot s$, for water $\eta \simeq 10^{-2} g/cm \cdot s$.

3.2 Energy conservation

The derivation of the energy conservation proceeds in the same manner as for an ideal fluid, except for the viscosity term which should be included; we get

$$\begin{aligned} \frac{\partial}{\partial t}\left(\tfrac{1}{2}\rho v^2 + \rho\varepsilon\right) = -div\left[\rho\boldsymbol{v}\left(\tfrac{1}{2}v^2 + w\right)\right] + \\ +\rho T\left(\tfrac{\partial s}{\partial t} + \boldsymbol{v}grads\right) + v_i\partial_j\sigma'_{ij} \ , \end{aligned} \tag{3.9}$$

where ε is the internal energy per unit mass, w is the enthalpy per unit mass, T is the temperature and s is the entropy per unit mass. In deriving equation (3.9) we use the continuity equation

$$\frac{\partial\rho}{\partial t} + div(\rho\boldsymbol{v}) = 0 \tag{3.10}$$

and the fundamental law of Thermodynamics

$$d\varepsilon = -pdV + Tds = -d(p/\rho) + \frac{dp}{\rho} + Tds \ , \tag{3.11}$$

[2] L. Landau and E. Lifshitz, *Course of Theoretical Physics*, vol. 6 (*Fluid Mechanics*), Elsevier, Oxford (1989). The transition to another system of coordinates for a tensor $t_{ij} = \partial_i v_j$ is performed by using $\partial_i v_j \boldsymbol{e}_i \boldsymbol{e}_j$, where $\boldsymbol{e}_i$ are the unit vectors of the system of coordinates; they are given by $\boldsymbol{e}_i = a_{i\alpha}\boldsymbol{e}_\alpha$, where $\boldsymbol{e}_\alpha$ are the unit vectors of the other system of coordinates and the matrix elements $a_{i\alpha}$ are functions of the new coordinates x_α. On the other hand, the coordinates $x_i = f_i(x_\alpha)$ are functions of the new coordinates x_α. A vector is written as $v_i\boldsymbol{e}_i = v_i a_{i\alpha}\boldsymbol{e}_\alpha = v_\alpha\boldsymbol{e}_\alpha$, where $v_\alpha = a_{i\alpha}v_i$, $v_i = b_{i\beta}v_\beta$ and $a_{i\alpha}b_{i\beta} = \delta_{\alpha\beta}$. Then, we have $\partial_i v_j \boldsymbol{e}_i\boldsymbol{e}_j = \frac{\partial x_\alpha}{\partial x_i}\partial_\alpha(b_{j\beta}v_\beta)a_{i\gamma}a_{j\delta}\boldsymbol{e}_\gamma\boldsymbol{e}_\delta$, whence the component $t_{\gamma\delta}$, all quantities expressed in the new coordinates.

where $V = 1/\rho$ is the specific volume per unit mass and $w = \varepsilon + p/\rho$ is the enthalpy per unit mass. The flow of an ideal fluid is adiabatic, *i.e.* $\partial s/\partial t + \boldsymbol{v} grad s = 0$. For a viscous fluid an additional term $v_i \partial_j \sigma'_{ij}$ appears in energy, which can be written as $v_i \partial_j \sigma'_{ij} = \partial_j (v_i \sigma'_{ij}) - \sigma'_{ij} \partial_j v_i$; and the energy conservation reads

$$\frac{\partial}{\partial t}\left(\tfrac{1}{2}\rho v^2 + \rho\varepsilon\right) = -\partial_j \left[\rho v_j \left(\tfrac{1}{2}v^2 + w\right) - v_i \sigma'_{ij}\right] + \\ + \rho T \left(\tfrac{\partial s}{\partial t} + \boldsymbol{v} grad s\right) - \sigma'_{ij} \partial_j v_i \; . \tag{3.12}$$

We can see that a transport of energy flux density $v_i \sigma'_{ij}$ appears, which arises from the momentum transfer during collisions, and, in addition, the term $\sigma'_{ij} \partial_j v_i$ is a source of heat. The viscosity has a double function: it produces mechanical collisions and also dissipates energy as heat. Consequently, diffusion of heat should be included in the energy conservation. The heat flux density reads $\boldsymbol{q} = -\kappa grad T$, where κ is the thermal conductivity, and the diffusion is given by $div \boldsymbol{q} = -\kappa \Delta T$ (q is energy per unit area of cross-section and unit time, and κ has dimensions $1/cm \cdot s$). We consider κ constant and uniform (like the viscosity coefficients; in general, κ is a function of temperature and pressure); in addition, $\kappa > 0$, in accordance with the fact that heat flows from hot to cold. Now, adding and subtracting the heat diffusion, we get

$$\frac{\partial}{\partial t}\left(\tfrac{1}{2}\rho v^2 + \rho\varepsilon\right) = -\partial_j \left[\rho v_j \left(\tfrac{1}{2}v^2 + w\right) - v_i \sigma'_{ij} - \kappa \partial_j T\right] + \\ + \rho T \left(\tfrac{\partial s}{\partial t} + \boldsymbol{v} grad s\right) - div(\kappa grad T) - \sigma'_{ij} \partial_j v_i \; , \tag{3.13}$$

or

$$\rho T \left(\frac{\partial s}{\partial t} + \boldsymbol{v} grad s\right) = div(\kappa grad T) + \sigma'_{ij} \partial_j v_i \tag{3.14}$$

and

$$\frac{\partial}{\partial t}\left(\frac{1}{2}\rho v^2 + \rho\varepsilon\right) = -\partial_j \left[\rho v_j \left(\frac{1}{2}v^2 + w\right) - v_i \sigma'_{ij} - \kappa \partial_j T\right] \; ; \tag{3.15}$$

the first equation is the equation of heat transfer, while the second equation is the equation of energy conservation. It is a remarkable fact that the equation of energy conservation is separated from the equation of heat conduction. The solution is determined in the same way as for ideal flows: we have three equations (the Navier-Stokes equation, the continuity equation and the equation of heat transfer) and three unknowns: the velocity, the density and the entropy. Instead of entropy we can use the temperature through $Tds = c_p dT$, where c_p is the specific heat for constant pressure (as shown in the Ideal Fluids chapter).

3.3 Production of entropy

By making use of the viscosity tensor (equation (3.3)), we get

$$\begin{aligned} \sigma'_{ij}\partial_j v_i &= \eta\left(\partial_i v_j + \partial_j v_i - \tfrac{2}{3}\delta_{ij}\partial_k v_k\right)\partial_j v_i + \zeta\delta_{ij}\partial_k v_k \partial_j v_i = \\ &= \tfrac{1}{2}\eta\left(\partial_i v_j + \partial_j v_i - \tfrac{2}{3}\delta_{ij}\partial_k v_k\right)^2 + \zeta(div\boldsymbol{v})^2 \ , \end{aligned} \tag{3.16}$$

such that the equation of heat transfer becomes

$$\begin{gathered} \rho T\left(\tfrac{\partial s}{\partial t} + \boldsymbol{v}grads\right) = div(\kappa gradT) + \\ + \tfrac{1}{2}\eta\left(\partial_i v_j + \partial_j v_i - \tfrac{2}{3}\delta_{ij}\partial_k v_k\right)^2 + \zeta(div\boldsymbol{v})^2 \ . \end{gathered} \tag{3.17}$$

We can see that the entropy increases as a result of the irreversible processes of thermal conduction and internal friction (viscosity). Indeed, let us calculate the change in the total entropy

$$\frac{d}{dt}\int dV\rho s = \int dV\frac{\partial(\rho s)}{\partial t} \ . \tag{3.18}$$

By making use of the continuity equation and the equation of heat transfer, we get

$$\begin{gathered} \tfrac{\partial(\rho s)}{\partial t} = \rho\tfrac{\partial s}{\partial t} + s\tfrac{\partial \rho}{\partial t} = -sdiv(\rho\boldsymbol{v}) - \\ -\rho\boldsymbol{v}grads + \tfrac{1}{T}div(\kappa gradT) + \\ + \tfrac{\eta}{2T}\left(\partial_i v_j + \partial_j v_i - \tfrac{2}{3}\delta_{ij}\partial_k v_k\right)^2 + \tfrac{\zeta}{T}(div\boldsymbol{v})^2 \ . \end{gathered} \tag{3.19}$$

The first two terms give $-div(\rho s\boldsymbol{v})$, which, integrated, gives the entropy flux through the surface at infinity; this flux is vanishing. Similarly, the diffusion term includes a vanishing surface term (for $gradT = 0$ at infinity); we are left with

$$\begin{gathered}\frac{d}{dt}\int dV\rho s = \int dV\frac{\kappa(gradT)^2}{T^2}+\\ +\int dV\frac{\eta}{2T}\left(\partial_i v_j+\partial_j v_i-\frac{2}{3}\delta_{ij}\partial_k v_k\right)^2+\\ +\int dV\frac{\zeta}{T}(div\boldsymbol{v})^2\ .\end{gathered}\tag{3.20}$$

Since the entropy can only increase, we can see that the viscosity coefficients and the thermal conductivity are all positive. This confirms Fourier's law of heat diffusion, according to which the heat flux $\boldsymbol{q}$ is proportional to $gradT$ and does not include $gradp$, for instance. We emphasize the different nature of the κ-term, which implies a surface contribution, and the η, ζ-terms, which are body contributions.

We can use $ds = (\partial s/\partial T)_p dT$ and $c_p = T(\partial s/\partial T)_p$ which transforms the heat transfer equation into

$$\begin{gathered}\rho c_p\left(\frac{\partial T}{\partial t}+\boldsymbol{v}gradT\right)=\kappa\Delta T+\\ +\frac{1}{2}\eta\left(\partial_i v_j+\partial_j v_i-\frac{2}{3}\delta_{ij}\partial_k v_k\right)^2+\zeta(div\boldsymbol{v})^2\ ;\end{gathered}\tag{3.21}$$

(for a compressible fluid the volume changes; the transport is performed at constant pressure). For the energy conservation we use the thermodynamic equilibrium relations $dw = d\varepsilon + dp/\rho$ and $d\varepsilon = -pd(1/\rho) + Tds$; the local equilibrium is attained, once the transport is realized. It is worth noting that the flow is a non-equilibrium (irreversible) process, which, for small velocities (much smaller than sound velocity), with small gradients, brings only small changes in equilibrium quantities; in particular, the change in entropy is a second-order effect, according to equation (3.17) (it is not so for the change in pressure, in general).

3.4 Incompressible fluid

Usually, the variations of the density and the entropy are very small, such that we may neglect them in many cases, while preserving the viscosity in the Navier-Stokes equation. In this chapter we assume an incompressible fluid with constant entropy. In a forthcoming chapter we shall see that under these conditions the viscosity may produce vorticity, though, very likely, the vortices are unstable.

For an incompressible fluid the dissipated energy can be calculated without resorting to internal energy and enthalpy. Indeed,

$$\begin{aligned} \frac{\partial}{\partial t}\left(\frac{1}{2}\rho v^2\right) = \rho\boldsymbol{v}\frac{\partial \boldsymbol{v}}{\partial t} = -\boldsymbol{v}grad p - \rho\boldsymbol{v}(\boldsymbol{v}grad)\boldsymbol{v} + v_i\partial_j\sigma^{'}_{ij} = \\ = -\rho\boldsymbol{v}grad\left(\frac{1}{2}v^2 + p/\rho\right) + \partial_j(v_i\sigma^{'}_{ij}) - \sigma^{'}_{ij}\partial_j v_i = \\ = -\partial_j\left[\rho v_j\left(\frac{1}{2}v^2 + p/\rho\right) - v_i\sigma^{'}_{ij}\right] - \sigma^{'}_{ij}\partial_j v_i \end{aligned} \tag{3.22}$$

(where we use $div\boldsymbol{v} = 0$; the viscosity coefficient ζ, called the second viscosity, does not contribute). This equation shows that the rate of change in time of the (kinetic) energy density is given by the energy flow (work done by pressure included), the flow of momentum transferred by collisions which account for viscosity and, finally, the dissipated energy. This later quantity can be written as

$$\begin{aligned} \dot{E}_{kin} = -\int dV\sigma^{'}_{ij}\partial_j v_i = -\frac{1}{2}\int dV\sigma^{'}_{ij}(\partial_j v_i + \partial_i v_j) = \\ = -\frac{1}{2}\eta\int dV(\partial_i v_j + \partial_j v_i)^2 \ ; \end{aligned} \tag{3.23}$$

therefore, by internal friction, the viscosity causes a loss of the kinetic energy, *i.e.* a dissipated energy, which is transformed in heat, which, in turn, may generate a change in entropy (ds), internal energy and enthalpy (Tds); the full set of equations are equations (3.14) and (3.15) (and $div\boldsymbol{v} = 0$).

In equation (3.22) $\boldsymbol{v}(\rho v^2/2)$ is the energy flux density associated with mass transfer (kinetic energy); $\boldsymbol{v}p$ is an energy flux density associated with the work done by pressure forces; $-v_i\sigma^{'}_{ij}$ is an energy flux density arising from momentum transfer in collisions, occurring in internal friction processes; finally, the last term on the right in equation (3.22),

$\sigma'_{ij}\partial_j v_i$, is the variation of the heat produced by viscosity; it is related to the variation of the entropy and the heat flux $\boldsymbol{q}$,

$$\sigma'_{ij}\partial_j v_i = \rho T \frac{ds}{dt} + div\boldsymbol{q} \ , \tag{3.24}$$

or

$$\begin{aligned} \rho T \tfrac{ds}{dt} &= \rho T \left(\tfrac{\partial s}{\partial t} + \boldsymbol{v} grads\right) = \\ &= -div\boldsymbol{q} + \sigma'_{ij}\partial_j v_i = div(\kappa gradT) + \sigma'_{ij}\partial_j v_i \ . \end{aligned} \tag{3.25}$$

For an incompressible fluid ($div\boldsymbol{v} = 0$) the viscosity tensor reduces to $\sigma'_{ij} = \eta(\partial_i v_j + \partial_j v_i)$, such that

$$\rho T \left(\frac{\partial s}{\partial t} + \boldsymbol{v} grads\right) = div(\kappa gradT) + \frac{\eta}{2}\left(\partial_i v_j + \partial_j v_i\right)^2 \ ; \tag{3.26}$$

since $\Delta T/T = div(gradT/T) + (gradT/T)^2$, we can see from equation (3.26) that the entropy increases in time.

3.5 Flow in a pipe

Let us consider a viscous steady flow of an incompressible fluid, with the Navier-Stokes equation

$$\rho(\boldsymbol{v} grad)\boldsymbol{v} = -gradp + \eta\Delta\boldsymbol{v} \tag{3.27}$$

(and $div\boldsymbol{v} = 0$). Let us assume that the flow takes place between two parallel infinite planes, separated by distance h along the y-axis and moving with relative constant velocity u along the x-axis. All the quantities depend only on y and the fluid velocity v is only in the x-direction. The inertial term does not contribute. We get $dp/dy = 0$ and $d^2v/dy^2 = 0$. Therefore, the pressure is uniform and $v = ay + b$ is a linear function of y, where a and b are constants. The boundary conditions are $v_{y=0} = 0$ and $v_{y=h} = u$, whence $v = \frac{u}{h}y$. The viscosity tensor is $\sigma'_{xy} = \eta\partial_y v_x = \eta\frac{u}{h}$, and the tangential force per unit area on the plane $y = 0$ is $\sigma'_{xy} = \eta\frac{u}{h}$ and $-\sigma'_{xy} = -\eta\frac{u}{h}$ on the plane $y = h$.

Let us assume now that the two planes are fixed, but we have a pressure gradient. Equation (3.27) gives

$$\frac{\partial p}{\partial x} = \eta \frac{\partial^2 v}{\partial y^2} \, , \; \frac{\partial p}{\partial y} = 0 \; . \tag{3.28}$$

Obviously, the solution is $\frac{\partial p}{\partial x} = p' = const$ and $v = \frac{1}{2\eta} p' y^2 + ay + b$, where a and b are constants. Making use of the boundary conditions we get

$$v = \frac{1}{2\eta} p' y(y - h) \; . \tag{3.29}$$

This is known as the Poiseuille flow.[3] The force of friction per unit area acting upon the plane $y = 0$ is $p'h/2$.

Let us consider now a flow through a pipe with constant cross-section, in the same conditions as above. The equations are $\partial p/\partial y = \partial p/\partial z = 0$ and

$$\eta \left(\frac{\partial^2 v}{\partial y^2} + \frac{\partial^2 v}{\partial z^2} \right) = \frac{dp}{dx} \; . \tag{3.30}$$

This means that the pressure gradient is uniform ($p' = const$). For a cylindrical cross-section with radius R equation (3.30) leads to

$$v = \frac{p'}{4\eta}(r^2 - R^2) \; . \tag{3.31}$$

A mass $2\pi r v \rho dr$ passes per unit time through the annular element $2\pi r dr$ of the cross-section. The mass per unit time (the discharge) is

$$Q = -\frac{\pi p'}{8\nu} R^4 \; , \tag{3.32}$$

where $\nu = \eta/\rho$ is called kinematic viscosity. The friction per unit area of the tube is $\sigma'_{rz} = -p'R/2$.

[3] G. H. L. Hagen, "Ueber die Bewegung des Wasers in engen cylindrischen Rochren", Ann. Physik Chem. (Poggendorf) **46** 423 (1839); J. L. M. Poiseuille, "Recherches experimentales sur le mouvement des liquides dans les tubes de tres petits diametres", C. R, Acad. Sci. **11** 961, 1041 (1840); G. G. Stokes, "On the theories of the internal friction of fluids in motion, and of the equilibrium and motion of elastic solids", Trans. Cambridge Phil. Soc. **8** 287 (1845) (in *Mathematical and Physical Papers by G. G. Stokes*, vol. 1, 75, Johnson Reprint Corp., NY (1966)).

Having known the velocity distribution we can compute the temperature distribution by using the equation of heat transfer. By making use of equation (3.23) we can get the rate of increase in temperature per unit time for the entire flow.

3.6 Flow between rotating cylinders

We assume two infinite coaxial cylinders with radii $R_{1,2}$ rotating about their axes with angular velocities $\Omega_{1,2}$. We consider the motion of an incompressible fluid comprised between these two cylinders. Obviously, only the velocity $v = v_\varphi$ is non-vanishing, and depends only on the radius r, $v_\varphi = v(r)$. Similarly, the pressure depends only on the radius r. In the Navier-Stokes equation we need to evaluate the term

$$(\boldsymbol{v}grad)\boldsymbol{v} = \frac{v}{r}\frac{\partial}{\partial\varphi}v\boldsymbol{e}_\varphi = -\frac{v^2}{r}\boldsymbol{e}_r \ , \tag{3.33}$$

where $\boldsymbol{e}_{\varphi,r}$ are the unit vectors along the angle and the radius (see Appendix). Therefore, we get the equation

$$\rho\frac{v^2}{r} = \frac{dp}{dr} \ . \tag{3.34}$$

The second equation is given by $(\Delta\boldsymbol{v})_\varphi$ from the viscosity term, which is

$$\frac{1}{r}\frac{d}{dr}(r\frac{dv}{dr}) - \frac{v}{r^2} = 0 \tag{3.35}$$

(where the vectorial laplacian is used, see Appendix). This equation has solution $v = ar + \frac{b}{r}$, where a and b are constants. The boundary conditions are $v = \Omega_{1,2}R_{1,2}$, such that we get

$$v = \frac{\Omega_2 R_2^2 - \Omega_1 R_1^2}{R_2^2 - R_1^2}r + \frac{(\Omega_1 - \Omega_2)R_1^2 R_2^2}{R_2^2 - R_1^2}\frac{1}{r} \tag{3.36}$$

($R_2 > R_1$). Equation (3.34) gives the pressure. This is known as the Couette flow.[4]

[4] M. Couette, *Etude sur le frottement des liquides*, Gauthiers-Villars, Paris (1890).

The force acting upon the unit area of the cylinders is the tangential force given by the tensor component $\sigma'_{r\varphi}$, which in this case is $\eta(\partial v/\partial r - v/r)$ (see Appendix). The moment of this force is obtained by multiplying it by the radius of the cylinder. If we multiply further by $2\pi R_{1,2}$, we get the moment of force per unit length of the cylinder.

We note that the viscous flow is very different from the ideal flow. In some cases the viscous velocity includes viscosity, but the force acting upon the boundary does not. In some other cases not even the velocity includes viscosity. Also, we note that in some cases the velocity is proportional to $1/\eta$, such that we cannot take the limit $\eta \rightarrow 0$.

3.7 Stokes' force

Let us consider a solid sphere with radius R moving with a constant velocity $-\boldsymbol{u}$ in an incompressible fluid at rest. Equally well, we may consider a sphere at rest and a fluid moving past the sphere with a constant and uniform velocity $\boldsymbol{u}$ at infinity. In the neighbourhood of the sphere the velocity of the fluid is modified, let it be $\boldsymbol{v}$; for a fixed sphere it is $\boldsymbol{v} - \boldsymbol{u}$. We are interested in steady flow.

In general, the Navier-Stokes equation reads

$$\rho \boldsymbol{V} grad \boldsymbol{V} = -grad p + \eta \Delta \boldsymbol{V} \ , \tag{3.37}$$

where $\boldsymbol{V}$ is the fluid velocity past the solid body at rest; the continuity equation is

$$div \boldsymbol{V} = 0 \ . \tag{3.38}$$

In the neighbourhood of the body the quadratic term in equation (3.37) is of the order of V^2/l, where l is the dimension of the body, while the viscosity term is of the order of $\eta V/l^2$. The ratio of these two terms is

$$R = \frac{(\rho \boldsymbol{V} grad \boldsymbol{V})}{(\eta \Delta \boldsymbol{V})} = \frac{Vl}{\nu} \ , \tag{3.39}$$

where $\nu = \eta/\rho$ is the kinematic viscosity. This ratio is a dimensionless number, which is called the Reynolds number.[5] We note that it is meaningful for finite dimensions of the body.

If $R \ll 1$, we may neglect the quadratic term in the Navier-Stokes equation (which is also called the inertial term), which becomes

$$\eta \Delta \boldsymbol{V} = grad p \ , \tag{3.40}$$

or

$$\Delta curl \boldsymbol{V} = 0 \ . \tag{3.41}$$

For the fluid flowing past the sphere we can write $\boldsymbol{v} - \boldsymbol{u} = curl \boldsymbol{A}$, since $div(\boldsymbol{v} - \boldsymbol{u}) = 0$. The vector $\boldsymbol{A}$ should be axial and it should depend linearly on $\boldsymbol{u}$, because the boundary conditions on the surface of the sphere are linear in $\boldsymbol{u}$. Therefore,

$$\boldsymbol{A} = f^{'}(r)\boldsymbol{n} \times \boldsymbol{u} = grad f \times \boldsymbol{u} \ ; \tag{3.42}$$

it gives

$$\boldsymbol{v} = \boldsymbol{u} + curl\, curl(f\boldsymbol{u}) \tag{3.43}$$

and

$$curl \boldsymbol{v} = -\Delta curl(f\boldsymbol{u}) \ . \tag{3.44}$$

Equation $\Delta curl(\boldsymbol{v} - \boldsymbol{u}) = 0$ (equation (3.41)) leads to

$$\Delta^2 grad f = 0 \ , \tag{3.45}$$

or $\Delta^2 f = const$. This constant must be zero, because the derivatives of f are vanishing at infinity. Therefore,

$$\Delta f = \frac{2a}{r} \ , \tag{3.46}$$

where a is a constant; hence $f = ar + b/r$, where b is another constant. These constants are determined from the boundary condition $\boldsymbol{v} = 0$ on the surface of the sphere ($a = 3R/4$, $b = R^3/4$). We get finally

$$\boldsymbol{v} = -\frac{3}{4}R\frac{\boldsymbol{u} + \boldsymbol{n}(\boldsymbol{un})}{r} - \frac{1}{4}R^3\frac{\boldsymbol{u} - 3\boldsymbol{n}(\boldsymbol{un})}{r^3} + \boldsymbol{u} \ , \tag{3.47}$$

[5] O. Reynolds, "An experimental investigation of the circumstances which determine whether the motion of water shall be direct or sinuous, and of the law of resistance in parallel channels", Proc. Roy. Soc. London **174** 935 (1883).

or

$$v_r = u\cos\theta\left(1 - \frac{3R}{2r} + \frac{R^3}{2r^3}\right) \ ,$$
$$v_\theta = -u\sin\theta\left(1 - \frac{3R}{4r} - \frac{R^3}{4r^3}\right) \qquad (3.48)$$

in spherical coordinates with respect to the $\boldsymbol{u}$-axis.

The pressure distribution is given by equation (3.40),

$$p = p_0 - \frac{3}{2}\eta\frac{\boldsymbol{un}}{r^2}R \ , \qquad (3.49)$$

where p_0 is the pressure at infinity.

The force acting upon the sphere is given by equation (3.8); it has only one component directed along the velocity $\boldsymbol{u}$,

$$F = \oint(-p\cos\theta + \sigma'_{rr}\cos\theta - \sigma'_{r\theta}\sin\theta)df \ , \qquad (3.50)$$

where the viscosity tensor is given by

$$\sigma'_{rr} = 2\eta\frac{\partial v_r}{\partial r} \ , \ \sigma'_{r\theta} = \eta\left(\frac{1}{r}\frac{\partial v_r}{\partial\theta} + \frac{\partial v_\theta}{\partial r} - \frac{v_\theta}{r}\right) \qquad (3.51)$$

(see Appendix). On the surface of the sphere $\sigma'_{rr} = 0$ and $\sigma'_{r\theta} = -\frac{3\eta u}{2R}\sin\theta$. The force is

$$F = \frac{3\eta u}{2R}\oint df = 6\pi\eta Ru \ . \qquad (3.52)$$

This is known as Stokes' viscosity (friction) force, or drag force.[6] In general, for bodies with other shapes, the force can be written as $F_i = \eta a_{ij}u_j$, where a_{ij} is a symmetric matrix,[7] not necessarily directed

[6] G. G. Stokes, "On the effect of the internal friction of fluids on the motion of pendulums", Trans. Cambridge Phil. Soc. **9** 8 (1851) (in *Mathematical and Physical Papers by G. G. Stokes*, vol. 3, 1, Johnson Reprint Corp., NY (1966)).

[7] This is a particular case of symmetry of the kinetical coefficients (reciprocal relations), L. Onsager, "Reciprocal relations in irreversible processes, I, II", Phys. Rev. **37** 405; **38** 2265 (1931); L. Landau and E. Lifshitz, *Course of Theoretical Physics*, vol. 5 (*Statistical Physics*), Pergamon Press, Oxford (1980).

along the velocity $\boldsymbol{u}$. Equation (3.52) can also be written as $F = 8\pi\eta a u$, where a is the constant in equation (3.46) (and the calculations are done with equation (3.47)).

It is worth noting that we have estimated the Reynolds number for distances of the order of the dimension of the body l. For a distance r we get $R(r) = Vr/\nu$, which is smaller than unity for $r < \nu/V \simeq \nu/u$. In general, for larger distances, we cannot neglect the quadratic term in the Navier-Stokes equation. Corrections to the Stokes force at large distances are known,[8] as well as corrections to velocity distribution at short distances.[9] Also, the force for an ellipsoid is known (for a disk moving perpendicular to its plane $F = 16\eta R u$, parallel to its plane $F = 32\eta R u/3$).[10] For an infinite cylinder we need to start directly from the equation with the quadratic term.

3.8 Fluid drop

Let us consider now a spherical drop of fluid with radius R and viscosity η' moving in gravitational field in a fluid with viscosity η. First we note that the drop preserves its spherical shape, which implies that the friction force should be much smaller that the surface tension; the former is of order $\eta u/R$ (per unit area), the latter is of order α/R, where u is the velocity of the drop and α is the surface-tension coefficient. We should have $\eta u/R \ll \alpha/R$. The viscosity force per unit volume is of order $\eta u/R^2$; it should equal the gravity force ρg, where ρ is the density of the drop and g is the gravitational acceleration; therefore, the velocity is of the order $u = \rho g R^2/\eta$, such that the inequality becomes $R \ll \sqrt{\alpha/\rho g}$. On the other hand, the condition of a small Reynolds number reads $R/\nu \ll 1/\sqrt{g}$.

---

[8] C. W. Oseen "Ueber die Stokes'sche Formel, und ueber eine verwandte Aufgabe in der Hydrodynamik", Ark. Math. Astron. Fys. **6** 154 (1910); H. Lamb, "On the uniform motion of a sphere through a viscous fluid", Phil. Mag. **21** 112 (1911).

[9] I. Proudman and J. R. A Pearson, "Expansions at small Reynolds numbers for the flow past a sphere and a circular cylinder", J. Fluid Mech. **2** 237 (1957); M. van Dyke, *Perturbation Methods in Fluid Mechanics*, Academic Press, Stanford (1964).

[10] H. Lamb, *Hydrodynamics*, Cambridge University Press, Cambridge (1879).

The velocity outside the drop is given by equation (3.43) where $f = ar + b/r$ ($\Delta^2 f = 0$). Inside the drop the equation $\Delta^2 f = const$ has solution

$$f = \frac{1}{4}Ar^2 + \frac{1}{8}Br^4 \tag{3.53}$$

and the velocity ($\boldsymbol{v} = curl\, curl(f\boldsymbol{u})$) is

$$v = -A\boldsymbol{u} + Br^2\left[\boldsymbol{n}(\boldsymbol{un}) - 2\boldsymbol{u}\right] \tag{3.54}$$

(we consider a reference frame where the drop is at rest). The normal velocities at the surface of the drop should be equal,

$$v_{i,r} = v_{e,r} \ , \tag{3.55}$$

where i/e means internal/external. Similarly, the tangential velocities should be equal, $v_{i,\theta} = v_{e,\theta}$. The component $\sigma'_{r\theta}$ gives the tangential friction force, therefore these internal/external components should be equal. We get four boundary conditions for the four coefficients a, b, A and B :

$$\begin{gathered} a = R\tfrac{2\eta+3\eta'}{4(\eta+\eta')} \ , \ b = R^3\tfrac{\eta'}{4(\eta+\eta')} \ , \\ A = -BR^2 = \tfrac{\eta}{2(\eta+\eta')} \ . \end{gathered} \tag{3.56}$$

The Stokes' force is $F = 8\pi\eta au$ (equation (3.52)), *i.e.*

$$F = 2\pi u\eta R\frac{2\eta + 3\eta'}{4(\eta + \eta')} \ . \tag{3.57}$$

For a solid sphere $\eta' \rightarrow \infty$, such that we recover the standard Stokes' formula. For a gas bubble $\eta' = 0$.
The force of gravity acting on the drop is $\frac{4\pi}{3}R^3(\rho - \rho')g$, where ρ' is the density of the drop, so we get the limiting velocity[11]

$$u = \frac{2R^2g(\rho - \rho')(\eta + \eta')}{3\eta(2\eta + 3\eta')} \ . \tag{3.58}$$

[11] W. Rybczynski, "Ueber die fortscheitende Bewegung einer fluessigen Kugel in einem zaehen Medium", Bull. Acad. Sci., Krakow, **A1** 40 (1911).

3.9 Spheres and disks

We consider two concentric spheres with radii $R_{1,2}$ ($R_1 < R_2$), and an incompressible fluid between them. Let the R_2-sphere be at rest and the R_1-sphere be rotating with the angular velocity $\boldsymbol{\Omega}_1$. The Reynolds number is $R = \Omega_1 R_1^2/\nu$. We consider it very small, so equations (3.38) and (3.40) apply. Obviously, the velocity is tangential in the plane perpendicular to the rotation axis, and the *gradp* in this direction is zero. Therefore, equation (3.40) becomes

$$\Delta \boldsymbol{v} = 0 \ . \tag{3.59}$$

From $div\boldsymbol{v} = 0$ the velocity derives from a $curl\boldsymbol{A}$, where $\boldsymbol{A}$ is an axial vector, like $\boldsymbol{\Omega}_1$. It follows $\boldsymbol{A} = f(r)\boldsymbol{\Omega}_1$ and

$$\boldsymbol{v} = curl(f\boldsymbol{\Omega}_1) = gradf \times \boldsymbol{\Omega}_1 \ . \tag{3.60}$$

From equation (3.59) we get $\Delta f = const$ and

$$f = ar^2 + \frac{b}{r} \ , \ \boldsymbol{v} = \left(\frac{b}{r^3} - 2a\right)\boldsymbol{\Omega}_1 \times \boldsymbol{r} \ . \tag{3.61}$$

The boundary conditions are $\boldsymbol{v} = 0$ for $r = R_2$ and $\boldsymbol{v} = \boldsymbol{\Omega}_1 \times \boldsymbol{r}$ for $r = R_1$. The result is

$$\boldsymbol{v} = \frac{R_1^3 R_2^3}{R_2^3 - R_1^3}\left(\frac{1}{r^3} - \frac{1}{R_2^3}\right)\boldsymbol{\Omega}_1 \times \boldsymbol{r} \ . \tag{3.62}$$

If the R_1-sphere is at rest and the R_2-sphere is rotating with $\boldsymbol{\Omega}_2 \times \boldsymbol{r}$, the velocity is

$$\boldsymbol{v} = \frac{R_1^3 R_2^3}{R_2^3 - R_1^3}\left(\frac{1}{R_1^3} - \frac{1}{r^3}\right)\boldsymbol{\Omega}_2 \times \boldsymbol{r} \ . \tag{3.63}$$

If both spheres rotate, the solution is the sum of these two solutions. The pressure is a constant p_0.

Let us assume that the outer sphere is absent, *i.e.* $R_2 \to \infty$, $\boldsymbol{\Omega}_2 = 0$ and $R_1 = R$, $\boldsymbol{\Omega}_1 = \boldsymbol{\Omega}$; then,

$$\boldsymbol{v} = \frac{R^3}{r^3}\boldsymbol{\Omega} \times \boldsymbol{r} \ . \tag{3.64}$$

In spherical polar coordinates with the polar axis along $\mathbf{\Omega}$ the velocity components are $v_r = v_\theta = 0$ and $v_\varphi = v = \frac{R^3}{r^2}\Omega \sin\theta$. There exists only one component of the viscosity tensor

$$\sigma'_{r\varphi} = \eta\left(\frac{\partial v}{\partial r} - \frac{v}{r}\right) = -3\eta\frac{R^3}{r^3}\Omega \sin\theta \ , \tag{3.65}$$

which is the friction force per unit area (equation (3.8)). The total moment of this force is

$$M = \int 3\eta\Omega \sin\theta \cdot R\sin\theta \cdot 2\pi R^2 \sin\theta \cdot d\theta = 8\pi\eta\Omega R^3 \ . \tag{3.66}$$

If the inner sphere is absent ($R_1 = 0$), the velocity is $\boldsymbol{v} = \mathbf{\Omega}_2 \times \boldsymbol{r}$, *i.e.* the fluid moves rigidly with the rotating sphere.

Let us consider now two parallel disks, comprising an incompressible fluid: one placed at $z = 0$ at rest, another rotating at height h with a constant angular velocity Ω. The velocity is tangential, depending on r and z. It is given by equation (3.61),

$$v_\varphi = v = \left(2a - \frac{b}{(r^2+z^2)^{3/2}}\right) r\Omega \ . \tag{3.67}$$

The boundary conditions are $v = 0$ for $z = 0$ and $v = r\Omega$ for $z = h$. For small values of h we get

$$v = \frac{rz^2}{h^2}\Omega \ . \tag{3.68}$$

The friction force per unit area is

$$-\sigma'_{z\varphi} = -\frac{2\eta rz}{h^2}\Omega \tag{3.69}$$

and its total moment acting upon the rotating disk is $M = -\pi\eta\Omega R^4/h$.

Let us suppose now that we have two parallel disks with radius R, one placed at $z = 0$, another at $z = h$. The upper disk moves vertically and uniformly with velocity u towards the lower disk, displacing the fluid which fills the room between the two disks. Since $h \gg R$, the velocity is mainly radial, so the equations are

$$\begin{gathered} \eta\frac{\partial^2 v_r}{\partial z^2} = \frac{\partial p}{\partial r} \ , \ \frac{\partial p}{\partial z} = 0 \\ \frac{1}{r}\frac{\partial(rv_r)}{\partial r} + \frac{\partial v_z}{\partial z} = 0 \ , \end{gathered} \tag{3.70}$$

with boundary conditions

$$\begin{aligned} & v_r = v_z = 0 \ , \ z = 0 \ , \\ & v_r = 0 \ , \ v_z = -u \ , \ z = h \ , \\ & p = p_0 \ , \ r = R \ , \end{aligned} \tag{3.71}$$

where p_0 is the external pressure. We get

$$v_r = \frac{1}{2\eta}\frac{dp}{dr}z(z-h) \tag{3.72}$$

and

$$u = \frac{1}{r}\frac{d}{dr}\int_0^h rv_r dz = -\frac{h^3}{12\eta r}\frac{d}{dr}\left(r\frac{dp}{dr}\right) \ . \tag{3.73}$$

Hence,

$$p = p_0 + \frac{3\eta u}{h^3}(R^2 - r^2) \ . \tag{3.74}$$

The total resistance force is $F = 3\pi\eta u R^4/2h^3$.

3.10 Kinetic-molecular drag

Let us consider a plane solid surface (solid body) placed vertically in the y, z-plane and moving horizontally with velocity $v > 0$ along the x-axis through an ideal classical gas. The energy and momentum conservation laws for an elastic collision of the surface with a gas molecule are

$$\begin{aligned} & Mv^2 + mv_x^2 = Mv'^2 + mv_x'^2 \ , \\ & Mv + mv_x = Mv' + mv_x' \ , \end{aligned} \tag{3.75}$$

where M is the mass of the body, m is the molecular mass, v' is the velocity of the body after collision and v_x, v_x' are the molecule velocities before and after collision, respectively. We assume that the collisions along the y, z-directions are balanced, such that they do not change the state of motion. Also, we note that the velocities in equations (3.75) are algebraic (with their sign). From equations (3.75) we get

$$v' = \frac{1-m/M}{1+m/M}v + \frac{2m/M}{1+m/M}v_x \ ,$$
$$v_x' = \frac{2}{1+m/M}v - \frac{1-m/M}{1+m/M}v_x \ . \tag{3.76}$$

The momentum (ΔP) and energy (ΔE) changes for the body are

$$\Delta P = M(v' - v) = -\frac{2m}{1+m/M}(v - v_x) \ ,$$
$$\Delta E = \frac{1}{2}M(v'^2 - v^2) = -\frac{2m(v-v_x)}{(1+m/M)^2}\left(v + \frac{m}{M}v_x\right) \ . \tag{3.77}$$

Let us consider now those molecules with velocity $v_x < 0$ along the x-axis. These molecules collide with the body, on the right side ($x > 0$). The number of collisions per unit time and unit area is $n(v - v_x)$, where n is the gas density. It follows that a force per unit area

$$p_1 = \Delta P \cdot n(v - v_x) = -\frac{2mn}{1 + m/M}(v - v_x)^2 \ , \ v_x < 0 \tag{3.78}$$

occurs on the right side of the body; it is directed along the negative x-axis. If we average this force over all velocities $v_x < 0$, we get the corresponding pressure. Let us consider now the molecules with velocity $v_x > 0$. For $v_x > v$, they collide with the body on the left side; the number of collisions per unit time and unit area is $n(v_x - v)$. If $v_x < v$, the body collides with these molecules on the right side; the number of these collisions per unit time and unit area is $n(v + v_x)$. It follows that we get the pressure

$$p_2 = \frac{2mn}{1+m/M}(v_x - v)^2 \ , \ v_x > v \ ,$$
$$p_2 = -\frac{2mn}{1+m/M}(v^2 - v_x^2) \ , \ 0 < v_x < v \ , \tag{3.79}$$

where the average must be taken over the values of v_x which are indicated in each row of equations (3.79). These averages imply truncated integrals of gaussians. It is convenient to estimate the pressure in two distinct cases: $v \ll v_{th}$ and $v \gg v_{th}$, where we take for the thermal velocity $v_{th} = \overline{v}_x = (T/2\pi m)^{1/2}$ (for $v_x > 0$, or $v_x < 0$), T being the temperature. For $v \ll v_{th}$ we may neglect the second raw in equations

(3.79), such that the total pressure is given by adding the first row of equations (3.79) and equation (3.78). We get

$$p \simeq -\frac{8mn}{1+m/M} v\overline{v}_x \ , \ v_x > 0 \ , \ v \ll v_{th} \ . \tag{3.80}$$

In the second case, $v \gg v_{th}$, we may neglect the first row in equations (3.79) and add the second row and equation (3.78) to get the total pressure

$$\begin{gathered} p \simeq -\frac{4mn}{1+m/M}\left(v^2 + v\overline{v}_x\right) \simeq \\ \simeq -\frac{4mn}{1+m/M} v^2 \ , \ v_x > 0 \ , \ v \gg v_{th} \ . \end{gathered} \tag{3.81}$$

The final result is[12]

$$\begin{gathered} p \simeq -\frac{8mn}{1+m/M} v v_{th} \ , \ v \ll v_{th} \ , \\ p \simeq -\frac{4mn}{1+m/M} v^2 \ , \ v \gg v_{th} \ . \end{gathered} \tag{3.82}$$

We can see that in both cases a drag force occurs (opposite to the direction of motion). An interpolation formula is

$$p \simeq -\frac{4mn}{1+m/M}\left(v^2 + 2vv_{th}\right) \ . \tag{3.83}$$

We may neglect the ratio m/M in the denominator of equation (3.83) and replace mn by the mass density ρ; for a macroscopic body the force implies the area A of its cross-section, such that, for small velocities the drag force is

$$F \simeq -8\rho v_{th} A v \ ; \tag{3.84}$$

for a small body with dimension r the effective area is reduced by transverse collisions to $\delta A \simeq r\Lambda$, where Λ is the mean freepath; the force becomes $F \simeq -8\rho\Lambda v_{th} r v$. The product $\rho\Lambda v_{th}$ is the viscosity η, such that the force is $\simeq -8r\eta v$, which is approximately Stokes'

[12]M. Apostol, "Drag force through gases and plasma", Ann. Math. Phys. **5** doi.org/10.17352/amp.000031 (2022).

formula $-6\pi r\eta v$ for a sphere of radius r.[13] For high velocities the formula was derived by Rayleigh by means of the fluid dynamics.[14] It is worth noting that the pressure $p \simeq -4\rho v^2$ given by equation (3.82) for $v \gg v_{th}$ is eigth times larger than the hydrodynamic pressure $\frac{1}{2}\rho v^2$. A similar result is obtained also for liquids, which, in this context, exhibit a molecular kinetics similar to gases.[15] The difference between the molecular drag force and the Stokes' force is that the force is not generated by individual molecular collisions, but it is caused by fluid particles collisions, where the molecular motion is averaged.

The drag force is derived here by using molecular kinetics (equivalent to transport equation with a macroscopic velocity v); the transported quantity is momentum. If we leave aside the molecular velocities, as for microscopic, but large, amounts of fluid, we may get Euler's equation of motion of fluids (Navier-Stokes equations). This was the original approach of Stokes (who used the empirical viscosity) and Rayleigh. Further on, for low velocities, from Euler's equation we may get the Navier-Cauchy equations of elastic motion. For low velocities these equations predict sound waves, in particular the hydrodynamic sound, with velocity $c = \sqrt{c_p T / c_v m}$ in an ideal classical gas, where $c_{p,v}$ are specific heats at constant pressure and volume, respectively. We note that c is close to the molecular thermal velocity $\simeq \sqrt{T/m}$. Sound is a collective excitation, which implies local thermal equilibrium, but not a global equilibrium: the pressure, the density and the temperature vary on a global scale (it is a mechanical motion compatible with the statistical motion; sound is not a transport phenomenon).

Let us examine first the case $v \ll v_{th}$. Let us assume a local displacement $\mathbf{u}$ in the gas; we assume that it is sufficiently small to produce a small pressure imbalance δp; since $u \ll \Lambda$, this imbalance is adiabatic, *i.e.* the molecules have not enough time to collide to each other. This displacement produces a small density imbalance $\delta\rho = -\rho div\mathbf{u}$.

---

[13] G. G. Stokes, "On the effect of the internal friction of fluids on the motion of pendulums", Trans. Cambridge Phil. Soc. **9** 8 (1851) (in *Mathematical and Physical Papers by G. G. Stokes*, vol. 3, 1, Johnson Reprint Corp., NY (1966)).

[14] Lord Rayleigh, "On the resistance of fluids", Phil. Mag. **2** 430 (1876).

[15] M. Apostol, "A model for the thermodynamics of simple liquids", Physica **B403** 3946 (2008); *Physical Kinetics*, Cambridge Scholars Publishing, Newcastle upon Tyne (2020).

According to Euler's equation $\ddot{\mathbf{u}}=-\frac{1}{\rho}grad\delta p$, we get

$$\frac{\partial^2\mathbf{u}}{\partial t^2}=-\frac{1}{\rho}\left(\frac{\partial p}{\partial\rho}\right)_s(-\rho grad\,div\mathbf{u})=\left(\frac{\partial p}{\partial\rho}\right)_s grad\,div\mathbf{u}\ , \qquad (3.85)$$

which is the equation of sound propagating with velocity given by $c^2=(\partial p/\partial\rho)_s$, where the suffix s means constant entropy (s). Since $(\partial p/\partial\rho)_s=(c_p/c_v)\,(\partial p/\partial\rho)_T$ (see Appendix), we get the sound velocity given above (where $(\partial p/\partial\rho)_T=T/m$). This is the hydrodynamic sound in an ideal gas. Since it is collisionless, it follows that it cannot be excited by the motion of the body through the gas, as long as $v\ll v_{th}$. In particular, the density variations produced by the sound are $\delta p=c^2\delta\rho=-\rho c^2 div\mathbf{u}=-\frac{\rho c_p T}{c_v m}div\mathbf{u}$, while the pressure variation given by equations (3.82) is $\delta p=-8\rho\sqrt{\frac{T}{2\pi m}}v$. This latter pressure does not produce variations of the density. The sound is a local-equilibrum phenomenon, while the drag force is caused by friction (viscosity), which, on its short scale, is a non-equilibrium phenomenon. The drag force dissipates energy, by mechanical work, which, per unit time, can be computed from the lost energy given by equations (3.77).

The situation is different in the regime $v\gg v_{th}$; in this case the body generates variations of the density, and, therefore, excites sound waves. It is worth noting that the pressure $-4\rho v^2$ given by equations (3.82) in this case does not depend on v_{th}, as if the molecular movement would be immaterial. This is the hydrodynamic regime. Indeed, let us assume that $Vh(\mathbf{r})$ is the characteristic function of the body, where V is the volume of the body; the characteristic function of the gas is $f(\mathbf{r})=1-Vh(\mathbf{r})$. For a pointlike body we can take $h(\mathbf{r})=\delta(\mathbf{r})$. We are interested in the variation δf of this function due to the movement of the body, *i.e.* the variation of the function $f(\mathbf{r}-\mathbf{v}t)=1-Vh(\mathbf{r}-\mathbf{v}t)$, where t denotes the time. The variation δF of any quantity F associated to the body is given by $\delta F=F\delta f$. The variation of this function can be written as $\delta f=V\mathbf{v}gradh(\mathbf{r}-\mathbf{v}t)\delta t$. As long as $v\ll v_{th}$ the time δt is much longer than the mean collision time τ, such that the equilibrium is restored rapidly, and the variation δf is zero. If $v\gg v_{th}$, the time δt is much shorter than τ, and we have an estimate $\delta f=V\tau\mathbf{v}gradh(\mathbf{r}-\mathbf{v}t)$ for the variation δf. Therefore, we have a variation of the pressure $\delta p = -(4\rho v^2)V\tau\mathbf{v}gradh(\mathbf{r}-\mathbf{v}t)$ and a variation of the density

$$\delta\rho = \rho V\tau\mathbf{v}gradh(\mathbf{r}-\mathbf{v}t) = \rho V\tau div[\mathbf{v}h(\mathbf{r}-\mathbf{v}t)] \ . \tag{3.86}$$

Since $\delta\rho = -\rho div\mathbf{u}$, where $\mathbf{u}$ is the displacement associated with the density variations, we get the displacement

$$\mathbf{u} = -V\tau\mathbf{v}h(\mathbf{r}-\mathbf{v}t) \ ; \tag{3.87}$$

obviously, this displacement satisfies the equation

$$\frac{\partial^2\mathbf{u}}{\partial t^2} - (\mathbf{v}grad)^2\,\mathbf{u} = 0 \ , \tag{3.88}$$

or

$$\frac{\partial^2 u}{\partial t^2} - v^2\frac{\partial^2}{\partial r^2}u = 0 \ , \tag{3.89}$$

where $\mathbf{u}$ and $\mathbf{r}$ are directed along the velocity $\mathbf{v}$ (and $\mathbf{u}$ is restricted to the cross-section of the body); this is a shock wave. Due to the equilibrium movement, the thickness of the tail increases in time with velocity v_{th}. The displacement $\mathbf{u}$ can be expanded in a Fourier series

$$\mathbf{u} = -V\tau\mathbf{v}\frac{1}{(2\pi)^3}\int d\mathbf{k}h(\mathbf{k})e^{i(\mathbf{kr}-\mathbf{kv}t)} \ , \tag{3.90}$$

which shows that the disturbance is a superposition of sound waves with frequency $\omega = vk = c(v/c)k$; since the wavevector is localized over a range $\Delta k \simeq 1/l$, where l is the dimension of the body, it follows that the sound extends up to a frequency of the order $\Delta\omega \simeq v/l$. The movement of the body in this case dissipates energy by exciting sound waves. It is worth noting that equations (3.88) and (3.89) are hydrodyamic equations, where there is no external force: the force exerted by the body on the gas is taken by the molecular movement, which determines (through τ) the amplitude of the free sound waves.

The above treatment can also be used for a plasma.[16] Let us assume an ionized gas, and consider only its ionized component. The density of the gas and the density of the plasma are very low. The electrons

[16]M. Apostol, *Physical Kinetics*, Cambridge Scholars Publishing, Newcastle upon Tyne (2020)

(e) have a temperature T_e, while the ions (i) have a temperature T_i, such that $T_e \gg T_i$. Both temperatures are much higher than the Coulomb energy q^2/a, where q is the electron (ion) charge and a is the mean separation distance between the particles. The electrons are correlated to the ions, through the Debye length, which in this case is very large. The corresponding cross-sections are $\sigma_{e-e} = \left(q^2/T_e\right)^2$, $\sigma_{i-i} = \left(q^2/T_i\right)^2$ and $\sigma_{e-i} = q^4/(T_e^2 + T_i^2)$; the mean freepaths are $\Lambda_{e-e} = (a^2/\sigma_{e-e})a = a^3T_e^2/q^4$, $\Lambda_{i-i} = a^3T_i^2/q^4$ and $\Lambda_{e-i} = a^3T_e^2T_i^2/ /q^4\left(T_e^2 + T_i^2\right)$; the mean lifetimes are $\tau_{e-e} = \Lambda_{e-e}/v_e = a^3T_e\sqrt{mT_e}/q^4$, $\tau_{i-i} = a^3T_i\sqrt{MT_i}/q^4$ and $\tau_{e-i} = a^3T_eT_i\sqrt{mMT_eT_i} /q^4\left(T_e\sqrt{mT_e} + T_i\sqrt{MT_i}\right)$, where m is the electron mass and M is the ion mass. If $T_e\sqrt{mT_e} \gg T_i\sqrt{MT_i}$ (which is the most realistic case), then $\tau_{e-i} \simeq \tau_{i-i} \ll \tau_{e-e}$, which shows that the electrons follow rapidly the ion motion, which ensures the equilibrium (electron-electron processes are immaterial for equilibrium); this is in accordance with the adiabatic hypothesis. In the unrealistic case $T_e\sqrt{mT_e} \ll T_i\sqrt{MT_i}$ the electrons lag behind the ions. In this case there exist an ionic displacement $\mathbf{u}_i$ and an electron displacement $\mathbf{u}_e$, as well as a restoring polarization force which is responsible for a plasma frequency given by $\omega_0^2 = 4\pi nq^2/\mu$, where μ is the reduced ion-electron mass. The ion-electron center of mass moves freely, with an ionic displacement coordinate which may be estimated as in equation (3.87); the electron component remains to be determined from boundary conditions. The ion-electron relative coordinate is subject to the polarization force and is a superposition of waves with frequencies given by $\omega^2 = \omega_0^2 + v^2k^2$. We note that this is a hydrodynamic regime, where the pressure force $\simeq -4\rho v^2$, although hydrodynamic, is equilibrated by the molecular movement.

3.11 Submerged jet

Let us consider a fluid flowing form a narrow tube in a mass of same fluid. This is a submerged jet. We are interested in its steady, viscous flow for a constant density. This is an exactly solvable problem.[17]

[17] L. Landau, "A new exact solution of the Navier-Stokes equations", C. R. (Dokl.) Acad. Sci. URSS **43** 286 (1944); L. Landau and E. Lifshitz, *Course of Theo-*

The Navier-Stokes equations read

$$\frac{\partial}{\partial t}(\rho v_i) = -\partial_j \Pi_{ij} \ , \tag{3.91}$$

where ρ is the density, $\boldsymbol{v}$ is the velocity and Π_{ij} is the tensor of the momentum flux density; it is given by

$$\begin{aligned} \Pi_{ij} &= -\sigma_{ij} + \rho v_i v_j \ , \\ \sigma_{ij} &= -p\delta_{ij} + \sigma'_{ij} \ , \\ \sigma'_{ij} &= \eta(\partial_i v_j + \partial_j v_i) \ , \end{aligned} \tag{3.92}$$

where p is the pressure, σ_{ij} is the stress tensor, σ'_{ij} is the viscosity tensor and η is the viscosity coefficient (a second contribution to the viscosity tensor is vanishing for an incompressible fluid). We assume that the mouth of the tube is a point, which is taken as the origin. We use spherical coordinates, with the polar axis directed along the tube. Due to the symmetry of the problem $v_\varphi = 0$ and $v_{r,\theta}$ do not depend on the azimuthal angle φ. The jet is generated by an external force, directed along the polar axis, which we denote by P; it is given by $\oint \Pi_{ij} dS_j$, where the integration is performed over the surface at infinity. This force may be called the "jet momentum". It follows that $v_{r,\theta}$ go like $1/r$, *i.e.* they have the form

$$v_r = \frac{F(\theta)}{r} \ , \ v_\theta = \frac{f(\theta)}{r} \ , \tag{3.93}$$

where θ is the zenithal angle, r is the radius and F, f are functions to be determined. The continuity equation

$$\frac{1}{r^2}\frac{\partial(r^2 v_r)}{\partial r} + \frac{1}{r\sin\theta}\frac{\partial(\sin\theta v_\theta)}{\partial\theta} = 0 \tag{3.94}$$

gives

$$F = -f' - f\cot\theta \ . \tag{3.95}$$

Due to the symmetry of this (steady) flow there is no momentum transfer, except the one brought about by the radial velocity along

---

retical Physics, vol. 6 (*Fluid Mechanics*), Elsevier, Oxford (1989).

the radius; *i.e.*, all the components of the momentum flux density are zero, except Π_{rr}. Its contribution is balanced by the external force. So, the Navier-Stokes equations are satisfied. It is useful to write down some components of the momentum flux density in spherical coordinates (see Appendix):

$$\begin{aligned} \Pi_{rr} &= p + 2\eta\tfrac{F}{r^2} + \rho\tfrac{F^2}{r^2} \ , \\ \Pi_{\theta\theta} &= p - \tfrac{2\eta}{r^2}(F + f') + \rho\tfrac{f^2}{r^2} \ , \\ \Pi_{\varphi\varphi} &= p - \tfrac{2\eta}{r^2}(F + f\cot\theta) \ , \\ \Pi_{r\theta} &= -\tfrac{\eta}{r^2}(F' - 2f) + \rho\tfrac{Ff}{r^2} \ . \end{aligned} \tag{3.96}$$

From these equations we can see that $\Pi_{\theta\theta} - \Pi_{\varphi\varphi} = 0 - 0 = 0$ leads to

$$f^2 - 2\nu f' + 2\nu f\cot\theta = 0 \ , \tag{3.97}$$

where $\nu = \eta/\rho$ is the kinematic viscosity. The solution of this equation is

$$f = -\frac{2\nu\sin\theta}{A - \cos\theta} \tag{3.98}$$

and

$$F = 2\nu\left[\frac{A^2 - 1}{(A - \cos\theta)^2} - 1\right] \ , \tag{3.99}$$

where $A > 1$. The constant A is determined from the external force

$$\begin{aligned} P &= \oint \Pi_{rr}\cos\theta dS = \\ &= 16\pi\nu^2\rho A\left[1 + \tfrac{4}{3(A^2-1)} - \tfrac{1}{2}A\ln\tfrac{A+1}{A-1}\right] \ , \end{aligned} \tag{3.100}$$

where the integration is performed over a spherical surface. This amounts to solving the Navier-Stokes equations. From $\Pi_{\theta\theta} = 0$ we get the pressure

$$p = p_0 - \frac{4\rho\nu^2}{r^2}\frac{A\cos\theta - 1}{(A - \cos\theta)^2} \ , \tag{3.101}$$

where p_0 is the pressure at infinity.

The streamlines are given by $dr/v_r = rd\theta/v_\theta$, whence we get

$$r = C\frac{A - \cos\theta}{\sin^2\theta} \ , \tag{3.102}$$

where C is a constant (for each streamline). The distance to the polar axis is

$$d = r\sin\theta = C\frac{A - \cos\theta}{\sin\theta} \ ; \tag{3.103}$$

it has a minimum $d_0 = C\sqrt{A^2 - 1}$ for $\cos\theta_0 = 1/A$. The jet sucks the surrounding fluid. A weak jet (small P) corresponds to $A \to \infty$; a strong jet (large P) corresponds to $A \to 1$.

The amount of mass flowing (per unit time) through a closed surface surrounding the origin can be obtained by integrating the velocity v_r over that surface; it is zero, as expected for a tube with zero thickness. It is different from zero for a tube with a finite thickness. In that case we need to consider an expansion in series of the ratio of the thickness of the tube to the distance. The solution given above corresponds to the leading term of this series. The mass of the fluid discharged per unit time may be viewed as another constant of integration (besides P). A non-zero angular momentum can also be considered.[18]

It is worth noting that the Reynolds number of this problem is $\sqrt{P/\rho\nu^2}$. We can see that it is not related explicitly to a length. For large velocities, *i.e.* large P, the dissipation is insufficient, discontinuity surfaces may appear, which are unstable. The turbulence appears in that case.

3.12 Laminar wake

Let us consider an incident flow with a constant velocity $\boldsymbol{U}$ at infinity, directed along the x-axis, past a solid body at rest. The fluid adheres to the surface of the body, so we have a slight rotational motion in a narrow region around the x-axis, in the neighbourhood of the body.

[18] L. Landau and E. Lifshitz, *Course of Theoretical Physics*, vol. 6 (*Fluid Mechanics*), Elsevier, Oxford (1989) and References therein; R. I. Mullyadzhanov and N. I. Yavorskii, "Solution of the problem of flow of a non-axisymmetric swirling submerged jet", J. Appl. Mech. Tech. Phys. **54** 207 (2013).

This happens at small Reynolds numbers, where the viscosity dominates the inertia of the fluid. At large distance, the inertial force dominates the viscosity, so we have a potential flow. This is the laminar wake. The velocity of the fluid is $\boldsymbol{U} + \boldsymbol{v}$, where $\boldsymbol{v}$ is small and vanishing at infinity. The inertial term in the Navier-Stokes equation can be approximated by $\rho(\boldsymbol{U} grad)\boldsymbol{v}$; this is called the Oseen approximation. This contribution is of the order of $\rho U v/x$ (since $\boldsymbol{v}$ varies slowly). The viscosity force is of the order of $\eta v/x^2$. For $Ux/\nu \gg 1$ we have a potential flow.

For small $\boldsymbol{v}$ the momentum flux density is $\simeq \rho U_i v_k$, where ρ (= *const*) is the fluid density. It follows that we have a drag force

$$F_x = -\rho U \int v_x dydz \tag{3.104}$$

and a lift force

$$F_{y,z} = -\rho U \int v_{y,z} dydz \ , \tag{3.105}$$

where the integration is performed over the narrow cross-section of the wake. These are known as the Joukowski-Kutta equations.[19]

Also, the viscosity term is of the order of $\eta v/d^2$, where d is the width of the wake; therefore, this width is of the order of

$$d \simeq \sqrt{\nu x/U} = \frac{x}{\sqrt{Ux/\nu}} \ll x \ . \tag{3.106}$$

3.13 Viscosity of suspensions

We consider many small and rare, identical solid particles in an incompressible fluid. This is a suspension. Obviously, the viscosity of the fluid changes, since the surface of the solid particles brings its own contribution. Therefore, the viscosity will increase. Although the viscosity acts at the surface, the change of the viscosity depends on the

[19]N. E. Joukowski, "Ueber die Konturen der Tragflaechen der Drachenflieger", Zeitschrift fuer Flugtechnik und Motorluftschiffahrt **1** 281 (1910); *Aerodynamique*, Gauthier-Villars, Paris (1931); W. M. Kutta, "Ueber eine mit den Grundlagen des Flugproblems in Beziehung stehende zweidimensionale Stroemung", Sitzungsber. Bayr. Akad. Wiss., math.-phys. Kl., Muenchen, 1 (1910).

displaced volume. Therefore, the relative change in viscosity will be proportional to the relative change in volume, $\Delta\eta/\eta \simeq \frac{\frac{4\pi}{3}R^3N}{V} \ll 1$, where R is the radius of the particles, assumed spherical, N is their number and V is the total volume; $n = N/V$ is the concentration of the suspension. The problem is to determine the numerical coefficient of this relation. For spherical particles the coefficient is 2.5.[20] For other shapes the coefficient varies in the neighbourhood of this value. The flow produces an orientational anisotropy for non-spherical particles, which competes with the Brownian motion. However, the orientational anisotropy is proportional to the gradient of the flow, which would introduce non-linear terms in the stress tensor, *i.e.* a second-order effect, which may be neglected.

Let us consider a solid sphere placed at the origin. The unperturbed (steady) fluid velocity can be written as

$$v_{0i} = \alpha_{ij}x_j \ , \tag{3.107}$$

where α_{ij} is a constant symmetric matrix, with $\alpha_{ii} = 0$ (continuity equation). In the vicinity of the spheres the fluid velocity is $\boldsymbol{v} = \boldsymbol{v}_0 + \boldsymbol{v}_1$. All these velocities are small, so we limit ourselves to the linear contributions. The unpertubed pressure is a constant, which we take zero. The velocity $\boldsymbol{v}_1$ is determined exactly as in equation (3.43), which now becomes

$$\boldsymbol{v}_1 = curl\, curl[(\boldsymbol{\alpha} grad)f] \ , \tag{3.108}$$

where $\boldsymbol{\alpha} grad$ is the vector with components $\alpha_{ij}\partial_j$ and $f = ar + b/r$ (equation (3.46)). The variation of the pressure is

$$p = \eta_0 \alpha_{ij}\partial_i\partial_j \Delta f \ , \tag{3.109}$$

where η_0 is the unperturbed viscosity (equation (3.40) $\eta_0\Delta\boldsymbol{v}_1 = gradp$). The boundary conditions on the surface of the sphere are $\boldsymbol{v} = 0$ (the sphere remains at rest, due to the symmetry of the flow). Making use of these conditions, we get

$$\begin{aligned} v_{1i} &= \frac{5}{2}\left(\frac{R^5}{r^4} - \frac{R^3}{r^2}\right)\alpha_{jk}n_in_jn_k - \frac{R^5}{r^4}\alpha_{ij}n_j \ , \\ p &= -5\eta_0\frac{R^3}{r^3}\alpha_{ij}n_in_j \ , \end{aligned} \tag{3.110}$$

[20] A. Einstein, "Eine neue Bestimmung der molekueldimensionen", Ann. Physik **19** 289 (1906).

where $\boldsymbol{n} = \boldsymbol{r}/r$.

In order to estimate the viscosity we use the stress tensor $\sigma_{ij} = -p\delta_{ij} + \eta(\partial_i v_j + \partial_j v_i)$. The equation of motion is $\partial_j \sigma_{ij} = 0$. For the unperturbed flow

$$\sigma_{ij} = 2\eta_0 \alpha_{ij} \ . \tag{3.111}$$

The perturbation is computed with $\boldsymbol{v}_1$ and p given by equations (3.87), multiplied by N. We are interested in an average effect, so we need to calculate the integral

$$\overline{\sigma}_{ij} = \frac{1}{V} \int dV \left[-p\delta_{ij} + \eta_0(\partial_i v_{1j} + \partial_j v_{1i})\right] \ , \tag{3.112}$$

where the integration is extended over a large volume V and the excluded small volume of the solid particles is neglected. The pressure contribution to equation (3.112) is a scalar, it should be proportional to $\alpha_{ii} = 0$. Since $\partial_j \sigma_{ij} = 0$, we can write

$$\sigma_{ij} = \partial_k(\sigma_{ik} x_j) \ , \tag{3.113}$$

such that the integral becomes

$$\overline{\sigma}_{ij} = n \oint \eta_0(\partial_i v_{1k} x_j + \partial_k v_{1i} x_j) df_k \ , \tag{3.114}$$

Only the $1/r^2$-term in equation (3.87) contributes to infinity; we get

$$\overline{\sigma}_{ij} = n\eta_0 \cdot 20\pi R^3 5\alpha_{kl}\overline{n_i n_j n_k n_l} - \alpha_{ik}\overline{n_j n_k} \ , \tag{3.115}$$

where the average is taken over directions. These averages are constructed with δ_{ij}; for instance, $\overline{n_i n_j} = \frac{1}{3}\delta_{ij}$ and

$$\overline{n_i n_j n_k n_l} = \frac{1}{15}\left(\delta_{ij}\delta_{kl} + \delta_{ik}\delta_{jl} + \delta_{il}\delta_{jk}\right) \ . \tag{3.116}$$

Finally we get

$$\overline{\sigma}_{ij} = 5n\eta_0\alpha_{ij} \cdot \frac{4\pi}{3} R^3 \ , \tag{3.117}$$

such that, on comparing with equation (3.111), we obtain the modified viscosity

$$\eta = \eta_0 \left(1 + \frac{5}{2} \cdot \frac{4\pi}{3} n R^3\right) \ . \tag{3.118}$$

This equation may serve to determine the dimension of the molecules, by measuring the viscosity. Although the molecular dimensions are much smaller than the dimension of the fluid particles, the calculations are valid because the viscosity occurs at the interface, and the presence of the molecules modifies the fluid viscosity. A similar result derived by molecular collisions would require an averaging over molecular motion (see the Kinetic-molecular drag section above).

3.14 Drift currents

Let us consider the motion of the plane-surface layers of a rotating fluid, like, for instance, the motion of the surface of seas and oceans under the action of winds. We introduce the local frame with coordinate axes oriented from North to South (x), West to East (y) and the vertical. The stationary motion for an incompressible fluid with small velocities is governed by the equation

$$-gradp + 2\rho\boldsymbol{v} \times \boldsymbol{\Omega} + \rho\boldsymbol{g} + \eta\Delta\boldsymbol{v} = 0 \ , \tag{3.119}$$

where p is the pressure, ρ is the density, $\boldsymbol{\Omega}$ is the angular velocity of the Earth, $\boldsymbol{g}$ is the gravitational acceleration and η is the viscosity coefficient. We include the Coriolis force, which depends on the velocity, and left aside the small centrifugal force, which may be included in pressure. For a point on the northern hemisphere $\boldsymbol{\Omega} = (-\Omega\cos\varphi, 0, \Omega\sin\varphi)$, where φ is the latitude. We assume $v_{x,y}(z)$, $p(z)$ and $v_z = 0$; the continuity equation is satisfied; equations (3.119) become

$$\begin{gathered} 2\Omega v_y \sin\varphi + \nu\frac{d^2v_x}{dz^2} = 0 \ , \\ -2\Omega v_x \sin\varphi + \nu\frac{d^2v_y}{dz^2} = 0 \ , \\ -\frac{dp}{dz} + 2\rho\Omega v_y\cos\varphi - \rho g = 0 \ , \end{gathered} \tag{3.120}$$

where $\nu = \eta/\rho$ is the kinematic viscosity. Since the Coriolis force is much smaller than the gravity force, the third equation (3.120) is satisfied by the hydrostatic-equilibrium equation $p = -\rho gz + const.$ The first two equations give

$$V = v_x + iv_y = Ce^{\sqrt{\frac{\Omega\sin\varphi}{\nu}}(1+i)z} \tag{3.121}$$

(for $z<0$), where C is an integration constant.
The fluid acts by a tangential viscosity force $-\eta\frac{dV}{dz}\mid_{z=0}$ per the unit area of surface; the wind, which generates the motion, acts by a pressure p_0 upon the same area. These forces should compensate each other, *i.e.* we should have $-\eta\frac{dV}{dz}\mid_{z=0}=p_0$, or

$$C=-\frac{|p_0|}{\sqrt{2\rho\eta\Omega\sin\varphi}}e^{-i\pi/4+i\delta} \ , \tag{3.122}$$

where we write $p_0=|p_0|\,e^{i\delta}$. It follows

$$V=v_0e^{\sqrt{\frac{\Omega\sin\varphi}{\nu}}z}e^{i\left(\sqrt{\frac{\Omega\sin\varphi}{\nu}}z-\pi/4+\delta\right)} \ , \tag{3.123}$$

where $v_0=-\,|p_0|\,/\sqrt{2\rho\eta\Omega\sin\varphi}$ for $\varphi\neq 0$.
We can see that the velocity of the fluid makes an angle $-\pi/4$ with the wind velocity and decreases with an attenuation length $\sqrt{\nu/\Omega\sin\varphi}$. This is Ekman's attenuation depth of the drift currents in oceans.[21] At depth $(3\pi/4)\sqrt{\nu/\Omega\sin\varphi}$ the current velocity is opposite to the wind velocity, and at depth $\pi\sqrt{\nu/\Omega\sin\varphi}$ the current is opposite to the surface current.

3.15 Oscillations in a viscous fluid

We consider an incompressible fluid bordered at $x=0$ by an yz-plane which executes an oscillatory motion along the y-direction, with velocity

$$u=u_0e^{-i\omega t} \ . \tag{3.124}$$

We are interested in the motion of the fluid.[22] The boundary conditions are $\boldsymbol{v}=\boldsymbol{u}$ for $x=0$. Obviously, all the quantities depend only on x, and $v_z=0$. From the continuity equation we get $v_x=0$. Also,

[21] V. W. Ekman, "On the influence of the Earth's rotation on ocean-currents", Ark. Mat. Astr. Phys. **2** 1 (1905).

[22] G. G. Stokes, "On the effect of the internal friction of fluids on the motion of pendulums", Trans. Cambridge Phil. Soc. **9** 8 (1851) (in *Mathematical and Physical Papers by G. G. Stokes*, vol. 3, 1, Johnson Reprint Corp., NY (1966)).

$(\boldsymbol{v}grad)\boldsymbol{v} = 0$. From the Navier-Stokes equations we get $dp/dx = 0$, *i.e.* $p = const$, and

$$\frac{\partial v}{\partial t} = \nu \frac{\partial^2 v}{\partial x^2} \ , \tag{3.125}$$

where $v = v_y$. This is the diffusion equation (heat conduction equation). Its solution is

$$v = u_0 e^{-\sqrt{\omega/2\nu}x} e^{-i(\omega t - \sqrt{\omega/2\nu}x)} \tag{3.126}$$

is a damped transverse wave; over a wavelength it decreases by the very large factor $e^{-2\pi}$. $\delta = \sqrt{2\nu/\omega}$ is the penetration depth; it is an estimate of the extension of the region where the fluid adheres to the solid plane. The friction force per unit area of the wall is the real part of

$$\sigma'_{xy} = \eta \partial_x v \mid_{x=0} = \sqrt{\frac{1}{2}\eta\omega\rho} u_0 (i-1) e^{-i\omega t} \ . \tag{3.127}$$

The average energy dissipated per unit area in unit time is $\overline{\sigma'_{xy}u} = \frac{1}{2}\sqrt{\eta\omega\rho/2} \mid u_0 \mid^2$.
For a $z > 0$-half-plane an additional contribution of force per unit length of the edge appears, from a velocity which is proportional to $\theta(z)$.
Let us consider now a body with finite dimension l, oscillating with frequency ω and small amplitude $a \ll l$, in an incompressible fluid. We write down the full Navier-Stokes equation

$$\frac{\partial \boldsymbol{v}}{\partial t} + (\boldsymbol{v}grad)\boldsymbol{v} = -\frac{1}{\rho} gradp + \nu \Delta \boldsymbol{v} \ . \tag{3.128}$$

Let us assume that the penetration depth is large, *i.e.* $\nu/\omega l^2 \gg 1$. The Reynolds number is $R = \omega a l/\nu = (a/l)/(\nu/\omega l^2) \ll 1$. Therefore, we may neglect the advection term in equation (3.128). In addition, the temporal term is of the order of $\omega^2 a$, so it can be neglected, since the frequency is very small. We are left with equation

$$-\frac{1}{\rho} gradp + \nu \Delta \boldsymbol{v} = 0 \ , \tag{3.129}$$

which shows that the flow is steady and non-inertial, and the drag force is the Stokes' force with the instantaneous velocity v.

Let us suppose now that the penetration depth is small, *i.e.* $\nu/\omega l^2 \ll 1$; the Reynolds number may not be small. The advection term is of the order $\omega^2 a^2/l = (a/l)(\omega^2 a)$, and we can see that it is much smaller than $\partial \boldsymbol{v}/\partial t$. Therefore, we can again neglect the advection term, and get

$$\frac{\partial \boldsymbol{v}}{\partial t} = -\frac{1}{\rho} grad p + \nu \Delta \boldsymbol{v} \ . \tag{3.130}$$

By taking the *curl* of this equation, we obtain

$$\frac{\partial}{\partial t} curl \boldsymbol{v} = \nu \Delta curl \boldsymbol{v} \ , \tag{3.131}$$

i.e. the *curl* satisfies a diffusion equation. Therefore, the flow is rotational in a small penetration depth around the body, and it becomes potential at large distance. In this far away region we have $curl \boldsymbol{v} = 0$ and $div \boldsymbol{v} = 0$, so $\Delta \boldsymbol{v} = 0$ and the viscosity is not important anymore, and the flow is governed by the Euler equation of ideal fluids. The average energy dissipated per unit area in unit time is calculated above.

3.16 Damping of gravity waves

Let us consider waves occurring near the free surface of a fluid in uniform gravitational field, *i.e.* gravity waves. They are similar to the oscillatory motion of a solid with dimension λ, where λ is the wavelength. This wavelength is large, so we assume $\nu/\omega\lambda^2 \ll 1$ and $a/\lambda \ll 1$, where ω is the frequency and a is the amplitude of motion. These assumptions are equivalent with a small penetration depth. Therefore, the flow is rotational in a "thin" surface layer, and potential in the rest of the fluid.[23]

At the free surface the force should be zero, *i.e.*

$$\sigma_{ij} n_j = -p n_i + \sigma'_{ij} n_j = 0 \ , \tag{3.132}$$

[23]The condition of a small penetration depth is different for gravity waves in a channel, where the role of the wavelength is played by the dimension of the channel (depth and width). The surface layer is then thin indeed, and the dissipation occurs within this layer only.

where $\boldsymbol{n}$ is the unit vector normal to the surface (equation (3.8)). The ideal flow does not satisfy this condition.

Since the wavelength is large, in calculating the dissipated energy we cannot limit ourselves to the surface layer, as in the case of a solid body; instead we need to integrate over the whole volume of the fluid, in equation (3.23),

$$\dot{E} = -\frac{1}{2}\eta \int dV(\partial_i v_j + \partial_j v_i)^2 \ . \tag{3.133}$$

In the dissipation equation we need to include the potential energy too, but this energy does not dissipate. In a potential flow $\boldsymbol{v} = grad\Phi$, such that this equation becomes

$$\dot{E} = -2\eta \int dV(\partial_i \partial_j \Phi)^2 \ , \tag{3.134}$$

where

$$\Phi = \Phi_0 \cos(kx - \omega t + \alpha)e^{kz} \tag{3.135}$$

is the velocity potential for the propagation along the x-direction; the coordinate z is along the depth. The result of integration in equation (3.134) is

$$\overline{\dot{E}} = -8\eta k^4 \int dV \overline{\Phi^2} \ , \tag{3.136}$$

where the upper bar indicates the time averaging.

The mean energy is the mean kinetic energy plus the mean potential energy. For small oscillations the two contributions are equal, so we get the mean energy

$$\overline{E} = \rho \int \overline{v^2} dV = 2\rho k^2 \int dV \overline{\Phi^2} \ . \tag{3.137}$$

The damping coefficient is given by the ratio

$$\gamma = \mid \overline{\dot{E}} \mid /2\overline{E} = 2\nu k^2 \ , \tag{3.138}$$

such that $\overline{E} \sim e^{-2\gamma t}$ (and the amplitude decreases as $e^{-\gamma t}$). Since $\omega^2 = gk$ for gravity waves, where g is the gravitational acceleration, the damping coefficient is $\gamma = 2\nu\omega^4/g^2$.

3.17 Helmholtz-Kelvin instability

Let us consider a steady flow of an incompressible fluid, governed by equations

$$(\boldsymbol{v}_0 grad)\boldsymbol{v}_0 = -\frac{1}{\rho} grad p_0 + \nu \Delta \boldsymbol{v}_0 \ , \ \ div \boldsymbol{v}_0 = 0 \ . \tag{3.139}$$

We add a small perturbation $\boldsymbol{v}_1$, which satisfies the linearized equations

$$\begin{aligned} &\tfrac{\partial \boldsymbol{v}_1}{\partial t} + (\boldsymbol{v}_0 grad)\boldsymbol{v}_1 + (\boldsymbol{v}_1 grad)\boldsymbol{v}_0 = \\ &= -\tfrac{1}{\rho} grad p_1 + \nu \Delta \boldsymbol{v}_1 \ , \ \ div \boldsymbol{v}_1 = 0 \ . \end{aligned} \tag{3.140}$$

Let us consider a surface of tangential discontinuity, along the x-direction (for $z = 0$). On one side of this surface the velocity $\boldsymbol{v}_0$ is a constant $\boldsymbol{v}$ along the x-direction. We leave the viscosity aside for the moment. Equation (3.140) becomes

$$\frac{\partial \boldsymbol{v}_1}{\partial t} + v \frac{\partial \boldsymbol{v}_1}{\partial x} = -\frac{1}{\rho} grad p_1 \ . \tag{3.141}$$

We take the div of this equation and use $div \boldsymbol{v}_1 = 0$,

$$\Delta p_1 = 0 \ . \tag{3.142}$$

Let $\zeta(x,t)$ be a small displacement of the surface along the z-direction; then, its velocity is

$$v_{1z} = \frac{\partial \zeta}{\partial t} + v \frac{\partial \zeta}{\partial x} \ . \tag{3.143}$$

Obviously, this equation has a solution of the form $v_{1z} \sim e^{i(kx-\omega t)}$, and the pressure p_1 has the same form. Equation (3.142) gives

$$\frac{d^2 p_1}{dz^2} - k^2 p_1 = 0 \ , \tag{3.144}$$

such that $p_1 \sim e^{i(kx-\omega t)} e^{-kz}$, for $z > 0$. From equation (3.141) we get

$$p_1 = -\frac{i\rho}{k} v_{1z} (\omega - vk) \ . \tag{3.145}$$

It is convenient to express v_{1z} by means of ζ, from equation (3.143), which has the same form. Therefore, we get

$$p_1 = -\frac{\rho}{k}\zeta(\omega - vk)^2 \ . \tag{3.146}$$

Let us consider now the other side of the surface, where $v = 0$. The calculations are the same, except for $\omega = 0$ and a changed sign for the derivative with respect to z in equation (3.141). Therefore, the pressure on this side is

$$p_1^{'} = \frac{\rho'}{k}\zeta\omega^2 \ , \tag{3.147}$$

where we use a different density, as for immiscible fluids. Since the two pressures must be equal, we get

$$\omega = vk\frac{\rho \pm i\sqrt{\rho\rho'}}{\rho + \rho'} \ . \tag{3.148}$$

We can see that the surface is unstable, either as a convective displacement, or just as a general instability (absolute instability). This is the Helmholtz-Kelvin instability.[24] The calculations above are for an ideal fluid. If we include the viscosity the instability is formally preserved. But for a small viscosity we cannot consider a sharp surface of discontinuity. In that case the discontinuity occurs over a narrow layer, and the instability arises as an effect of transport of momentum across this layer by collisions. This is the origin of turbulence, as we shall see in a forthcoming chapter. The Helmholtz-Kelvin instability shows rather that we need to include viscosity in the description of fluids.

[24]Unstable discontinuity surfaces in fluids are known since long, H. Helmholtz, "Ueber discontinuirliche Flussigkeitsbewegungen", Monatsberichte Akad. Wiss. zu Berlin 215 (1868); Lord Kelvin (W. Thomson), "Hydrokinetic solutions and observations", Phil. Mag. **42** 362 (1871).

4 Turbulence. Boundary Layer

4.1 Reynolds number

The Navier-Stokes equation (for incompressible fluids, $div\boldsymbol{v} = 0$),

$$\partial_t \boldsymbol{v} + (\boldsymbol{v}grad)\boldsymbol{v} = -\frac{1}{\rho}gradp + \nu\Delta\boldsymbol{v} \ , \tag{4.1}$$

where $\nu = \eta/\rho$ is the (kinematic) viscosity and η is the (dynamic) coefficient of viscosity, has a remarkable similarity property: by making the scale transformations

$$\boldsymbol{r} = L\boldsymbol{r}' \ , \ t = \frac{L}{V}t' \ , \tag{4.2}$$

where L is a length scale and V is a velocity scale ($\boldsymbol{v} = V\boldsymbol{v}'$), the Navier-Stokes equation becomes

$$\partial_{t'}\boldsymbol{v}' + (\boldsymbol{v}'grad')\boldsymbol{v}' = -grad'(p'/\rho') + \frac{\nu}{LV}\Delta\boldsymbol{v}' \ , \tag{4.3}$$

or

$$\partial_{t'}\boldsymbol{v}' + (\boldsymbol{v}'grad')\boldsymbol{v}' = -grad'(p'/\rho') + \frac{1}{R}\Delta\boldsymbol{v}' \ , \tag{4.4}$$

where

$$R = \frac{LV}{\nu} \tag{4.5}$$

is known as the Reynolds number (and $p/\rho = V^2(p'/\rho')$; ν is measured in cm^2/s). We can see that R is a length times velocity divided by kinematic viscosity. The Reynolds number was introduced by Stokes,[1] used intensively by Reynolds[2] and named by Sommerfeld.[3]

If L/V is distinct from the temporal scale factor T in equation (4.2), then the solution of the Navier-Stokes equation depends on two numbers: the Reynolds number R and another number $S = VT/L$, called the Strouhal number. Another number in the Navier-Stokes equation may appear when external forces are present; it is $F = V^2/Lf$, where f is the external force per unit mass. F is called the Froude number. All these numbers are helpful in studying laboratory models of fluid flows.

4.2 Stability of the flow

The Navier-Stokes equations, inertial term included, can be solved exactly, to some extent numerically, in a few cases: flow through a channel,[4] a rotating disk,[5] or submerged jet.[6] Such solutions are not of much importance, because, when the inertial term matters, instabilities and turbulence may set in.

With usual notations the equations of the fluid mechanics for an in-

[1] G. G. Stokes, "On the effect of the internal friction of fluids on the motion of pendulums", Trans. Cambridge Phil. Soc. **9** 8 (1851) (in *Mathematical and Physical Papers by G. G. Stokes*, vol. 3, 1, Johnson Reprint Corp., NY (1966)).

[2] O. Reynolds, "An experimental investigation of the circumstances which determine whether the motion of water shall be direct or sinuous, and of the law of resistance in parallel channels", Proc. Roy. Soc. London **174** 935 (1883).

[3] A. Sommerfeld, "Ein Beitrag zur hydrodynamischen Erklaerung der turbulenten Fluessigkeitsbewegungen", Proc. Int. Math. Congress **3** 116 (1908).

[4] G. Hammel, "Spiralfoermige Bewegungen zaeher Fluessigkeiten", Jahresbericht der Deutchen Mathematiker-Vereinigung **25** 34 (1917).

[5] Th. von Karman, "Ueber laminare und turbulente Reibung", Z. angew. Math. Mech. **1** 233 (1921).

[6] L. Landau and E. Lifshitz, *Course of Theoretical Physics*, vol. 6 (*Fluid Mechanics*), Elsevier, Oxford (1989).

compressible viscous fluid are

$$\rho\frac{\partial \boldsymbol{v}}{\partial t} + \rho(\boldsymbol{v}grad)\boldsymbol{v} = -gradp + \eta\Delta\boldsymbol{v} \ ,$$

$$div\boldsymbol{v} = 0 \ , \tag{4.6}$$

$$\rho c_p\left(\frac{\partial T}{\partial t} + \boldsymbol{v}gradT\right) = \kappa\Delta T + \frac{1}{2}\eta\left(\partial_i v_j + \partial_j v_i\right)^2 \ .$$

The inertial (advection) term $\rho(\boldsymbol{v}grad)\boldsymbol{v}$ is a transport of velocity (momentum). The terms on the right in the first equation above (the Navier-Stokes equation) generate a force and a viscosity force. The η-term in the third equation is a source of heat, which determines a temperature T. This temperature is transported with the fluid velocity ($\boldsymbol{v}gradT$) and the molecular velocity through the thermoconductivity κ (collisions, interaction). In addition, the change in temperature causes a change in the pressure p, such that the first and the third equations above are coupled.

By multiplying the Navier-Stokes equation by $\boldsymbol{v}$, we arrive straightforwardly at

$$\begin{gathered}\frac{\partial}{\partial t}\left(\frac{1}{2}\rho v^2\right) + div\left[\boldsymbol{v}\left(\frac{1}{2}\rho v^2 + p\right) - \frac{1}{2}\eta grad(v^2)\right] + \\ +\eta\left(\partial_j v_i\right)^2 = 0 \ .\end{gathered} \tag{4.7}$$

The term $\eta grad(v^2)$ represents a transfer of energy by collisions (viscosity) energy; the term $\eta\left(\partial_j v_i\right)^2$ represents the heat produced by viscosity. We integrate this equation over a volume V enclosed by a surface f,

$$\begin{gathered}\frac{\partial}{\partial t}\int dV\left(\frac{1}{2}\rho v^2\right) + \oint df\left[v_n\left(\frac{1}{2}\rho v^2 + p\right) - \frac{1}{2}\eta\partial_n(v^2)\right] + \\ +\eta\int dV\left(\partial_j v_i\right)^2 = 0 \ ,\end{gathered} \tag{4.8}$$

where v_n is the velocity component normal to the surface and ∂_n is the derivative along the normal to the surface.

Let us assume that the flow is "internal" (or "closed"), *i.e.* the surface is closed and the boundary conditions are $v_n = 0$ (both for ideal or viscous fluids). Equation (4.8) becomes

$$\frac{\partial}{\partial t}\int dVv^2 - \nu\oint df\partial_n(v^2) + 2\nu\int dV\left(\partial_j v_i\right)^2 = 0 \ , \tag{4.9}$$

where $\nu = \eta/\rho$ is the kinematic viscosity. In this equation the variables can be separated; we write the velocity as

$$\boldsymbol{v} = f(t)\boldsymbol{u}(\boldsymbol{r}) \ ; \tag{4.10}$$

Equation (4.9) becomes

$$k\frac{df^2}{dt} + \nu(2h - c)f^2 = 0 \ , \tag{4.11}$$

where we introduce the notations

$$\begin{gathered} k = \int dV u^2 \ , \ c = \oint df \partial_n(u^2) \ , \\ h = \int dV \left(\partial_j u_i\right)^2 \end{gathered} \tag{4.12}$$

(k from kinetic, c from collisions, h from heat). Outside the surface u^2 is zero, inside the surface $u^2 \neq 0$; on the surface $u^2 = 0$. The derivative $\partial_n(u^2)$ in c is the limit of the inside derivatives, so $c < 0$; and $k > 0$, $2h - c > 0$ in equation (4.11). The solution of this equation is

$$f = f_0 e^{-\frac{\nu(2h-c)}{2k}t} \ . \tag{4.13}$$

We can see that the flow is stable. The initial perturbation f_0 is transformed by collisions into heat, so it is dissipated in time, as expected. We note that the orders of magnitude are $c \simeq Su^2/l$, $h \simeq Vu^2/l^2$, where S is the area of the surface, V is the volume and l is the length over which the velocity varies; their ratio is $c/h \simeq Sl/V$, *i.e.* the coefficient c is a surface contribution. The time constant in equation (4.13) is the ratio of the kinetic energy to the dissipated energy ($k/\nu h$). If the fluid is ideal the initial perturbation is preserved, as expected.

In equation (4.8) we may neglect the surface term, as long as the flow is macroscopic. We are left with

$$k\frac{df^2}{dt} + 2\nu h f^2 = 0 \ , \tag{4.14}$$

which has stable solutions. By using this equation we can approach an "open" flow, *i.e.* a flow bordered by a surface which is not closed, as well as a fluid with a free surface, with or without an external force potential. In all these cases the flow is stable. For a free surface in gravitational field, for instance, the free surface in equation (4.8) may oscillate; the resulting gravitational waves are damped by viscosity. We note that the above analysis can be extended to compressible fluids, for $\rho v^2/2+\rho\varepsilon$, where ε is the internal energy per unit mass, entropy included. The main role in the above arguments is played by terms with derivatives normal to the surface, such that we will have instabilities for large variations of both the velocity and the temperature across the surface.

Let us focus now on a small element of fluid with volume V, with linear dimension l, enclosed by a surface with area S, such that $V \simeq lS$. The energy (e) and the heat term in equation (4.8) are of the order of $e \simeq S\rho v^3$ and $heat \simeq \eta V\frac{v^2}{l^2}$, such that $\frac{e}{heat} = \frac{vl}{\nu} = R$ is of the order of the Reynolds number; the pressure term is of the order Spv. We note that all these terms are positive. Let us assume that the velocity acquires a high increase on crossing the surface, over a very small distance $\delta \ll l$. The collision-term in equation (4.8) becomes of order $coll \simeq S\eta\frac{v^2}{\delta}$. Since δ is very small, we have $\frac{coll}{e} \simeq \frac{\nu}{v\delta} \gg 1$, $\frac{coll}{heat} \simeq \frac{l}{\delta} \gg 1$ and $\frac{coll}{Spv} \simeq \frac{\eta v}{p\delta} \gg 1$. Such a discontinuity surface is energetically favourable, because it decreases appreciably the energy. Under these conditions equation (4.11) becomes

$$k\frac{df^2}{dt} - \nu cf^2 = 0 \ , \tag{4.15}$$

where $c > 0$; its solution is unstable. We conclude that such small discontinuities may appear, and disappear, spontaneously and continuously.[7] The corresponding Reynolds number is $R = \frac{vl}{\nu} = \frac{v\delta}{\nu}\frac{l}{\delta} \gg \frac{v\delta}{\nu}$. This latter quantity plays the role of a critical Reynolds number.[8] In addition, we note that such discontinuites are associated with a non-vanishing vorticity. This is why they are called vortices, eddies.[9]

[7]Unstable discontinuity surfaces in fluids are known since long, see, for instance, H. Helmholtz, "Ueber discontinuirliche Flussigkeitsbewegungen", Monatsberichte Akad. Wiss. zu Berlin 215 (1868); Lord Kelvin (W. Thomson), "Hydrokinetic solutions and observations", Phil. Mag. **42** 362 (1871).

[8]O. Reynolds, "On the dynamical theory of incompressible viscous fluids and the determination of the criterion", Proc. Roy. Soc. London **56** 40 (1894).

[9]We note that a similar analysis is valid for independent time dependencies of

From the above analysis we can see that the flow is stable as long as the heat term (h) dominates the inertial term (e), *i.e.* for small Reynolds numbers. On increasing the Reynolds number the heat production becomes insufficient, and the energy begins to be more and more mechanical. It may be so high, that the flow cannot take it, so it is transferred to the collision mechanical energy (c), as being more favourable. Noteworthy, this term ($coll$) is governed by viscosity. This means fluctuations in velocity, which take place by discontinuous small surfaces, with vorticity. As expected, they are unstable, in the sense that they appear and disappear continuously, and spontaneously. This is the turbulence, which, essentially, reflects the incapacity of the flow to sustain large Reynolds numbers; even the local thermodynamic equilibrium is damaged in such a situation. This is the physical picture of the occurrence of turbulence. It is worth noting that the increase of Reynolds numbers occurs by increasing the velocity (*i.e.* a dominance of the inertial term), or a decrease of viscosity, over large distances. The fluid responds by abrupt changes in velocity over small distances, such as to absorb the exceeding mechanical energy by collisions (and to reduce the Reynolds number). Unfortunately, this response is unstable. The turbulence is the incapacity of fluids to dissipate energy as heat. Basically, it is a non-equilibrium problem, where large velocities at one point cannot be dissipated as heat, not transported as energy, but they are transferred to the adjacent points, across a surface of velocity discontinuities, by molecular collisions; and such a transfer is fluctuating. The transfer is maximized by returning velocities, or a motion in finite spatial regions, which gives the phenomenon an aspect of fluctuating eddies. It is worth noting that this picture holds for large discontinuities in velocities occurring at surfaces.

Large velocity gradients may be achieved also by an external source of heat, thus leading to larger Reynolds numbers. The instability occurs again, by the same mechanism as the one described above. Then we have convective instabilities.[10]

---

the components of the velocity.

[10]H. Benard, "Les tourbillons cellulaires dans une nappe liquide", Revue Generale des Sciences Pures and Appliquees **11** 1261, 1309 (1900); Lord Rayleigh, "On convection currents in a horizontal layer of fluid, when the higher temperature is on the under side", Phil. Mag. **32** 529 (1916); J. R. A. Pearson, "On convection cells induced by surface tension", J. Fluid Mech. **4** 489 (1958).

Formally, the instability of a steady flow can also be seen by linearizing the Navier-Stokes equations in the neighbourhood of a fixed steady solution. We get a system of first-order differential equations with coefficients depending on position. The boundary conditions determine the frequency ω of the solution, which may exhibit an imaginary part.[11]

4.3 Large Reynolds numbers

For small Reynolds numbers in the Navier-Stokes equation (4.4) the spatial derivatives of the velocity are small; consequently, we may treat the non-linear term $(\boldsymbol{v}grad)\boldsymbol{v}$ as a perturbation, which can even be ignored in the first approximation; such a treatment greatly simplifies the solving of the Navier-Stokes equation. For small values of viscosity ν the Reynolds numbers may still be small, for small characteristic velocities and lengths; then, the viscosity term in the Navier-Stokes equation preserves its original meaning of friction, which causes the damping of the motion.

Quite a different situation may appear for large Reynolds numbers. The small contribution of the factor $1/R$ in the viscosity term $\Delta\boldsymbol{v}/R$ is compensated by large spatial variations of the velocity, over small spatial regions.[12] There appear regions in the fluid flow where the velocity exhibit swirls, called eddies, or locally closed streamlines, called vortices. For instance, suppose that we have a solid obstacle in a fluid flow, as shown in Fig. 4.1. On collision with the obstacle the flow will be reflected back, and the adherence force (denoted by f in Fig. 4.1) bends the reflected streamlines in small eddies. Similar situations occur for a fluid which flows past a sphere, a cylinder, wedges, past bodies of various forms, a plane surface, etc, as shown in Figs. 4.2, 4.3. In all these cases the velocity undergoes large variations over small spatial distances, such that a solution of the Navier-Stokes equation

[11] L. Landau and E. Lifshitz, *Course of Theoretical Physics*, vol. 6 (*Fluid Mechanics*), Elsevier, Oxford (1989). An example was given by E. N. Lorenz, "Deterministic non-periodic flow", J. Atmos. Sci. **20** 130 (1963), which exhibits an irregular solution known usually as chaotic behaviour.

[12] W. Heisenberg, "Ueber Stabilitaet und Turbulenz von Fluessigkeitsstroemen", Ann. Physik **74** 577 (1924).

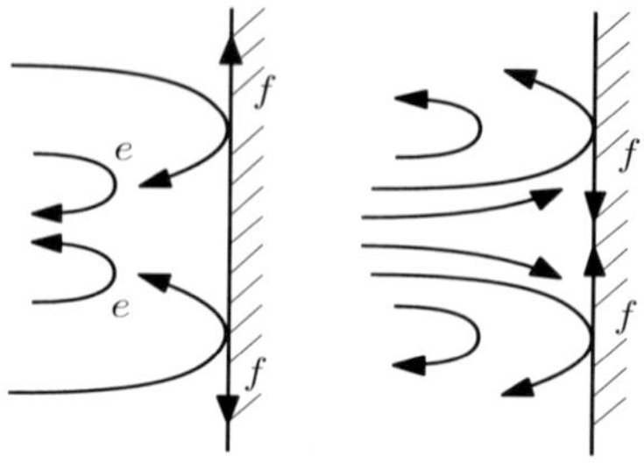

Figure 4.1: Flow colliding with an obstacle, developing eddies (e); note the adherence force f.

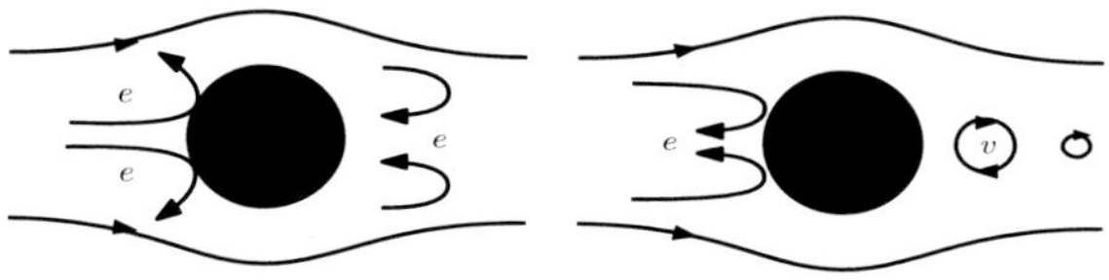

Figure 4.2: Turbulent flow past a sphere, exhibiting eddies (e) and vortices (v).

in these regions is practically impossible, at least analytically;[13] a solution could only be obtained for certain distinct, separate spatial regions. We may call such a highly-varying solution of the Navier-Stokes equation an irregular solution, in order to distingush it from regular solutions which belong to distinct, finite regions where such high spatial variations are absent. An irregular solution looks like exhibiting "ruptures" (discontinuities) between regions with regular solutions.

Flows which exhibit eddies or vortices are called turbulent flows. Instances of turbulent flows are shown in Fig. 4.2, where a turbulent wake is depicted in flow past a sphere, or in Fig. 4.3, where a turbulent boundary layer occurs in flow past a plane surface.[14] It seems that turbulence appears for Reynolds numbers higher than a critical value R_c, specific to each situation.

[13]It is not only impossible, but it is also useless.

[14]Many instances of turbulence and turbulent flows are shown in M. Van Dyke, *An Album of Fluid Motion*, The Parabolic Press, Stanford, CA (1982). A turbulent wake reduces the drag force ("drag crisis").

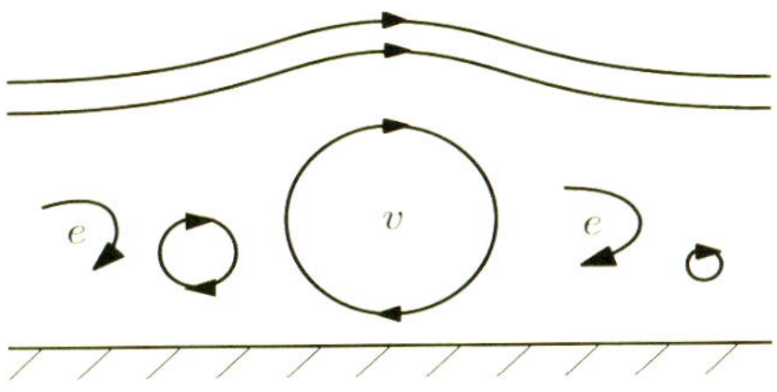

Figure 4.3: Eddies (e) and vortices (v) in a turbulent boundary layer past a plane surface.

It is worth noting that large Reynolds numbers, actually infinite Reynolds numbers, may be associated with ideal fluids ($\nu = 0$); however, such an association does not appear by necessity, because there may happen to be no characteristic length or velocity, such that, in such situations, the Reynolds number is in fact zero. This may explain the fact that ideal fluids may sustain non-turbulent, *i.e.* regular solutions of the Navier-Stokes equation, at least over limited spatial regions. It follows that the Reynolds numbers are not global, but rather local parameters of the fluid flow. On the other hand, in the presence of small perturbations, as those generated by small inhomogeneities for instance, or those caused by the adherence force at the contact with other bodies, the motion of the ideal fluids may become unstable, leading, very likely, to turbulence; in such a situation, the Reynolds number seems to jump from zero to large values. Turbulent regions in fluid flows, like the turbulent boundary layer, or the turbulent wake, etc, are separated from regular regions, exhibiting practically a regular flow; such a separation surface is well known in turbulent flows.

Apart from regular and irregular spatial regions, and apart from local instabilities, the turbulent flow exhibits another characteristic: the velocity in turbulent regions fluctuates (in time). Very likely, these fluctuations are related to the unstable character of the flow. After the flow develops an instability, as caused by a perturbation, it may find new ways to accommodate itself to that perturbation, the effect of the latter in the new state being thus reduced to fluctuations only. The fluctuations may be small, or they may be very large (in the latter case the turbulence looks like a chaotical motion); when the fluctuations are small, a mean velocity can be used. In any case, the surface separating the turbulent flow from the regular one fluctuates, a phenomenon which is known as the intermitency of the turbulence. Finally, a fully developed turbulence exhibits a large number of small eddies and vortices, separated by a large number of small regular regions, such that the whole ensemble may be viewed statistically.

4.4 Reynolds equations

The Navier-Stokes equation

$$\frac{\partial v_i}{\partial t} = -\frac{1}{\rho}\partial_i p - v_j \partial_j v_i + \nu \Delta v_i \tag{4.16}$$

written for an incompressible fluid with the continuity equation $\partial_i v_i = 0$ can also be written as

$$\frac{\partial v_i}{\partial t} = -\frac{1}{\rho}\partial_i p - \partial_j (v_i v_j) + \nu \Delta v_i \ . \tag{4.17}$$

We make the decomposition $v_i = \overline{v}_i + v_i^{'}$, where $\overline{v}_i$ is a time average (so, $\overline{v}_i$ does not depend on the time) and $v_i^{'}$ is a fluctuating velocity ($\overline{v_i^{'}} = 0$); we get

$$\begin{gathered}\frac{\partial v_i}{\partial t} = \frac{\partial \overline{v}_i}{\partial t} + \frac{\partial v_i^{'}}{\partial t} = \frac{\partial v_i^{'}}{\partial t} = -\frac{1}{\rho}\partial_i \overline{p} - \frac{1}{\rho}\partial_i p^{'} - \\ -\partial_j(\overline{v}_i\overline{v}_j + v_i^{'}\overline{v}_j + \overline{v}_i v_j^{'} + v_i^{'} v_j^{'}) + \nu \Delta \overline{v}_i + \nu \Delta v_i^{'} \ ,\end{gathered} \tag{4.18}$$

where the pressure is also decomposed into its mean value $\overline{p}$ and its fluctuating part $p^{'}$. Taking here the time average we get

$$0 = -\frac{1}{\rho}\partial_i \overline{p} - \partial_j(\overline{v}_i\overline{v}_j + \overline{v_i^{'} v_j^{'}}) + \nu \Delta \overline{v}_i \ ; \tag{4.19}$$

and the continuity equations

$$\partial_i \overline{v}_i = 0 \ , \ \partial_i v_i^{'} = 0 \ . \tag{4.20}$$

Equation (4.19) represents a system of three equations with three unknown mean velocities $\overline{v}_i$ and six unknown correlation functions

$\overline{v_i'v_j'}$; the continuity equation for mean velocities should be added. This system of equations is under-determined. It is known as Reynolds equations; $\overline{v_i'v_j'}$ are the components of the Reynolds stress tensor.[15] It is thought that the Reynolds tensor provides an additional force in the fluid equation of motion (of equilibrium, in fact).

Now we multiply equations (4.18) by v_i, sum over i, take the time average and use the continuity equations (4.20), to get the energy conservation equation

$$\begin{gathered}0=\overline{\frac{\partial}{\partial t}\left(\frac{1}{2}v_i^2\right)}=\overline{v}_i\left[-\frac{1}{\rho}\partial_i\overline{p}-\partial_j(\overline{v}_i\overline{v}_j+\overline{v_i'v_j'})+\nu\Delta\overline{v}_i\right]-\\ -\overline{v_i'v_j'}\partial_j\overline{v}_i-\frac{1}{2}\partial_j\left(\overline{v_i'^2}\right)\overline{v}_j+\nu\overline{v_i'\Delta v_i'}\ ,\end{gathered}\tag{4.21}$$

where we have dropped out third-order terms involving products of three fluctuating velocities and their derivatives, as well as the contribution of the fluctuating part of the pressure; we assume that the velocity fluctuations are much smaller than the mean flow velocities.

First, we note that $\overline{v}_i\partial_j\overline{v_i'v_j'}$ and $\overline{v_i'v_j'}\partial_j\overline{v}_i$ can be written as $\partial_j\left(\overline{v}_i\overline{v_i'v_j'}\right)$ in equation (4.21), which becomes

$$\begin{gathered}0=\overline{\frac{\partial}{\partial t}\left(\frac{1}{2}v_i^2\right)}=-\frac{1}{\rho}\partial_i\overline{v}_i\overline{p}-\frac{1}{2}\partial_j\left[\overline{v}_j\left(\overline{v}_i^2+\overline{v_i'^2}\right)\right]-\\ -\partial_j(\overline{v}_i\overline{v_i'v_j'})+\nu\left(\overline{v}_i\Delta\overline{v}_i+\overline{v_i'\Delta v_i'}\right)\ .\end{gathered}\tag{4.22}$$

The term on the left in this equation (multiplied by density) is the rate of change in time of the energy density; the first term on the right is the work done by pressure forces per unit time (divided by density); the second term implies the energy flux density $\frac{1}{2}\rho\overline{v}_j(\overline{v}_i^2+\overline{v_i'^2})$ caused by mass transfer, both in the mean and turbulent flow; its divergence (∂_j) indicates the transport of this energy by the mean flow (velocity $\overline{v}_j$); the third term on the right is the transport of the energy flux density $\rho\overline{v}_i\overline{v_i'v_j'}$ by the turbulent flow (velocity v_j'). The last term on

[15] O. Reynolds, "On the dynamical theory of incompressible viscous fluids and the determination of the criterion", Phil. Trans. Roy. Soc. Lond. **A186** 123 (1895).

the right in equation (4.22) can be written as

$$\nu\left(\overline{v}_i\Delta\overline{v}_i+\overline{v_i'\Delta v_i'}\right)=$$
$$=\tfrac{1}{2}\nu\Delta\left(\overline{v}_i^2+\overline{v_i'^2}\right)-\nu\left(\partial_j\overline{v}_i\right)^2-\nu\overline{\left(\partial_j v_i'\right)^2}\ . \tag{4.23}$$

The first term on the right in this equation is related to the transport of the energy flux density caused by the internal friction (viscosity, collisions; momentum transfer), while the last term is different: it represents the heat source generated by viscosity. This heat production is partly conducted (transported) thermally and partly it increases the entropy; therefore, we have

$$T\frac{ds}{dt}=T\left(\frac{\partial s}{\partial t}+v_i\partial_i s\right)=-\frac{1}{\rho}div\boldsymbol{q}+\nu\left(\partial_j\overline{v}_i\right)^2+\nu\overline{\left(\partial_j v_i'\right)^2}\ , \tag{4.24}$$

or

$$T\left(\frac{\partial s}{\partial t}+v_i\partial_i s\right)=\frac{\kappa}{\rho}\Delta T+\nu\left(\partial_j\overline{v}_i\right)^2+\nu\overline{\left(\partial_j v_i'\right)^2}\ , \tag{4.25}$$

where s is the entropy per unit mass, T is the temperature, $\boldsymbol{q}=-\kappa gradT$ is the heat flux density and κ (uniform) is the thermal conductivity (in general it depends on pressure and temperature). We note that the total change in entropy

$$\tfrac{d}{dt}\int dV s=\int dV\tfrac{\partial s}{\partial t}=\tfrac{\kappa}{\rho}\int d\mathbf{r}\tfrac{(gradT)^2}{T^2}+$$
$$+\nu\int dV\left[\left(\partial_j\overline{v}_i\right)^2+\overline{\left(\partial_j v_i'\right)^2}\right] \tag{4.26}$$

is positive (κ and ν are positive). In equation (4.25) the entropy and the temperature should be decomposed into their mean and fluctuating values and the average should be taken:

$$\overline{T}\left(\overline{v}_i\partial_i\overline{s}+\overline{v_i'\partial_i s'}\right)+\overline{T'\tfrac{\partial s'}{\partial t}}+\overline{v}_i\overline{T'\partial_i s'}+\overline{T'v_i'}\partial_i\overline{s}=$$
$$=\tfrac{\kappa}{\rho}\Delta\overline{T}+\nu\left(\partial_j\overline{v}_i\right)^2+\nu\overline{\left(\partial_j v_i'\right)^2}\ , \tag{4.27}$$

whence we can see that the change in entropy implies fluctuations; in view of our assumptions of small fluctuations we may neglect their contributions to equation (4.27), which becomes

$$\overline{T}\overline{v}_i \partial_i \overline{s} = \frac{\kappa}{\rho} \Delta \overline{T} + \nu \left(\partial_j \overline{v}_i \right)^2 \; ; \tag{4.28}$$

here we may use $T d\overline{s} = c_p dT$, where c_p is the specific heat at constant pressure, in order to get the temperature distribution. Equations (4.19) to (4.28) are the general equations of equilibrium, energy conservation and heat transfer in turbulent flows.[16]

Now we go back to equation (4.22) of energy conservation. We note that it splits into two distinct equations; the square bracket in equation (4.22) is the "equation of motion" (of equilibrium) (4.17); it can be obtained from the basic equation (4.17) by multiplying by $\overline{v}_i$ and taking the average; therefore, we have the equation of conservation of the mean flow

$$0 = \overline{\frac{\partial}{\partial t}\left(\frac{1}{2}\overline{v}_i^2\right)} = \overline{v}_i \left[-\frac{1}{\rho}\partial_i \overline{p} - \partial_j (\overline{v}_i \overline{v}_j + \overline{v_i' v_j'}) + \nu \Delta \overline{v}_i \right] \; ; \tag{4.29}$$

the remaining part of equation (4.22) can be obtained from the basic equation (4.17) by multiplying by v_i' and taking the average; we get the equation of conservation of the turbulence energy

$$0 = \overline{\frac{\partial}{\partial t}\left(\frac{1}{2}v_i'^2\right)} = -\overline{v_i' v_j'} \partial_j \overline{v}_i - \frac{1}{2}\partial_j \left(\overline{v}_j \overline{v_i'^2} \right) + \nu \overline{v_i' \Delta v_i'} \; . \tag{4.30}$$

It is worth noting that in equations (4.29) and (4.30) the transport term $-\partial_j(\overline{v}_i \overline{v_i' v_j'})$ is split into its components $-\overline{v}_i \partial_j \overline{v_i' v_j'}$ and $-\overline{v_i' v_j'} \partial_j \overline{v}_i$; the first component is viewed as a source term for the mean flow energy (a production rate of the mean flow energy), while the second term is viewed as a production rate of turbulence energy; both terms would indicate a coupling (an interaction) between the mean flow and the turbulent flow.

[16] L. Landau and E. Lifshitz, *Course of Theoretical Physics*, vol. 6 (*Fluid Mechanics*), Pergamon Press, Oxford (1987).

For a homogeneous turbulence the fluctuations have a small, local (finite) variation;[17] if we take a spatial average in equation (4.30) we get

$$\overline{v_i^{'}v_j^{'}}\partial_j\overline{v}_i=-\nu\overline{\left(\partial_j v_i^{'}\right)^2} \ . \tag{4.31}$$

Let

$$v_i^{'}=\sum_{\omega,\boldsymbol{k}}v_i^{'}(\omega,\boldsymbol{k})e^{-i\omega t+i\boldsymbol{k}\boldsymbol{r}} \tag{4.32}$$

be a Fourier decomposition of a fluctuating velocity, where t is the time and $\boldsymbol{r}$ is the position vector; we have

$$\begin{gathered}\left(\partial_j v_i^{'}\right)^2=\\ =-\sum\nolimits_{\omega,\boldsymbol{k},\omega^{'},\boldsymbol{k}^{'}}k_jk_j^{'}v_i^{'}(\omega,\boldsymbol{k})v_i^{'}(\omega^{'},\boldsymbol{k}^{'})e^{-i(\omega+\omega^{'})t+i(\boldsymbol{k}+\boldsymbol{k}^{'})\boldsymbol{r}} \ ;\end{gathered} \tag{4.33}$$

a time average implies $\omega=-\omega^{'}$, and a space average implies $\boldsymbol{k}=-\boldsymbol{k}^{'}$; we get

$$\overline{\left(\partial_j v_i^{'}\right)^2}=\sum_{\omega\boldsymbol{k}}k_j^2\left|v_i^{'}(\omega,\boldsymbol{k})\right|^2=C=const>0 \ , \tag{4.34}$$

(since $v_i^{'}$ is a real quantity). Equation (4.31), which is the equation of the conservation of the turbulence energy spatially averaged, is an additional equation to the Reynolds equation (4.30); it establishes a connection between correlation functions (Reynolds stress tensor) and the mean velocities, through an undetermined constant C; combined with the symmetries of the problem it may bring a certain simplification towards reducing the number of unknowns in the Reynolds equations. It is worth noting that for bounded flows the constant C may retain a dependence on certain coordinates, along which the turbulence is not homogeneous; if such a dependence is slow, we may treat it perturbationally, and view the slow spatial dependence as a correction to the constant C.

[17] G. I. Taylor, "Statistical Theory of Turbulence", Proc. Roy. Soc. Lond. **A151** 421 (1935); *ibid*, 444 (1935); G. K. Batchelor, *The theory of homogeneous turbulence*, Cambridge University Press, Cambridge (1953).

It is also worth noting that νC is the heat source, so we may write equation (4.30) as

$$0 = \overline{\frac{\partial}{\partial t}\left(\frac{1}{2}v_i'^2\right)} = -\overline{v_i'v_j'}\partial_j\overline{v}_i - \nu C \ , \tag{4.35}$$

after spatial averaging; we can see that the production rate of turbulence energy $-\overline{v_i'v_j'}\partial_j\overline{v}_i$ is a constant heat source. In a homogeneous turbulence all the turbulence energy is transformed in heat; and its rate of transformation (dissipation) is constant.

The spatial scale seems to play an important role in turbulence, such that a spatial Fourier analysis looks promising. In the equation of energy conservation the inertial (non-linear) term from the Navier-Stokes equation connects, by energy transfer, various wavelengths. By making use of assumptions which look reasonable, and by using dimensional analysis, this term is simplified, and the energy spectrum can be derived. This is known as the Kolmogorov spectrum.[18] A similar decoupling of the hierarchical equations of correlation functions can be achieved, with similar results.[19]

4.5 Boundary layer

An ideal fluid may flow laminary over a flat, unbounded surface. A laminar flow consists of parallel flowing layers. The surface friction

[18]A. N. Kolmogorov, "Dissipation of energy in locally isotropic turbulence", Dokl. Akad. Nauk SSSR **32** 16 (1941) (Proc. Roy. Soc. Lond. **A434** 15 (1991)); "Precission sur la structure locale de la turbulence dans un fluide visqueux aux nombres de Reynolds eleves", in *La turbulence en Mecanique des Fluides*, 447, eds. A. Favre, L. S. G. Kovasznay, R. Dumas, J. Gaviglio and M. Coantic, Gauthier-Villars, Paris (1961); A. M. Obukhov, "On the distribution of energy in the spectrum of turbulent flow", Dokl. Akad. Nauk SSSR **32** 22 (1941).

[19]S. Chandrasekhar, *The Theory of Turbulence*, ed. E. A. Spiegel, Springer, Dordrecht (2011); W. Heisenberg, "Zur statistischen Theorie der Turbulenz", Z. Phys. **124** 628 (1948); "On the theory of statistical and isotropic turbulence", Proc. Roy. Soc. London **A195** 402 (1948); C. F. von Weizsaecker, "Das Spektrum der Turbulenz bei grossen Reynoldsschen", Z. Phys. **124** 614 (1948); "The evolution of galaxies and stars", Astrophys. J. **114** 165 (1951); R. H. Kraichnan, "The structure of isotropic turbulence at very high Reynolds numbers", J. Fluid Mech. **5** 497 (1959).

imposes the surface velocity to be zero (including the parallel-to-the surface component). The velocity increases gradually in the bulk of the fluid. We have a laminar boundary layer. The boundary layer was discovered by Prandtl.[20] At higher velocities the surface friction may cause turbulence, which extends in the whole fluid by viscosity; then, we have a turbulent boundary layer. The ideal fluid is defined usually by a vanishing viscosity. However, in contact with a solid surface the viscosity is never zero, such that the fluids close to a solid surface are never ideal.

Let us consider a steady flow with velocity u along the x-direction over an unbounded flat surface; the Navier-Stokes equation reduces to

$$0 = -\frac{1}{\rho}\frac{\partial p}{\partial x} + \nu\frac{\partial^2 u}{\partial z^2} ; \tag{4.36}$$

with a constant pressure-gradient the solution is

$$u(z) = u_1 z - \frac{1}{2\nu}\left(-\frac{1}{\rho}\frac{\partial p}{\partial x}\right) z^2 , \tag{4.37}$$

where $u\mid_{z=0}= 0$ and $\frac{\partial u}{\partial z}\mid_{z=0}= u_1$; for u_1 and $(1/\rho)\frac{\partial p}{\partial x}$ of opposite signs the velocity exhibits a swirl towards negative values for increasing z; this is an eddy, pointing out an instability of the flow towards turbulence.[21]

For a vanishing pressure gradient (constant pressure) the velocity is $u(z) = az$, where the constant a is determined from the boundary conditions. An unbounded surface is unphysical, so the linear law of velocity is valid within a height much smaller than the finite extension, say d, of the surface. Therefore, we have a natural cutoff of the order

[20] L. Prandtl, "Ueber Flussigkeitsbewegung bei sehr kleiner Reibung", Proc. III Intern. Math. Congr., Heidelberg 1904, A. Krazer, ed., Teubner, Leipzig, Germany (1904), p.484 ("Vier Abhdlg. zur Hydro- u. Aerodynamik", Gottingen (1927); Nat. Adv. Comm. Aeronautics (NACA) Tech. Memo. #452 (1928)); see also *Early Developments of Modern Aerodynamics*, J. A. K. Ackroyd, B. P. Axcell and A. I. Ruban, eds., Butterworh-Heinemann, Oxford, UK (2001), p.77; H. Blasius, *Grenzschichten in Fluessigkeiten mit kleiner Reibung*, Z. angew. Math. und Physik, **56** 1 (1908).

[21] S. Talpos and M. Apostol, "Displaced logarithmic profile of the velocity distribution in the boundary layer of a turbulent flow over an unbounded flat surface", Phys. Lett. **A379** 3102 (2015).

of d, which allows the boundary condition $u(z=d)=u_0$, where u_0 is the velocity of the main flow. Therefore, $u(z)=\frac{u_0}{d}z$. The friction force per unit area is $dP_x/df=-\sigma'_{xz}=-\eta\partial_z u=-\eta u_0/d$. The heat transfer equation reads $\kappa\frac{d^2T}{dz^2}+\eta(du/dz)^2=0$, whence

$$T=T_0(z/d)+(T_s/d+\eta u_0^2 z/2\kappa d^2)(d-z) \ , \tag{4.38}$$

where T_s is the temperature of the surface and T_0 is the temperature of the main flow.

4.6 Logarithmic profile of the velocity

Let us consider an incompressible fluid with density ρ flowing turbulently over a flat unbounded surface. There is a friction with the surface, as described by the force per unit area $\sigma=\rho u_*^2$, where u_* is a characteristic velocity. It is assumed that this velocity and the distance z from the surface are sufficient to determine the flow velocity u along the surface; then, the only possibility is $du/dz=u_*/z$, whence

$$u=\frac{u_*}{k}\ln\frac{z}{z_0} \ ; \tag{4.39}$$

where k is a numerical constant and z_0 is a cutoff distance. Since the Reynolds number is $R=z_0u_*/\nu$, and since $R\simeq 1$ is the critical value, then $z_0=\nu/u_*$ and the law becomes

$$u=\frac{u_*}{k}\ln\frac{zu_*}{\nu} \ . \tag{4.40}$$

If the surface is rough, z_0 is of the order of the roughness height. Sometimes, the logarithmic law is written as $u\sim\ln(z+z_0)/z_0$, to allow for $z\ll z_0$, where the flow is laminar. Below z_0 the flow, though "turbulent", is almost laminar (sub-layer, with a separation surface). Various schemes of approximation have been suggested in turbulence, like Prandtl's mixing length l in $u=ldu/dz$, or von Karman's mixing length l in $du/dz=ld^2u/dz^2$; then, l is made z, and a logarithmic law is obtained.

The logarithmic law is universal and very popular; it is based on the work of von Karman, Taylor and Prandtl (k is called the von

Karman constant).[22] The derivation given here seems to belong to Landau.[23] It is also called the "the wall law". It seems that, in spite of being universal, it is fufilled by many experiments in a wide range $10 < z/z_0 < 10^4$ (actually in the neighbourhood of 10^2 with good accuracy), both in pipes and on plates.[24] The nature of the surface is included in the parameter u_* (in general, z_0 is very low and $\nu = 0.125cm^2/s$ for air and $\nu = 0.01cm^2/s$ for water; in fact, the experimental absolute range of z is rather tight).

The surface friction is localized at the surface, and its effects should be included in the boundary conditions. The boundary conditions are missing in the derivation of the logarithmic profile. Turbulence is caused by viscosity (free wakes may be turbulent). It may, at most, be initiated by the bounding surfaces, and may be influenced by them, but not essentially governed by them. Viscosity is absent in the derivation of the logarithmic profile.

Once initiated, the turbulence is characterized by viscosity and fluctuations. The bounding surfaces are introduced only by boundary conditions. The fluctuations and viscosity, as well as the boundary conditions are absent in the derivation of the logarithmic profile.

4.7 Singular wind

We show in this section that the Reynolds equations for a turbulent flow over an unbounded flat surface in the presence of a constant pressure-gradient lead to a displaced logarithmic profile of the velocity distribution.[25] The displaced logarithmic profile is obtained by

[22]Th. von Karman, "Mechanische Ahnlichkeit und Turbulenz", Nachrichten von der Gesellschaft der Wissenschaften zu Gottingen, Fachgruppe 1 (Mathematik) **5** 58 (1930) (also as "Mechanical Similitude and Turbulence", Nat. Adv. Comm. Aeronautics Tech. Mem. #611 (1931)); G. I. Taylor, "The transport of vorticity and heat through fluids in turbulent motion", Proc. Roy. Soc. **A135** 685 (1932); "Note on the distribution of turbulent velocities in a fluid near a solid wall", Proc. Roy. Soc. Lond. **A135** 678 (1932); L. Prandtl, "Bemerkungen zur Theorie der freien Turbulenz", Z. angew. Math. Mech. **22** 241 (1942).

[23]L. Landau and E. Lifshitz, *Course of Theoretical Physics*, vol. 6 (*Fluid Mechanics*), Pergamon Press, Oxford (1987).

[24]H. Schlichting, *Boundary-Layer Theory*, McGraw-Hill, NY (1979).

[25]S. Talpos and M. Apostol, *loc. cit.*

assuming a constant production rate of turbulence energy. The displacement height measured on the (vertical) axis perpendicular to the surface is either positive or negative. For a positive displacement height the boundary layer exhibits an inversion, while for a negative displacement height the boundary layer is a direct one. In an inversion boundary layer the logarithmic velocity profile is disrupted into two distinct branches separated by a logarithmic singularity. The viscosity transforms this logarithmic singularity into a sharp edge, governed by a generalized Reynolds number. The associated temperature distribution is calculated, and the results are discussed in relation to meteorological boundary-layer jets and stratified layers. The effects of gravitation and atmospheric thermal or fluid-mixture concentration gradients ("external forcings") are also considered; it is shown that such circumstances may lead to various modifications of the boundary layers. Also, a brief presentation of a similar situation is described for a circular pipe.

Usually, the velocity logarithmic profile in the boundary layer of a turbulent flow over an unbounded flat surface is derived by using dimensional or similarity arguments.[26] We examine here the implications of the Reynolds equations for the turbulent boundary layer in the presence of a pressure gradient. It is shown that a constant pressure-gradient leads to a linear dependence of the Reynolds shear stress on the distance from the surface. Such a dependence, combined with the assumption of a constant production rate of turbulence energy, yields a displaced logarithmic profile of velocity. The displacement height, measured along the (vertical) axis perpendicular to the surface, is either positive or negative, corresponding to an inversion or a direct boundary layer, respectively. The difference between the two types of boundary layers arises from the boundary conditions. In the inversion layer the fluid flows in the direction opposite to the main flow, and the logarithmic law of the velocity profile is split into two branches separated by a logarithmic singularity. The viscosity transforms this singularity into a sharp edge, governed by a generalized Reynolds number. The temperature distribution associated with such boundary layers is calculated, and the results are discussed in connection with the meteorological boundary-layer jets and stratified layers.

[26]H. Schlichting, *loc. cit.*; L. Landau and E. Lifshitz, *loc. cit.*

Gravitation and atmospheric thermal or fluid-mixture concentration gradients ("external forcings") are also considered; it is shown that such circumstances may lead to various modifications of the boundary layers. A similar situation is presented for a circular pipe.

Specifically, we are interested in a turbulent flow of an incompressible fluid along an infinite plane surface. The coordinates x and y lie on the surface and the coordinate z is perpendicular to the surface (vertical coordinate). The velocity components (u, v, w) correspond to the (x, y, z)-directions. The fluid flows along the x axis with velocity u. As usually, we introduce the mean velocities $\overline{u}$, $\overline{v}$ and $\overline{w}$ and the fluctuating velocities $u^{'}$, $v^{'}$ and $w^{'}$, by $u \rightarrow \overline{u} + u^{'}$, etc (we consider time averaging; also, spatial averaging will be discussed below). We assume $\overline{v} = \overline{w} = 0$ and $\overline{u}(z) \neq 0$ depending only on z (a uniform flow along the x-direction). Under these conditions the Navier-Stokes equations lead to the Reynolds equations[27]

$$\begin{aligned} 0 &= -\frac{1}{\rho}\frac{\partial p}{\partial x} - \frac{\partial}{\partial x}\overline{u'^2} - \frac{\partial}{\partial y}\overline{u'v'} - \frac{\partial}{\partial z}\overline{u'w'} + \nu\frac{\partial^2\overline{u}}{\partial z^2} , \\ 0 &= -\frac{\partial}{\partial x}\overline{u'v'} - \frac{\partial}{\partial y}\overline{v'^2} - \frac{\partial}{\partial z}\overline{v'w'} , \\ 0 &= -\frac{\partial}{\partial x}\overline{u'w'} - \frac{\partial}{\partial y}\overline{v'w'} - \frac{\partial}{\partial z}\overline{w'^2} , \end{aligned} \tag{4.41}$$

where ρ is the fluid density, p is the (mean) pressure (depending only on x) and ν is the viscosity coefficient. We note the occurrence in equation (4.41) of the correlation functions $\overline{u'^2}$, $\overline{u'v'}$, etc (also called variances, like $\overline{u'^2}$, and covariances, like $\overline{u'v'}$); these quantities are the components of the Reynolds stress tensor (which, multiplied by ρ, is a momentum flux density). We assume a constant, negative pressure-gradient $\partial p/\partial x = const < 0$. We note that the Reynolds stress tensor generates forces which may compete with the pressure-gradient force in equation (4.41) (and with the viscosity "force"); therefore, equation (4.41) is in fact an equilibrium equation (corresponding to a steady flow), as expected.

[27]H. Schlichting, *loc. cit.*; O. Reynolds, "On the dynamical theory of incompressible viscous fluids and the determination of the criterion", Phil. Trans. Roy. Soc. Lond. **A186** 123 (1895); H. Lamb, *Hydrodynamics*, Cambridge University Press, Cambridge (1916).

Equations (4.41) represent a system of three equations with seven unknowns: the components of the Reynolds tensor and the mean velocity $\overline{u}$; it is an under-determined system of equations. We are interested in the first equation (4.41), where we assume $\partial(\overline{u'v'})/\partial y = 0$, $\overline{u'w'}(z) \neq 0$ depending only on z and $\partial\overline{u'^2}/\partial x = const$. In general, constant Reynolds stress components with respect to a coordinate amount to a homogeneous turbulence along that axis.[28] Under these conditions, the first equation (4.41) reads

$$0 = A - \frac{d}{dz}\overline{u'w'} + \nu\frac{d^2\overline{u}}{dz^2} \ , \tag{4.42}$$

where $A = -(1/\rho)dp/dx - \partial\overline{u'^2}/\partial x$; for the sake of generality we keep for the moment $\partial\overline{u'^2}/\partial x = const$ in equation (4.42), corresponding to an inhomogeneous turbulence along the x-axis.

We leave aside for the moment the viscosity term in equation (4.42); then, the integration of this equation gives

$$\overline{u'w'} = Az - \beta u_*^2 \ , \tag{4.43}$$

where $\beta = \pm 1$ and the parameter u_* is a surface friction velocity; for the sake of generality we keep both signs in the boundary condition $\overline{u'w'} \mid_{z=0} = -\beta u_*^2$; ρu_*^2 is the friction force per unit area of the surface and $\rho\overline{u'w'}$ is the xz-component of the momentum flux density (Reynolds shear stress). Equation (4.43) can also be written as

$$\overline{u'w'} = \frac{\beta u_*^2}{h}(z-h) \ , \ h = \frac{\beta u_*^2}{A} \ , \ A \neq 0. \tag{4.44}$$

Such a linear dependence of the shear stress is known in the atmospheric turbulence of the boundary layers[29] and in turbulent flow on flat plates or in channels.[30] We note that the displacement height h may have both signs.

[28] G. K. Batchelor, *The theory of homogeneous turbulence*, Cambridge University Press, Cambridge (1953); G. I. Taylor, "Statistical Theory of Turbulence", Proc. Roy. Soc. Lond. **A151** 421 (1935); "Statistical Theory of Turbulence. II", Proc. Roy. Soc. Lond. **A151** 444 (1935).

[29] S. J. Caughey, J. C. Wyngaard and J. C. Kaimal, "Turbulence in the evolving stable boundary layer", J. Atmos. Sci. **36** 1041 ((1979).

[30] H. Schlichting, *loc. cit.*

Multiplying the Navier-Stokes equations by $\overline{u}$ and using the same procedure ($u \rightarrow \overline{u} + u'$, etc), we get the conservation law for the mean-flow energy

$$0 = \overline{\frac{\partial}{\partial t}\left[\frac{1}{2}\overline{u}^2\right]} =$$
$$= \overline{u}\left[-\frac{1}{\rho}\frac{\partial p}{\partial x} - \frac{\partial}{\partial x}\overline{u'^2} - \frac{\partial}{\partial y}\overline{u'v'} - \frac{\partial}{\partial z}\overline{u'w'} + \nu\frac{\partial^2 \overline{u}}{\partial z^2}\right] \quad , \tag{4.45}$$

which is the first equation (4.41) multiplied by $\overline{u}$. Similarly, multiplying the Navier-Stokes equations by the fluctuating velocities and taking the average we get the conservation equation for the turbulence energy

$$0 = \overline{\frac{\partial}{\partial t}\left[\frac{1}{2}(u'^2 + v'^2 + w'^2)\right]} =$$
$$= -\overline{u'w'}\frac{\partial \overline{u}}{\partial z} - \frac{1}{2}\overline{u}\frac{\partial}{\partial x}\left(\overline{u'^2} + \overline{v'^2} + \overline{w'^2}\right) + \tag{4.46}$$
$$+\nu\left(\overline{u'\Delta u'} + \overline{v'\Delta v'} + \overline{w'\Delta w'}\right) \quad ,$$

where third-order terms involving products of three fluctuating velocities and velocity derivatives, as well as the contribution of the fluctuating part of the pressure have been dropped out; in addition, in deriving equation (4.46) the continuity equation $\partial u'/\partial x + \partial v'/\partial y + \partial w'/\partial z = 0$ has been used. The main assumption made here is that the fluctuations are small in comparison with the mean flow. Adding the two equations (4.45) and (4.46), we get the conservation law of the total energy

$$0 = \overline{\frac{\partial}{\partial t}\left[\frac{1}{2}(u^2 + v^2 + w^2)\right]} =$$
$$= -\frac{1}{\rho}\frac{\partial p}{\partial x}\overline{u} - \frac{\partial}{\partial x}\left(\overline{u}\overline{u'^2}\right) - \frac{\partial}{\partial y}\left(\overline{u}\overline{u'v'}\right) -$$
$$-\frac{\partial}{\partial z}\left(\overline{u}\overline{u'w'}\right) - \frac{\partial}{\partial x}\left[\frac{1}{2}\overline{u}\left(\overline{u'^2} + \overline{v'^2} + \overline{w'^2}\right)\right] + \tag{4.47}$$
$$+\nu\left(\overline{u}\frac{\partial^2 \overline{u}}{\partial z^2} + \overline{u'\Delta u'} + \overline{v'\Delta v'} + \overline{w'\Delta w'}\right) \quad .$$

The first term on the right in equation (4.47) is related to the work done by the pressure forces per unit time; the next four terms are related to the energy flux density due to the fluid mass transfer; the last term, involving the viscosity coefficient, can be written as

$$\nu\left(\overline{u}\frac{\partial^2\overline{u}}{\partial z^2}+\overline{u'\Delta u'}+\overline{v'\Delta v'}+\overline{w'\Delta w'}\right)=$$

$$=\nu\frac{\partial^2}{\partial z^2}\left(\tfrac{1}{2}\overline{u}^2\right)+\tfrac{1}{2}\nu\Delta\left(\overline{u'^2}+\overline{v'^2}+\overline{w'^2}\right)-$$

$$-\nu\left(\frac{\partial\overline{u}}{\partial z}\right)^2-\nu\left[\overline{\left(\frac{\partial u'}{\partial x}\right)^2}+\overline{\left(\frac{\partial u'}{\partial y}\right)^2}+\overline{\left(\frac{\partial u'}{\partial z}\right)^2}\right]- \tag{4.48}$$

$$-\nu\left[\overline{\left(\frac{\partial v'}{\partial x}\right)^2}+\overline{\left(\frac{\partial v'}{\partial y}\right)^2}+\overline{\left(\frac{\partial v'}{\partial z}\right)^2}\right]-$$

$$-\nu\left[\overline{\left(\frac{\partial w'}{\partial x}\right)^2}+\overline{\left(\frac{\partial w'}{\partial y}\right)^2}+\overline{\left(\frac{\partial w'}{\partial z}\right)^2}\right]\ ;$$

the first two terms on the right in equation (4.48) are related to the energy flux density due to internal friction (momentum transfer through collisions caused by viscosity); the remaining terms imply heat production; for an adiabatic flow they are equal to $-(1/\rho)div\mathbf{q}$, where $\mathbf{q}$ is the heat flux density.[31]

A similar analysis can be done for each of the equations (4.45) and (4.46) separately. The terms $-\overline{u}\frac{\partial}{\partial z}\overline{u'w'}$ on the right in equation (4.45) and $-\overline{u'w'}\frac{\partial\overline{u}}{\partial z}$ in equation (4.46) (which together give an energy flux density $-\frac{\partial}{\partial z}\left(\overline{u}\overline{u'w'}\right)$), have a special meaning when taken separately: each of them represents a coupling between the mean flow and the turbulent flow. In particular, $-\overline{u'w'}\frac{\partial\overline{u}}{\partial z}$ is a production rate of turbulence energy (while $-\overline{u}\frac{\partial}{\partial z}\overline{u'w'}$ is a production rate of mean-flow energy; "production" means here either a positive or a negative contribution).

We focus now on equation (4.46). The components of the Reynolds stress tensor have a small, local (finite) variation, at least for a homogeneous turbulence; assuming a homogeneous turbulence along the

[31] L. Landau and E. Lifshitz, *Course of Theoretical Physics*, vol. 6, (*Fluid Mechanics*), Pergamon Press, Oxford (1987).

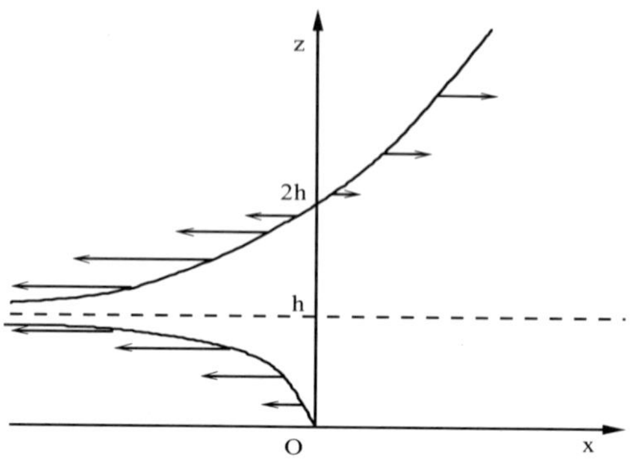

Figure 4.4: Logarithmic singularity (velocity "rupture") in an inversion layer, which may occur in the turbulent boundary layer. The arrows indicate the velocity (equation (4.52)).

x, y-coordinates and taking a spatial averaging with respect to these coordinates we get from equation (4.46)

$$\begin{gathered}\overline{u'w'}\frac{d\overline{u}}{dz} = \nu\left(\overline{u'\frac{d^2}{dz^2}u'} + \overline{v'\frac{d^2}{dz^2}v'} + \overline{w'\frac{d^2}{dz^2}w'}\right) - \nu C = \\ = \frac{1}{2}\nu\frac{d^2}{dz^2}\left(\overline{u'^2} + \overline{v'^2} + \overline{w'^2}\right) - \\ -\nu\left[\overline{\left(\frac{du'}{dz}\right)^2} + \overline{\left(\frac{dv'}{dz}\right)^2} + \overline{\left(\frac{dw'}{dz}\right)^2}\right] - \nu C \ ,\end{gathered} \tag{4.49}$$

where C is a positive constant arising from spatial averages of the type $\overline{u'\frac{\partial^2}{\partial x^2}u'}$ (for the sake of simplicity, in equation (4.49) the spatial averaging is not indicated explicitly by an additional averaging sign). The first term on the right in equation (4.49) involves the turbulence energy, while the second term is the heat produced by internal friction (viscosity) along the z-axis. For small fluctuations these terms have a slow variation with z, at least for not too small distances from the surface; therefore, we may assume that the main contribution to the *rhs* of equation (4.49) is a constant. The deviations of this term from a constant are viewed as small corrections, which are neglected here; this implies a limited, intermediate range of validity for the coordinate z.

Therefore, we assume a constant production rate of turbulence energy

$$\overline{u'w'}\frac{d\overline{u}}{dz} = const = -\beta u_*^2 u_1 \ , \qquad (4.50)$$

where $u_1 = \frac{d\overline{u}}{dz}\mid_{z=0} = \overline{u}' \mid_{z=0}$. This scheme of approximation amounts to a perturbation-theoretical treatment, where the correlation functions of the fluctuating quantities (including correlation functions of derivatives of fluctuating quantities), like those involved in equation (4.49), are assumed to vary slowly along the z-axis. The sign of the constant in equation (4.50) (the sign of the product βu_1) remains undetermined. A constant production rate of turbulence energy is suggested by experimental studies, at least for not too small distances from the surface.[32]

Equation (4.50) is a closure assumption, which amounts to a turbulence model; using equation (4.44) we get

$$\frac{d\overline{u}}{dz} = -\frac{hu_1}{z-h} \qquad (4.51)$$

and

$$\overline{u}(z) = -hu_1 \ln\left|\frac{z-h}{h}\right| \qquad (4.52)$$

for a vanishing velocity on the surface ($\overline{u}\mid_{z=0} = 0$). This law is different from the well-known logarithmic profile $\overline{u}(z) = (u_*/k)\ln(z/z_0)$,[33] where $k \simeq 0.4$ is the von Karman constant and z_0 is a cut-off parameter; in particular equation (4.52) includes a characteristic length h and the slope of the velocity at $z = 0$. The difference arises from

[32] S. J. Kline, W. C. Reynolds, F. A. Schraub and P. W. Runstadler, "The structure of turbulent boundary layer", J. Fluid Mech. **30** 741 (1967); P. S. Klebanoff, "Characteristics of turbulence in a boundary layer with zero pressure gradient", Nat. Adv. Comm. Aeronautics (NACA) Tech. Note #3178 (1954).

[33] Th. von Karman, "Mechanische Ahnlichkeit und Turbulenz", Nachrichten von der Gesellschaft der Wissenschaften zu Gottingen, Fachgruppe 1 (Mathematik) **5** 58 (1930) (also as "Mechanical Similitude and Turbulence", Nat. Adv. Comm. Aeronautics (NACA) Tech. Mem. #611 (1931); G. I. Taylor, "Note on the distribution of turbulent velocities in a fluid near a solid wall", Proc. Roy. Soc. Lond. **A135** 678 (1932); "The transport of vorticity and heat through fluids in turbulent motion", Proc. Roy. Soc. **A135** 685 (1932); L. Prandtl, "Bemerkungen zur Theorie der freien Turbulenz", Z. angew. Math. Mech. **22** 241 (1942).

the fact that the governing quantities of the velocity distribution are not only the velocity u_* and the coordinate z, but the Reynolds tensor too. Various logarithmic laws have been suggested for various z-regions in turbulent boundary layers, including similarity analysis in the presence of a pressure gradient. Equation (4.52) gives a displaced logarithmic profile of velocity, with h the displacement height. For large values of z ($z \gg h$) we recover a logarithmic law

$$\overline{u}(z) \simeq -hu_1 \ln |z/h| \ , \ z \gg |h| \ , \tag{4.53}$$

while for small values of z (comparable with $|h|$) equation (4.52) gives a linear z-dependence $\overline{u}(z) \simeq u_1 z$; this linear dependence corresponds also to the absence of the pressure-gradient ($A = 0$ in equation (4.44)), where $h \to \infty$. For $h > 0$ ($\beta/A > 0$) the displaced logarithmic profile given by equation (4.52) corresponds to an inversion boundary layer, while for $h < 0$ ($\beta/A < 0$) the displaced logarithmic profile indicates a direct boundary layer. A positive displacement height (in an inversion layer) disrupts the logarithmic velocity profile $\ln z$ in two branches, separated by a logarithmic singularity at $z = h$, as shown by equation (4.52) and in Fig. 4.4. For $z < 2h$ the direction of the flow is opposite (negative, for $u_1 < 0$) to the direction of the main flow for $z > 2h$. In the vicinity of the surface the velocity is linear in z, $\overline{u}(z) \simeq u_1 z$ for $z \ll h$, and negative ($u_1 < 0$).

Records are known of inversion layers in Meteorology ("counter-gradient" layers).[34] The inversion layer may also be viewed as a "sub-layer", though it is different from what is usually called a sub-layer in the logarithmic profile;[35] rather, the range of small values of z where the velocity is linear in z can be considered a sub-layer. The "linear sub-layer" is well documented, both experimentally and theoretically.[36] An offset to the z-coordinate of the h-type inside the logarithm in equation (4.52) has been pointed out in atmospheric turbulence, especially in connection with self-similarity symmetries of the

[34] R. B. Stull, *An Introduction to Boundary Layer Meteorology*, Kluwer, Dordrecht (1988).

[35] L. Landau and E. Lifshitz, *Course of Theoretical Physics*, vol. 6, (*Fluid Mechanics*), Pergamon Press, Oxford (1987).

[36] S. J. Kline, W. C. Reynolds, F. A. Schraub and P. W. Runstadler, *loc. cit.*; H. Eckelmann, "The structure of the viscous sublayer and the adjacent wall region in a turbulent channel flow", J. Fluid Mech. **65** 439 (1974).

equations of motion. Local self-similarity arguments used in getting a consistent organization of the experimental data in the turbulent structure of the atmospheric boundary layer indicated the existence of a finite parameter similar to h (the boundary-layer depth), a linear dependence on the coordinate z of the shear stress $\overline{u'w'}$, a connection between the shear stress $\overline{u'w'}$ and the gradient of the mean velocity as given in equations (4.46) and (4.49), and, especially, a constant rate of production of the turbulence energy, as expressed by equation (4.49), at least for not too small distances from the surface (usually for $z/|h| > 0.2$; the Coriolis force and the gravitational force are neglected in our considerations).[37] A parametrization model of the planetary boundary layer in conditions of general atmospheric circulation suggests similar conclusions.[38]

It is also worth noting that assuming an inhomogeneous turbulence along the x-axis with $\partial\overline{u'^2}/\partial x = const$ within the same procedure as above, equation (4.46) leads to

$$\overline{u'w'}\frac{d\overline{u}}{dz} = C_1\overline{u} + C_2 \ , \tag{4.54}$$

where $C_{1,2}$ are two constants; making use of $\overline{u'w'}$ given by equation (4.44), we get $\overline{u} \sim (z-h)^{C_3} + const$, where C_3 is another constant. Therefore, in this case the velocity profile is changed into a power law. Consequently, it seem that a (displaced) logarithmic profile of velocity is obtained only for homogeneous turbulence along the in-plane directions and a constant pressure-gradient. In the equilibrium equation (4.42) this pressure gradient is compensated by the gradient (along the vertical axis) of the Reynolds shear stress (for vanishing viscosity).

We turn now to equation (4.42) and include the viscosity coefficient; we are interested in the limit $\nu \rightarrow 0$. We assume a homogeneous turbulence, *i.e.* $A = -(1/\rho)dp/dx > 0$ $(\partial\overline{u'^2}/\partial x = 0)$. For a direct boundary layer we have $\beta = -1$ and $u_1 > 0$; changing h into $-h$, we

[37] A. L. M. Grant, "The structure of the near-neutral atmospheric boundary layer", J. Atmos. Sci. **49** 226 (1992); J. C. Wyngaard, "Atmospheric Turbulence", Annu. Rev. Fluid Mech. **24** 205 (1992).

[38] J. W. Deardorff, "Parametrization of the planetary boundary layer for use in general circulation models", Monthly Weather Rev. **100** 93 (1972).

get from equations (4.42) and (4.49)

$$0=\frac{u_*^2}{h}-\frac{d}{dz}\frac{u_*^2u_1}{\overline{u}'}+\nu\overline{u}'' \tag{4.55}$$

(with the notation $A=-\frac{1}{\rho}\frac{dp}{dx}=u_*^2/h$). A first integration leads to

$$\overline{u}'=\frac{u_*^2}{2h\nu}\left[-(z+\overline{h})\pm\sqrt{(z+\overline{h})^2+z_h^2}\right] \ , \tag{4.56}$$

where

$$\overline{h}=h\left(1-\frac{1}{R}\right) \ , \ z_h^2=\frac{4h^2}{R} \tag{4.57}$$

and $R=u_*^2/\nu u_1$ is a generalized Reynolds number; $R_c=1$ plays the role of a critical value for the existence of a positive parameter $\overline{h}$ ($\overline{h}>0$ for $R>R_c=1$). Equation (4.51) is recovered in the limit $\nu\rightarrow 0$ ($R\rightarrow\infty$) by the sign $+$ in equation (4.56). The solution $\overline{u}$ can be obtained by integrating this equation with the boundary condition $\overline{u}\mid_{z=0}=0$. We get

$$\begin{aligned}\overline{u}(z)=hu_1\{\ln\frac{z+\overline{h}+\sqrt{(z+\overline{h})^2+z_h^2}}{\overline{h}+\sqrt{\overline{h}^2+z_h^2}}-\frac{z_h^2}{2\left[z+\overline{h}+\sqrt{(z+\overline{h})^2+z_h^2}\right]^2}+\\ +\frac{z_h^2}{2\left(\overline{h}+\sqrt{\overline{h}+z_h^2}\right)^2}\} \ ;\end{aligned} \tag{4.58}$$

by comparing this equation with the displaced logarithmic equation (4.52), we can see that the only effect of the viscosity is to bring corrections of the order $1/R$ in the limit $R\gg 1$.

The modification brought by viscosity is more appreciable in the inversion boundary layer. For an inversion layer we assume $A=-(1/\rho)dp/dx>0$, $\beta=1$ and $u_1<0$; equation (4.42) becomes

$$0=\frac{u_*^2}{h}+\frac{d}{dz}\frac{u_*^2u_1}{\overline{u}'}+\nu\overline{u}'' \tag{4.59}$$

with $A=-\frac{1}{\rho}\frac{dp}{dx}=u_*^2/h>0$. A first integration leads to

$$\overline{u}'=-\frac{u_*^2}{2h\nu}\left[z-\overline{h}\pm\sqrt{(z-\overline{h})^2+z_h^2}\right] \ , \tag{4.60}$$

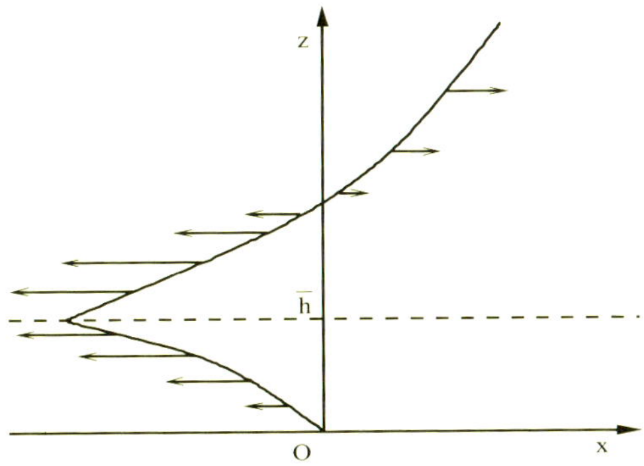

Figure 4.5: A sharp-edged inversion layer (equation (4.63)), shown schematically.

where $\overline{h} = h(1 - 1/R)$, $z_{\overline{h}}^2 = 4h^2/R$ and $R = u_*^2/\nu\,|u_1|$; a positive displacement height $\overline{h}$ exists for $R > R_c = 1$. Equation (4.51) is recovered in the limit $\nu \to 0$ ($R \to \infty$) by

$$\overline{u}' = \begin{cases} -\frac{u_*^2}{2h\nu}\left[z - \overline{h} + \sqrt{(z-\overline{h})^2 + z_{\overline{h}}^2}\right] & , \ z < \overline{h} \ , \\ -\frac{u_*^2}{2h\nu}\left[z - \overline{h} - \sqrt{(z-\overline{h})^2 + z_{\overline{h}}^2}\right] & , \ z > \overline{h} \ ; \end{cases} \tag{4.61}$$

we can see that the derivative $\overline{u}'$ has a discontinuity at $z = \overline{h}$, given by

$$\overline{u}'(\overline{h}) = sgn(z - \overline{h})\,|u_1|\,\sqrt{R} \ ; \tag{4.62}$$

the functi $\overline{u}(z)$ acquires a "sharp-edge" profile in the vicinity of $z = \overline{h}$, ctrolled by the Reynolds number R; for $R \to \infty$ ($\nu \to 0$) the logarithmic singularity is recovered. The soluti $\overline{u}$ can be obtained by integrating equatis (4.61) with the boundary cditi $\overline{u}\,|_{z=0} = 0$. We get

$$\overline{u}(z) = -h\,|u_1|\,\{\ln\left|\frac{\sqrt{(z-\overline{h})^2+z_h^2}-|z-\overline{h}|}{\sqrt{\overline{h}^2+z_h^2}-\overline{h}}\right| +$$

$$+\tfrac{1}{2z_h^2}\left[\sqrt{(z-\overline{h})^2+z_h^2}-\left|z-\overline{h}\right|\right]^2 - \tag{4.63}$$

$$-\tfrac{1}{2z_h^2}\left(\sqrt{\overline{h}^2+z_h^2}-\overline{h}\right)^2\}$$

and

$$\overline{u}(\overline{h}) = -\frac{1}{2}h\,|u_1|\left(\ln R + 1 - \frac{1}{R}\right) . \tag{4.64}$$

By comparing with equation (4.52) (with h negative) we can see that $\sqrt{R}$ scales as $h/|z-h|$ for z in the vicinity of h and $R \to \infty$. For a viscous flow the inversion layer is preserved for $R > R_c = 1$ ($\nu < u_*^2/|u_1|$), while acquiring an edge at $\overline{h}$, which gets sharper with increasing R; an estimate for the width of the sharpness of the inversion layer is the parameter z_h, while $2\overline{h}$ can be taken as an estimate of the width of the inversion layer for large values of R. A sketch of a sharp-edged inversion layer is shown in Fig. 4.5. Such turnovers or cusps in wind profile have been recorded in nocturnal jets in turbulent atmospheric boundary layers.[39] We note also that for large values of the viscosity ($R \ll 1$) the sign of the "renormalized" height $\overline{h}$ may change in the above equations, and a transition direct-inversion layer may occur. The inverse wind described herein may be viewed as an example of a large turbulent vortex.

We turn now to the temperature distribution associated with the turbulent flow in the boundary layer. We assume a homogeneous turbulence along the x, y-directions, according to the discussion above.

[39] L. Mahrt and J.-C. Andre, "On the stratification of turbulent mixed layer", J. Geophys. Res. **88** 2662 (1983); L. Mahrt, "Vertical stratification and turbulence in the very stable boundary layer", J. Atmos. Sci. **42** 2333 (1985); J. Malcher and H. Krauss, "Low-level jet phenomena described by an integrated dynamic PBL model", Bound. Layer Meteor. **27** 327 (1983); J. R. Garratt, "Inland boundary layer at low altitudes. Part 1, The nocturnal jet", Bound. Layer Meter. **32** 307 (1985); H. Krauss, J.Malcher and E. Schaller, "Nocturnal low-level jet during PUKK", Bound. Layer. Meteor. **31** 187 (1985); R. R. Brook, "Koorin nocturnal low-level jet", Bound. Layer Meteor. **32** 133 (1985).

According to our hypothesis the production rate of turbulence energy is a constant, $-\overline{u'w'}\frac{d\overline{u}}{dz}=\beta u_*^2 u_1$; for both a direct and an inversion layer the product βu_1 is negative. We can see that the turbulence energy is diminished in fact (the "production" rate of turbulence energy is in fact a "loss" rate of turbulence energy). Equation (4.50) gives a representation of the main contribution to the sum of the correlation functions involved in the *rhs* of equation (4.49), spatially averaged, in terms of our notations $-\beta u_*^2 u_1$. We note that for the opposite sign $\beta u_1 > 0$, the flow direction is reversed both in the direct and inversion layer, corresponding to an increase in the turbulence energy (this situation can also be obtained for $-(1/\rho)dp/dx < 0$). In addition, it is worth noting that there is a non-vanishing energy flux transported along the z-coordinate for large z, due to the fact that the surface is unbounded (as seen from an increasing velocity $\overline{u}(z)$ with increasing z); this is why the validity of the present results is limited to finite values of z.

According to equation (4.48), the equation of heat transfer is

$$T\frac{ds}{dt}=-\frac{1}{\rho}div\mathbf{q}+\nu\left(\frac{\partial\overline{u}}{\partial z}\right)^2+ \\ +\nu\left[\overline{\left(\frac{\partial u'}{\partial z}\right)^2}+\overline{\left(\frac{\partial v'}{\partial z}\right)^2}+\overline{\left(\frac{\partial w'}{\partial z}\right)^2}\right]+\nu C\ , \tag{4.65}$$

where s is the entropy per unit mass, T is the temperature, $\mathbf{q}$ is the heat flux density and C is the same constant that appears in equation (4.49); the thermal conduction is given by $\mathbf{q}=-\kappa gradT$, where κ is the thermal conductivity. Equation (4.65) shows the increase in time of the entropy by internal friction and heat conduction. In general, in the *lhs* of equation (4.65) we take both a time averaging and a spatial averaging with respect to the in-plane coordinates; it implies both mean and fluctuating temperatures and entropies (and velocities). Since the fluctuating quantities are assumed to be small in comparison with the mean quantities, we may leave aside these contributions to equation (4.65). In addition, the *lhs* of equation (4.65) is vanishing (as for an adiabatic flow) for our particular problem (since we have only one component $\overline{u}$ of the mean velocity along the homogeneous-turbulence x-axis). Then the temperature distribution is given by

$$\kappa\frac{d^2T}{dz^2}=-\rho\nu\left(\frac{d\overline{u}}{dz}\right)^2 . \tag{4.66}$$

For $\nu\rightarrow 0$ we may use equation (4.51) for $d\overline{u}/dz$; we get

$$T=\frac{\rho\nu u_1^2h^2}{\kappa}\ln\left|\frac{z-h}{h}\right|+\frac{\rho\nu u_1^2h}{\kappa}z \tag{4.67}$$

with the boundary conditions $T\mid_{z=0}=\frac{dT}{dz}\mid_{z=0}=0$. We can see that the temperature decreases with the distance from the surface; for an inversion layer the temperature has a logarithmic singularity at $z=h$; in general, the temperature distribution for an inversion layer exhibits a non-monotonous dependence on z. For the evolution of the temperature in time, we use $ds=c_pdT$ in equation (4.65), where c_p is the specific heat at constant pressure; the solution is specific to a diffusion process. Models of temperature distributions in atmospheric boundary layers, or associated with stratified air-flow inversion layers (jets), or in convective atmospheric boundary layers have been discussed.[40]

A more general situation may appear for boundary layers in the presence of so-called "external forcings", which may generate local gradients of temperature or of fluid-mixture concentration along the z-coordinate.[41] Under such circumstances we may expect a z-dependence of the pressure p; consequently, the parameter $A=(-1/\rho)(\partial p/\partial x)$ is not a constant anymore, as assumed heretofore, but, instead, it becomes a function of z (the gravitation produces a similar effect). For instance, for a simple linear dependence of the form $A=(-1/\rho)(\partial p/\partial x)$ $=a+2bz$, where a and b are two (positive) constants, the procedure described herein (equations (4.42) and (4.49)) leads to a quadratic z-dependence of $\overline{u'w'}$ and $1/(d\overline{u}/dz)$, of the form $\overline{u'w'}=az+bz^2-u_*^2$, for $\beta=1$ (inversion layer). The velocity distribution is $\overline{u}\sim\ln\left[\left|z/h_1-1\right|/(z/h_2+1)\right]$, where $h_2>h_1>0$ are two parameters related to the roots of the equation $az+bz^2-u_*^2=0$. We can see that the

[40]R. B. Stull, *An Introduction to Boundary Layer Meteorology*, Kluwer, Dordrecht (1988); J. W. Deardorff, "The counter-gradient heat flux in the lower atmosphere and in laboratory", J. Atmos. Sci. **23** 503 (1966); H. Tennekes, "A model for the dynamics of the inversion above a convection boundary layer", J. Atmos. Sci. **30** 558 (1973).

[41]R. B. Stull, *loc. cit.*

inversion layer is preserved, but the logarithmic velocity profile suffers now an appreciable change for large z, where the velocity tends to a constant $\sim \ln(h_2/h_1)$ (we recall that the derivations made here are not valid for large z). A similar result is obtained for $b<0$ (gravitation effect), where the velocity profile $\overline{u} \sim \ln|(z/h_1-1)/(z/h_2-1)|$, $h_2 > h_1 > 0$, exhibits a "scissors-like", double inversion layer (stratified layers).

The logarithmic singularity of the inversion layer at $z=h$ or the discontinuity in the slope of the function $\overline{u}(z)$ at $z=\overline{h}$ might be viewed as unsatisfactory features of the present model. However, $\overline{u'w'} \sim z-h$ given by equation (4.44) can be considered as a first-order contribution in powers of z; an additional, next-order z^2-term could lead to $\overline{u'w'}$ of the form $\overline{u'w'} \sim (z-h)^2+\varepsilon^2$, where ε is a small parameter. Indeed, it seems that the condition of vanishing $\overline{u'w'}$ at $z=h$ is too strong. Similarly, the constant production rate of turbulence energy $\overline{u'w'}\frac{d\overline{u}}{dz} = -u_*^2u_1$ (equation (4.49)) may acquire an additional z-term, leading to $\overline{u'w'}\frac{d\overline{u}}{dz}$ of the form $\overline{u'w'}\frac{d\overline{u}}{dz} \sim z-h'$ (where h' is, in general, different from h). Then, $\overline{u}'$ behaves essentially as $\overline{u}' \sim (z-h)/[(z-h)^2+\varepsilon^2]$ and $\overline{u} \sim \ln[(z-h)^2+\varepsilon^2]$, which might be a more acceptable functional dependence of the velocity on z. However, the derivation of higher-order terms in the shear stress and the production rate of turbulence energy require further, additional model assumptions.

In conclusion, we may say that a displaced logarithmic profile of velocity has been derived here for the boundary layer in the turbulent flow over an unbounded flat surface by assuming a constant pressure-gradient and a constant production rate of turbulence energy, associated with a homogeneous turbulence along the in-plane directions. For a direct boundary layer the displacement height is negative, while for an inversion boundary layer the displacement height is positive. The inversion layer splits the logarithmic profile in two branches, separated by a logarithmic singularity (a "rupture" in the velocity distribution). In the inversion layer the fluid flows in direction opposite to the main flow. In a viscous flow the singularity in the inversion layer becomes a sharp edge controlled by a generalized Reynolds number R (small values of the viscosity bring only corrections of the order $1/R \ll 1$ for the direct boundary layer). The associated temperature distribution has also been estimated for such boundary layers. It has been shown that gravitation and atmospheric, local, thermal or fluid-mixture concentration gradients ("external forcings") can bring various modifications to the inversion layers. The procedure employed in this paper of treating the turbulent boundary layer is based on the Reynolds equations and on certain assumptions regarding the components of the Reynolds stress tensor, derived from the assumption of homogeneous turbulence, or the symmetry of the problem. The Reynolds equations remain, in general, undetermined, as well as some features of the boundary conditions; for instance, the distinction criterion between a direct boundary layer and an inversion one (the choice of the parameter $\beta = \pm 1$), or the choice of the sign of the product βu_1 (as well as the parameter u_*^2) remain undetermined; there appears to be no obvious, physical, quantitative criteria to discriminate between the two types of boundary layers, the difference being made by the boundary conditions, and, probably, the (undetermined) preparation conditions of the turbulence.

Finally, we include here a brief description of a similar situation for a pipe with a circular cross-section of radius R. For a uniform flow with mean velocity $u(r)$ along the z-axis of the pipe the Reynolds shear stress is

$$2r\overline{v_r' v_z'} = \frac{\beta u_*^2}{h}\left(r^2 - R^2 - 2hR\right) \ , \tag{4.68}$$

where $h = \beta u_*^2/A$, $A = -(1/\rho)(dp/dz) = const > 0$ (r being the radial coordinate). The constant production rate of turbulence energy is $-\overline{v_r' v_z'}(du/dr) = -\beta u_*^2 u_1$, where $u_1 = -\frac{du}{dr}\mid_{r=R}$ and the velocity distribution is given by

$$u(r) = hu_1 \ln\left|\frac{r^2 - R^2 - 2hR}{2hR}\right| \tag{4.69}$$

for $u(R) = 0$ (and zero viscosity). For $\beta = +1$ and $u_1 > 0$ the boundary layer is a direct one, and the viscosity brings only small corrections in the limit $\nu \to 0$. For $\beta = -1$ and $u_1 < 0$ the parameter h is negative and we have an inversion layer; the velocity exhibits a logarithmic singularity at $r = (R^2 - 2\,|h|\,R)^{1/2}$ (the flow is possible for $R > 2\,|h|$), which is made finite by viscosity.

4.8 Turbulence

4.8.1 Introduction

We represent the fully developed turbulence as a fluctuating, homogeneous and isotropic vorticial liquid, with singular velocities in a discrete set of points. On average, the Navier-Stokes equation is trivially satisfied by such a vorticial liquid, in the sense that the mean velocity field and the mean inertial force are zero everywhere except at the singularities. We are left with a fluid of discrete, pointlike singularities, whose Navier-Stokes equation is reduced to Newton's equation of motion. Such sets of singularities can be treated as a (quasi) ideal classical gas in thermodynamic equilibrium (or a solution of vortices in the background fluid). Apart from the general case, we discuss also three particular examples of vorticial liquids: a two-dimensional liquid of parallel, rectilinear filaments (coulombian liquid in two dimensions), a coulombian vorticial liquid in three dimensions and a dipolar vorticial liquid with pointlike vorticities.

As it is well known, for high velocities with large variations over small distances, the solutions of the Navier-Stokes equation may become unstable, and the fluid may develop turbulence. In its extreme manifestation the turbulent flow displays very irregular, disordered velocities, fluctuating in time at each point in space. This is known as a fully developed turbulence. By using such fluctuating velocities, besides a steady mean velocity, the Navier-Stokes equation becomes an infinite hierarchy of equations for velocity correlation functions, known as Reynolds's equations,[42] which need closure assumptions. According to the experimental observations, it was realized that such irregular movements of the fluid exhibit distributions of swirls (eddies, vortices), of various magnitudes; it is likely that the large eddies transfer energy to the small eddies, which dissipate it.[43] Statistical

[42]O. Reynolds, "On the dynamical theory of the incompressible viscous fluids and the determination of the criterion", Phil. Trans. Roy. Soc. London **A186** 123 (1895).

[43]L. F. Richardson, "The supply of energy from and to atmospheric eddies", Proc. Roy. Soc. London **A97** 354 (1920); *Weather Prediction by Numerical Process*, Cambridge University Press, Cambridge (1922); L. Prandtl, "The generation of vortices in fluids of small viscosity", J. Roy. Aeronaut. Soc. **31** 720 (1927).

concepts like correlations, homogeneity and isotropy have been introduced in the theory of turbulence,[44] and dimensional analysis and similarity arguments allowed the derivation of the energy spectrum of the turbulent eddies.[45]

Meanwhile, the relation of this statistical turbulence with the Navier-Stokes equation remained unclear. Could the Navier-Stokes equation describe a turbulent motion? To what extent and in what sense? Has the Navier-Stokes equation smooth and stable solutions? What is the appropriate representation of a turbulent field of velocities?[46]

According to experimental evidence, we adopt for the extreme manifestation of the turbulent flow a representation consisting of a fluctuating, homogeneous and isotropic vorticial liquid, with singular velocities in a discrete set of points. We think that such a representation is appropriate for what is called a fully developed turbulence. Intermediate regimes of less-developed, turbulent flows, which do not fulfil such conditions, remain indefinite to a large extent.

In general, a vorticial liquid develops internal forces, which make it unstable; in order to be satisfied, the Navier-Stokes equation needs very special external forces, which, physically, is not always attainable in practice. For fluctuating, homogeneous and isotropic distributions of vorticities the mean velocity field and the mean inertial force are zero, such that, on average, the Navier-Stokes equation is satisfied trivially. Consequently, we are left with a set of discrete singularities, for which the Navier-Stokes equation is reduced to Newton's equation of pointlike particles. The ensemble of pointlike singularities may be viewed as a (quasi) ideal classical gas of particles in thermodynamic equilibrium (or a solution of vortices in the background fluid).

---

[44] G. I. Taylor, "Statistical theory of turbulence", Proc. Roy. Soc. London **A151** 421; II. 444 (1935); G. K. Batchelor, *The Theory of Homogeneous Turbulence*, Cambridge University Press, Cambridge (1953).

[45] A. N. Kolmogorov, "The local structure of turbulence in incompressible viscous fluid for very large Reynolds numbers", Dokl. Akad. Nauk SSSR **30** 299 (1941) (in Russian) (Proc. Roy. Soc. London **A434** 9 (1991)); "Dissipation of energy in locally isotropic turbulence", Dokl. Akad. Nauk SSSR **32** 19 (1941) (in Russian) (Proc. Roy. Soc. Lond. **A434** 15 (1991)); A. M. Obukhov, "On the distribution of energy in the spectrum of turbulent flow", Dokl. Akad. Nauk SSSR **32** 22 (1941) (in Russian).

[46] J. Leray, "Sur le mouvement d'un liquide viscueux emplissant l'espace", Acta Mathematica **63** 193 (1934).

We discuss also three particular examples of vorticial liquids: a distribution of parallel, rectilinear filaments (coulombian liquid in two dimensions), a coulombian vorticial liquid in three dimensions and a dipolar vorticial liquid with pointlike vorticities.

4.8.2 A scale problem

Let us assume a viscous fluid which flows with large velocity, and large variations of velocity in space and time. We expect for such variations to appear in the vicinity of an inhomogeneity, *e.g.* a solid surface, because fluids adhere to the solid surfaces. We recall that the equations of the fluid mechanics involve space and time derivatives of velocity, pressure, density, temperature, etc, *i.e.* variations of these quantities. We can get a feeling of how large a change δp in pressure could be, by using a Bernoulli-type equation $\delta p \simeq \rho v^2$, where ρ is the density and v is the velocity. For instance, a velocity $v = 100km/h$, which is a high velocity, produces a change $\delta p \simeq 10^4 dyn/cm^2$ in air ($\rho = 10^{-3} g/cm^3$), whose normal pressure is $10^6 dyn/cm^2$; therefore, $\delta p/p \simeq 10^{-2}$. Such a velocity ($\simeq 3 \times 10^3 cm/s$) is close to the mean thermal velocity $\simeq 10^4 cm/s$ (for normal air), and close to the sound velocity in normal air $c \simeq 3.5 \times 10^4 cm/s$. Therefore, as high as such a velocity is, we still expect local thermal equilibrium. The change in density is given by $\delta p = K(\delta\rho/\rho)$, where $K = -V(\partial p/\partial V)$ is the (say, isothermal) modulus of compression. For air $K \simeq 10^6 dyn/cm^2$, for water $K \simeq 10^{10} dyn/cm^2$, such that we get $\delta\rho/\rho \simeq 10^{-2}, 10^{-6}$. The change in temperature is obtained from $\delta p = \alpha T(\delta T/T)$, where $\alpha = (\partial p/\partial T)_V$ is the thermal pressure coefficient (at constant volume V). For water $\alpha \simeq 10^{22}/cm^3$, for gases it is much higher; for normal temperature $T = 300K$ we get $\delta T/T \simeq 10^{-4}$, or much lower. Consequently, we expect an almost ideal, incompressible flow, the velocity being determined solely by the external force which acts in the fluid. This was the original formulation of the fluid mechanics. Only later second-order phenomena, related to temperature variations, thermoconductivity and viscosity, have been considered. The sound, which implies compressibility, though discovered earlier, is rather a propagation of interaction; the shock waves are propagation of shocks; the relativistic fluid mechanics relates, mainly, to the propagation of whole volumes of fluid, though variations of velocity are allowed (assuming local thermodynamic equilibrium). Capillarity involves an adherence force, superfluidity is the flow of a quantum fluid.

Therefore, let us consider an incompressible (classical) fluid. This is a widely applicable assumption, except for some cases, like, for instance, the sound, but, as it is said above, the sound is rather a "flow" of interaction (compression waves), where the amplitude of the fluid motion is small. For an incompressible fluid the continuity equation

$$\frac{\partial \rho}{\partial t} + div(\rho \boldsymbol{v}) = 0 \tag{4.70}$$

becomes

$$div\boldsymbol{v} = 0 \; . \tag{4.71}$$

Similarly, for an incompressible fluid the Navier-Stokes equation

$$\rho \frac{\partial \boldsymbol{v}}{\partial t} + \rho(\boldsymbol{v} grad)\boldsymbol{v} = -gradp + \left(\tfrac{1}{3}\eta + \zeta\right) grad\, div\boldsymbol{v} + \eta \Delta \boldsymbol{v} \tag{4.72}$$

can be written as

$$\frac{\partial \boldsymbol{v}}{\partial t} + (\boldsymbol{v} grad)\boldsymbol{v} = -\tfrac{1}{\rho} gradp + \nu \Delta \boldsymbol{v} \; , \tag{4.73}$$

where η, ζ are the viscosity coefficients and $\nu = \eta/\rho$ is the kinematical viscosity. Since the variations of the density and the temperature are small, the pressure in equation (4.73) is the external pressure.

For large velocities with large variations over small distances the main contribution to this equation arises from the term quadratic in $\boldsymbol{v}$,

$$(\boldsymbol{v} grad)\boldsymbol{v} = 0 \; , \tag{4.74}$$

which is the inertial force, or

$$\boldsymbol{v} \times curl\boldsymbol{v} = grad\left(\frac{1}{2}v^2\right) \; . \tag{4.75}$$

This equation indicates in fact an equilibrium, which equalizes the Euler forces $grad(v^2/2)$ and $\boldsymbol{v} \times curl\boldsymbol{v}$. We note that equation (4.74) holds as if $\nu = 0$. Together with a high velocity with high spatial variations over small distances, this amounts to large Reynolds numbers.

The continuity equation may require that $\boldsymbol{v} = curl\boldsymbol{A}$, where $\boldsymbol{A}$ is a vector potential.

Also, it is worth noting that the time variations of the velocity are a second-order effect, such that, from equation (4.73), we may write

$$\frac{\partial \boldsymbol{v}}{\partial t} = -\frac{1}{\rho} gradp + \nu \Delta \boldsymbol{v} \ , \tag{4.76}$$

a diffusion equation with sources.

The heat-transfer equation

$$\begin{gathered} \rho T \tfrac{ds}{dt} = \kappa \Delta T + \sigma'_{ij} \partial_j v_i \ , \\ \sigma'_{ij} = \eta \left(\partial_i v_j + \partial_j v_i - \tfrac{2}{3} \delta_{ij} \partial_k v_k \right) \ , \\ \sigma'_{ij} \partial_j v_i = \tfrac{1}{2} \eta \left(\partial_i v_j + \partial_j v_i - \tfrac{2}{3} \delta_{ij} \partial_k v_k \right)^2 + \zeta \left(\partial_k v_k \right)^2 \ , \end{gathered} \tag{4.77}$$

where κ is the thermoconductivity, becomes

$$\rho T \frac{ds}{dt} = \kappa \Delta T + \frac{1}{2} \eta \left(\partial_i v_j + \partial_j v_i \right)^2 \tag{4.78}$$

for an incompressible fluid, where part of the entropy variation may be vanishing by local equilibrium; for the remaining part we get

$$\rho c_p \left(\frac{\partial T}{\partial t} + \boldsymbol{v} gradT \right) = \kappa \Delta T + \frac{1}{2} \eta \left(\partial_i v_j + \partial_j v_i \right)^2 \ , \tag{4.79}$$

where c_p is the specific heat at constant pressure, or

$$\frac{\partial T}{\partial t} + \boldsymbol{v} gradT = \chi \Delta T + \frac{\nu}{2c_p} \left(\partial_i v_j + \partial_j v_i \right)^2 \ , \tag{4.80}$$

where $\chi = \kappa / \rho c_p$ (thermometric conductivity).

The parameter ν / c_p is extremely small. For air $\eta \simeq 10^{-4} g/cm \cdot s$, for water $\eta \simeq 10^{-2} g/cm \cdot s$; for both c_p is of the order $c_p \simeq 10^{23}/g$; we get $\nu / c_p \simeq 10^{-24}$, $10^{-25} g \cdot cm^2/s$. For fairly high velocity gradients, the last term in equation (4.80) may be neglected. We can see that the production of heat due to viscosity is practically zero, as expected. The thermometric conductivity is of the order $\chi \simeq 10^{-1} cm^2/s$ for air and $\chi \simeq 10^{-3}cm^2/s$ for water. In the remaining part of equation (4.80) the main contribution may arise from the $\boldsymbol{v}$-term, for high velocities with high variations over small distances. Therefore, we have

$$\boldsymbol{v}gradT = 0 \tag{4.81}$$

and a diffusion equation

$$\frac{\partial T}{\partial t} = \chi \Delta T \tag{4.82}$$

for temperature. These equations determine the temperature as a second-order effect.

We can see that the motion of the fluids is a scale problem. For various regimes of flow some terms in the equations of the fluid mechanics may prevail over others. In our case, as discussed above, where the velocity is large, with large variations over small distances, the equations of the fluid mechanics are split in

$$\begin{gathered} div\boldsymbol{v} = 0 \ , \ \ \boldsymbol{v}grad\boldsymbol{v} = 0 \ , \\ \frac{\partial \boldsymbol{v}}{\partial t} = -\frac{1}{\rho}gradp + \nu\Delta\boldsymbol{v} \ , \\ \boldsymbol{v}gradT = 0 \ , \ \frac{\partial T}{\partial t} = \chi\Delta T \ . \end{gathered} \tag{4.83}$$

In the opposite case, where the velocity is small, with small variations, we may neglect the non-linear terms in equations (4.73) and (4.80), which become

$$\begin{gathered} \frac{\partial \boldsymbol{v}}{\partial t} = -\frac{1}{\rho}gradp + \nu\Delta\boldsymbol{v} \ , \\ \frac{\partial T}{\partial t} = \chi\Delta T \ . \end{gathered} \tag{4.84}$$

If the quantities involved in these equations have larger variations, preserving the small variations of the velocity, then we may keep the full continuity equation and the variations of the internal pressure; also, we may keep the second coefficient of viscosity. This is the realm of the usual equations of fluid mechanics. In the potential flow, or in the Couette flow, we may retain the non-linear term in the Navier-Stokes equation, which, however, in this context is associated with second-order corrections. If the involved quantities have large variations, other phenomena than the fluid flow may appear; for instance, a large temperature gradient may lead to a convective instability. Solutions of Navier-Stokes equations for larger variations of the quantities are, very likely, unstable. We consider below large velocities with large variations over small distances.

4.9 Vorticity

The Euler equation for an ideal fluid can be written as

$$\frac{d\boldsymbol{v}}{dt} = -grad w \ , \tag{4.85}$$

where w is the enthalpy ($dw = \frac{1}{\rho}dp$); noteworthy, we use the total time derivative in this equation. By taking the *curl*, we get

$$curl\frac{d\boldsymbol{v}}{dt} = 0 \ . \tag{4.86}$$

On the other hand,

$$\frac{d\boldsymbol{v}}{dt} = \frac{\partial \boldsymbol{v}}{\partial t} + (\boldsymbol{v} grad)\boldsymbol{v} = \frac{\partial \boldsymbol{v}}{\partial t} + \left(\frac{\partial \boldsymbol{v}}{\partial t}\right)_f \ , \tag{4.87}$$

where the suffix f indicates that the derivative is taken along the flow. Equation (4.86) becomes

$$\frac{\partial}{\partial t}curl\boldsymbol{v} + \left(\frac{\partial}{\partial t}curl\boldsymbol{v}\right)_f = 0 \tag{4.88}$$

and, since the two variations of the $curl\boldsymbol{v}$ are independent, we get

$$curl\boldsymbol{v} = 0 \ , \tag{4.89}$$

i.e. the vorticity $curl\boldsymbol{v}$ is conserved along the flow. Therefore, we cannot create, or destroy, vorticity $curl\boldsymbol{v}$ in the flow of an ideal fluid. Equation (4.89) is valid in the absence of special external force, which do not derive from a gradient. This is Helmholtz's circulation law.

For a real, viscid fluid the Navier-Stokes equation is

$$\rho\frac{d\boldsymbol{v}}{dt} = -gradp + \eta\Delta\boldsymbol{v} + \left(\frac{1}{3}\eta + \zeta\right) grad\, div\boldsymbol{v} \ , \tag{4.90}$$

where ρ is the density, p is the pressure and η, ζ are the viscosity coefficients. By taking the *curl*, we get

$$curl\,(\rho\tfrac{d\boldsymbol{v}}{dt}) = grad\rho \times \tfrac{d\boldsymbol{v}}{dt} + \rho curl\tfrac{d\boldsymbol{v}}{dt} = \\ = \eta\Delta curl\boldsymbol{v}\ ; \tag{4.91}$$

we can see that the viscosity η can generate vorticity. In general, the velocity $\boldsymbol{v}$ rotates about the vorticity $curl\boldsymbol{v}$. This is a vortex.

Equation (4.91) is in conflict with the continuity equation

$$\frac{d\rho}{dt} + \rho div\boldsymbol{v} = 0\ . \tag{4.92}$$

Indeed, if $curl\boldsymbol{v} \neq 0$, the velocity should be derived from a *curl* (not from a *grad*!), *i.e.* we should have $\boldsymbol{v} = curl\boldsymbol{A}$, where $\boldsymbol{A}$ is a vector potential. Consequently, the fluid should be incompressible ($div\boldsymbol{v} = div\,curl\boldsymbol{A} = 0$). The density should be constant both in time and space (along the flow). This indicates that in a compressible fluid we cannot have vortices. The density may have longitudinal variations, along the velocity of the flow, as in sound waves, where the term with $grad\rho$ disappears from equation (4.91), and $curl\boldsymbol{v} = 0$. Usually, the variations of the density are small (and the amplitude of the sound waves are small), such that they may be neglected for the present purpose.

Therefore, we may limit to an incompressible fluid ($div\boldsymbol{v} = 0$), for which equation (4.91) becomes

$$curl\frac{d\boldsymbol{v}}{dt} = \nu\Delta curl\boldsymbol{v}\ , \tag{4.93}$$

or

$$\frac{\partial}{\partial t}curl\boldsymbol{v} - curl\,(\boldsymbol{v} \times curl\boldsymbol{v}) = \nu\Delta curl\boldsymbol{v}\ , \tag{4.94}$$

where $\nu = \eta/\rho$ is the kinematical viscosity and we have used the identity $(\boldsymbol{v}grad)\boldsymbol{v} = -\boldsymbol{v} \times curl\boldsymbol{v} + grad(v^2/2)$. This is the equation of vorticity; it can also be written as

$$\frac{\partial}{\partial t}curl\boldsymbol{v} + curl\,[(\boldsymbol{v}grad)\boldsymbol{v}] = \nu\Delta curl\boldsymbol{v}\ . \tag{4.95}$$

In principle, this equation gives the velocity. The Navier-Stokes equation gives the pressure. By taking the *curl* in equation (4.95), we get

$$\begin{aligned} grad\,div\left[(\boldsymbol{v}grad)\boldsymbol{v} + \tfrac{1}{\rho}gradp\right] &= 0 \ , \\ grad\,div\left(\tfrac{\partial \boldsymbol{v}}{\partial t} - \nu\Delta\boldsymbol{v}\right) &= 0 \ . \end{aligned} \tag{4.96}$$

Actually, these equations can be simplified. If we write the Navier-Stokes equation as

$$\frac{\partial}{\partial t}curl\boldsymbol{A} + (\boldsymbol{v}grad)\boldsymbol{v} = -\frac{1}{\rho}gradp + \nu\Delta curl\boldsymbol{A} \ , \tag{4.97}$$

we get

$$div\left[(\boldsymbol{v}grad)\boldsymbol{v} + \frac{1}{\rho}gradp\right] = 0 \tag{4.98}$$

and, of course,

$$div\left(\frac{\partial \boldsymbol{v}}{\partial t} - \nu\Delta\boldsymbol{v}\right) = 0 \ , \tag{4.99}$$

which is an identity ($div\boldsymbol{v} = 0$, $\boldsymbol{v} = curl\boldsymbol{A}$). In some cases a particular solution of these equations is provided by

$$\begin{aligned} (\boldsymbol{v}grad)\boldsymbol{v} &= -grad(p/\rho) \ , \\ \tfrac{\partial \boldsymbol{v}}{\partial t} &= \nu\Delta\boldsymbol{v} \ . \end{aligned} \tag{4.100}$$

We can see that the Navier-Stokes equation is split into (the derivatives of) a diffusion equation and an equilibrium equation, between the pressure force ($-grad(p/\rho)$) and Euler's force ($(\boldsymbol{v}grad)\boldsymbol{v}$). Actually, the diffusion equation is a diffusion equation for the vorticity, because the above equations are valid for a non-vanishing vorticity. Also, it is easy to see that these equations are a generalization of the equations for the Couette flow.

However, we have also the heat-transfer equation. For an incompressible fluid it reads

$$\rho c_p\frac{dT}{dt} = \kappa\Delta T + \frac{1}{2}\eta\left(\partial_i v_j + \partial_j v_i\right)^2 \ , \tag{4.101}$$

or

$$\frac{dT}{dt} = \chi \Delta T + \frac{1}{2} \frac{\nu}{c_p} \left(\partial_i v_j + \partial_j v_i \right)^2 \quad , \tag{4.102}$$

where c_p is the specific heat per unit mass at constant pressure, κ is the thermoconductivity, and $\chi = \kappa / \rho c_p$ is the thermometric conductivity. For an incompressible fluid this equation can be transformed into an equation for the derivatives of the pressure,

$$\frac{dp}{dt} = \chi \Delta p + \frac{1}{2} \frac{\alpha \nu}{c_p} \left(\partial_i v_j + \partial_j v_i \right)^2 \quad , \tag{4.103}$$

where $\alpha = (\partial p / \partial T)_v$ is the thermal pressure coefficient. In general, this equation is not compatible with the Navier-Stokes equation (or equation (4.98)). Therefore, rigorously speaking, we cannot have vorticity in an incompressible fluid either. Usually, the coefficient ν / c_p is very small (of the order $10^{-24} - 10^{-25} g \cdot cm^2/s$), such that, for small gradients of velocity, we have a low rate of entropy production (though, a factor of the order $10^{23} K/erg$ should be included). Under these conditions, we may assume that the flow is isentropic and the heat-transfer equation may be neglected.

Therefore, within these approximations (incompressibility and constant entropy), we are left with equation (4.95) and the Navier-Stokes equation (or equation (4.98)) for a vorticial flow. But, in general, an external pressure which satisfies the Navier-Stokes equation is very special. We may conclude that, if they exist, the vortices are, in fact, unstable. They develop an Euler's force, which is difficult to be equilibrated by an external force.

We can create a vortex by (special) external means, due to viscosity; for instance, a uniform angular momentum $L = vr$ (per unit mass), for $\boldsymbol{v}$ perpendicular to $\boldsymbol{r}$, gives a singular distribution of velocities $v \sim 1/r$; this is a filamentary vortex. If we have a uniform angular velocity ω, the velocity is $v = \omega r$ and the angular momentum is $L = \omega r^2$, as for a uniform rotation. Other types of vortices can be generated, either singular or non-singular, by various external means, all by the presence of the viscosity. The creation of such a vorticity amounts to generate a vortex from a delocalized vorticity, *i.e.* from many smaller vortices. On the other hand, such a vortex can diffuse, according to equation (4.93), *i.e.* it can be transformed in smaller vortices, which also means generating vorticity, besides destroying vorticity at the same time. In general, the vortices are unstable.

We show below that the viscosity, apart from generating vorticity, may also lead to instabilities.

For small variations of the velocity we may neglect the inertial term in the vorticity equation (4.95), which becomes

$$\frac{\partial}{\partial t} curl\boldsymbol{v} = \nu\Delta curl\boldsymbol{v} \ . \tag{4.104}$$

By making use of $\boldsymbol{v} = curl\boldsymbol{A}$, we get

$$(\Delta - grad\, div)\left(\frac{\partial \boldsymbol{A}}{\partial t} - \nu\Delta\boldsymbol{A}\right) = 0 \ . \tag{4.105}$$

A solution of this equation is provided by

$$\frac{\partial \boldsymbol{A}}{\partial t} - \nu\Delta\boldsymbol{A} = 0 \ , \tag{4.106}$$

which leads to

$$\boldsymbol{A} = \boldsymbol{A}_0 e^{-\lambda\nu t} e^{\pm i\sqrt{\lambda}r}/r \ , \tag{4.107}$$

where $\boldsymbol{A_0}$ and λ are two constants. The velocity acquires the form

$$\boldsymbol{v} = -\boldsymbol{A}_0 \times grad\left(e^{-\lambda\nu t} e^{\pm i\sqrt{\lambda}r}/r\right) \ , \tag{4.108}$$

and the pressure is uniform within this approximation. We note that, although the spatial dependence of the solution does not depend on viscosity, it is generated by the viscosity term $\nu\Delta\boldsymbol{v}$.

4.9.1 Instabilities

With usual notations the equations of the fluid mechanics for an incompressible viscous fluid are

$$\rho\frac{\partial v}{\partial t} + \rho(\boldsymbol{v}grad)\boldsymbol{v} = -gradp + \eta\Delta\boldsymbol{v} \ ,$$

$$div\boldsymbol{v} = 0 \ , \tag{4.109}$$

$$\rho c_p\left(\frac{\partial T}{\partial t} + \boldsymbol{v}gradT\right) = \kappa\Delta T + \frac{1}{2}\eta\left(\partial_i v_j + \partial_j v_i\right)^2 \ .$$

The inertial term $\rho(\boldsymbol{v}grad)\boldsymbol{v}$ is a transport of velocity (momentum). The terms on the right in the first equation above (the Navier-Stokes equation) generate a force and a viscosity force. The η-term in the third equation is a source of heat, which determines a temperature T. This temperature is transported with the fluid velocity ($\boldsymbol{v}gradT$) and with the molecular velocity through the thermoconductivity κ (collisions, interaction). In addition, the change in temperature causes a change in the pressure p, such that the first and the third equations above are coupled.

On multiplying the Navier-Stokes equation by $\boldsymbol{v}$ we arrive straightforwardly to

$$\begin{gathered}\frac{\partial}{\partial t}\left(\tfrac{1}{2}\rho v^2\right)+div\left[\boldsymbol{v}\left(\tfrac{1}{2}\rho v^2+p\right)-\tfrac{1}{2}\eta grad(v^2)\right]+\\+\eta\left(\partial_j v_i\right)^2=0\ .\end{gathered}\tag{4.110}$$

The term $\eta grad(v^2)$ represents an energy flux associated with collisions (viscosity); the term $\eta\left(\partial_j v_i\right)^2$ represents the heat produced by viscosity. We integrate this equation over a volume V enclosed by a surface f,

$$\begin{gathered}\frac{\partial}{\partial t}\int dV\left(\tfrac{1}{2}\rho v^2\right)+\oint df\left[v_n\left(\tfrac{1}{2}\rho v^2+p\right)-\tfrac{1}{2}\eta\partial_n(v^2)\right]+\\+\eta\int dV\left(\partial_j v_i\right)^2=0\ ,\end{gathered}\tag{4.111}$$

where v_n is the velocity component normal to the surface and ∂_n is the derivative along the normal to the surface.

We compare the orders of magnitude of the surface terms and the η-volume term, and get ratios of the form $\frac{Sl}{V}R$, $\frac{Sl}{V}$, $\frac{Sl}{V}(p/\rho v^2)R$, where S is the area of the surface, V is the volume, l is the distance over which the velocity varies, v is the velocity and $R=vl/\nu$ is the Reynolds number, $\nu=\eta/\rho$ being the kinematic viscosity. Since $Sl/V\ll 1$ we can neglect the surface contributions in comparison with the heat term. By writing

$$\boldsymbol{v}=f(t)\boldsymbol{u}(\boldsymbol{r})\ ,\tag{4.112}$$

the above equation reads

$$\frac{\partial}{\partial t}f^2\cdot\int dV\left(\frac{1}{2}\rho u^2\right)+\eta f^2\int dV\left(\partial_j u_i\right)^2=0\ .\tag{4.113}$$

We can see that the time dependence of the velocity is a damped exponential. The flow is stable, as a consequence of the dissipated heat. The η-term in equation (4.113) gives, in fact, the increase of entropy.

Let us assume that the integration domains are sufficiently small, such that Sl/V is of the order of unity; the velocity varies over a distance l inside the domains, but we assume that it suffers a large discontinuity across the surface, over a small distance $\delta \ll l$. Then, it is easy to see that the dominant term in equation (4.111) is the collision term, such that equation (4.111) becomes

$$\frac{\partial}{\partial t}f^2 \cdot \int dV \left(\frac{1}{2}\rho u^2\right) - \frac{1}{2}\eta f^2 \oint df \partial_n(v^2) = 0 \ . \tag{4.114}$$

We can see that for a positive normal derivative the flow is unstable. The viscosity is insufficient to disipate the energy as heat, and the energy is transferred by molecular collisions (viscosity) through surfaces of discontinuities. The process occurs in small domains, with large discontinuities of velocity across their surface, and the instabilities imply returning, *i.e.* swirling, and fluctuating, velocities. This is the turbulence phenomenon. We note that the instabilities are governed by viscosity, as expected, which, also, gives turbulence. Turbulence cannot appear in an ideal fluid, where swirls ($curl\boldsymbol{v}$) are absent. Moreover, we note that the inertial term does not appear in instabilities, but it plays an important role in turbulence.

We note that the above arguments can be extended to compressible fluids. The main role in these arguments is played by terms with derivatives normal to the surface. Consequently, we can have instabilities brought by large variations of both the velocity and the temperature across the surface. However, the fluctuating part of the density is small, such that it may be neglected. On the other hand, since these instabilities are governed by viscosity, and the viscosity produces vorticity, the fluid should be incompressible, as shown in the preceding section.

4.9.2 Vortex

We consider an incompressible fluid flow with velocity $\boldsymbol{v}$, governed by the Navier-Stokes equation

$$\frac{\partial \boldsymbol{v}}{\partial t} + (\boldsymbol{v} grad)\boldsymbol{v} = -\frac{1}{\rho} gradp + \boldsymbol{F} + \nu \Delta \boldsymbol{v} \ , \tag{4.115}$$

where ρ is the density of the fluid, p denotes the (internal) pressure, $\boldsymbol{F}$ is an external force (per unit mass) and ν is the kinematical viscosity. We add the heat-transfer equation of

$$\frac{\partial T}{\partial t} + \boldsymbol{v} gradT = \chi \Delta T \ , \tag{4.116}$$

where T is the temperature and χ is the thermometric conductivity ($\chi = \kappa/\rho c_p$, where κ is the thermoconductivity and c_p is the specific heat per unit mass at constant pressure); the viscosity term in this equation is very small, such that it can be left aside.[47]

Let us consider a velocity field given by

$$\boldsymbol{v} = \boldsymbol{\omega} \times gradf(r) \ , \tag{4.117}$$

where $\boldsymbol{\omega}$ is a vector which may depend only on the time and the function $f(r)$ is smooth everywhere, except, possibly, at the origin $\boldsymbol{r} \neq 0$, and vanishing rapidly at infinity. Obviously, the velocity $\boldsymbol{v}$ rotates about $\boldsymbol{\omega}$; such a velocity field is a vortex.

We can ckeck that $div\boldsymbol{v} = 0$ and

$$\boldsymbol{v} = -curl\left[\boldsymbol{\omega} f(r)\right] \ , \tag{4.118}$$

such that we can define a vector potential $\boldsymbol{A} = -\boldsymbol{\omega} f(r)$ ($\boldsymbol{v} = curl\boldsymbol{A}$). We note that $div\boldsymbol{A} \neq 0$ and the vorticity

$$curl\boldsymbol{v} = \Delta\left[\boldsymbol{\omega} f(r)\right] - grad\, div\left[\boldsymbol{\omega} f(r)\right] \neq -\Delta \boldsymbol{A} \ . \tag{4.119}$$

If the velocity given by equation (4.118) satisfies the second equation (4.100), we have $\boldsymbol{\omega} \sim e^{-\nu\lambda t}$ and $\Delta f + \lambda f = 0$, *i.e.* $f \sim e^{\pm i\sqrt{\lambda} r}/r$, where

[47] L. Landau and E. Lifshitz, *Course of Theoretical Physics*, vol. 6 (*Fluid Mechanics*), Elsevier, Oxford (1989).

λ is, in general, complex. In two dimensions f is a Bessel function. The Navier-Stokes equation gives the pressure.

The most common example of a velocity field given by equation (4.117) is the coulombian filamentary vortex, with $\boldsymbol{\omega} = const$, $f(r) = -\ln r$ and the two-dimensional position vector $\boldsymbol{r}$ perpendicular to $\boldsymbol{\omega}$. In this case $\boldsymbol{v} = \boldsymbol{r} \times \boldsymbol{\omega}/r^2$, $\boldsymbol{A} = \boldsymbol{\omega} \ln r$ and $curl\boldsymbol{v} = -2\pi\boldsymbol{\omega}\delta(\boldsymbol{r})$. The velocity is singular at $\boldsymbol{r} = 0$. For the coulombian filamentary vortex the Navier-Stokes equation (4.115) can be satisfied for $p = -\rho\omega^2/2r^2$. We need an external potential $\varphi = p/\rho = -\omega^2/2r^2$, which may be provided by the gravitational field for a fluid with a free surface.

The terms in the Navier-Stokes equation for the vortex are

$$\begin{aligned} \frac{\partial \boldsymbol{v}}{\partial t} &= (\dot{\boldsymbol{\omega}} \times \boldsymbol{r})\frac{f'}{r} \ , \\ (\boldsymbol{v}grad)\boldsymbol{v} &= \left[\boldsymbol{\omega}(\boldsymbol{\omega}\boldsymbol{r}) - \omega^2\boldsymbol{r}\right]\frac{f'^2}{r^2} \ , \\ \Delta\boldsymbol{v} &= (\boldsymbol{\omega} \times \boldsymbol{r})\frac{(f'' + 2f'/r)'}{r} \ , \end{aligned} \tag{4.120}$$

where the primes denote the derivatives of the function f; the second equation (4.120) is derived by using the identity

$$(\boldsymbol{v}grad)\boldsymbol{v} = -\boldsymbol{v} \times curl\boldsymbol{v} + grad(v^2/2) \tag{4.121}$$

for the inertial force $(\boldsymbol{v}grad)\boldsymbol{v}$. For a filamentary vortex, the above expressions become

$$\begin{aligned} (\boldsymbol{v}grad)\boldsymbol{v} &= -\omega^2\boldsymbol{r}\frac{f'^2}{r^2} \ , \\ \Delta\boldsymbol{v} &= (\boldsymbol{\omega} \times \boldsymbol{r})\frac{(f'' + f'/r)'}{r} \ . \end{aligned} \tag{4.122}$$

In general, the vortex given by equation (4.117) does not satisfy the Navier-Stokes equation; it develops internal forces which cannot be fully compensated; the vortex is unstable. In particular, according to equation (4.121), there exists a force $-\boldsymbol{v} \times curl\boldsymbol{v}$, an energy $v^2/2$ and another force $grad(v^2/2)$ (per unit mass); we call these forces Euler's forces.

4.9.3 Vorticial liquid

A set of vectors $\boldsymbol{\omega}_i$ placed at $\boldsymbol{r}_i$ form a vorticial liquid. The velocity field is

$$\boldsymbol{v} = \sum_i \boldsymbol{\omega}_i \times gradf_i(R_i) \ , \tag{4.123}$$

where $\boldsymbol{R}_i = \boldsymbol{r} - \boldsymbol{r}_i$. This velocity field is a superpositon of independent vortices, which do not satisfy the Navier-Stokes equation, according to the above discussion. Each i-th vortex develops an internal force, while the inertial term generates an interaction which brings an additional force; this additional force depends on all the other j-th vortices, $j \neq i$. Therefore, the Navier-Stokes equation is not satisfied by the velocity field given by equation (4.123), as expected.

We assume randomly fluctuating vectors $\boldsymbol{\omega}_i$, as a distinctive feature of a fully developed turbulence. It is likely that the fluid develops fluctuations as a reaction to its uncompensated internal forces. Specifically, we assume $\overline{\boldsymbol{\omega}}_i = 0$ and $\overline{\omega_i^\alpha \omega_j^\beta} = \frac{1}{3}\overline{\omega_i^2}\delta_{ij}\delta^{\alpha\beta}$, where α, $\beta = 1,2,3$ are the cartesian labels of the components of the vectors $\boldsymbol{\omega}_i$, and the overbar indicates the average over time. Then, the average velocity is zero ($\overline{\boldsymbol{v}} = 0$) and we are left with the inertial term

$$\overline{(\boldsymbol{v}grad)\boldsymbol{v}} = \overline{v_j \partial_j v_i} \ . \tag{4.124}$$

The calculation of this term is straightforward; we get

$$\begin{aligned}\overline{(\boldsymbol{v}grad)\boldsymbol{v}} = \tfrac{1}{3}\textstyle\sum_i \omega_i^2 [\tfrac{1}{2}grad\,(gradf_i(R_i))^2 - \\ -gradf_i(R_i)\cdot\Delta f_i(R_i)] \ .\end{aligned} \tag{4.125}$$

The averaged Euler forces are

$$\begin{aligned}-\overline{\boldsymbol{v} \times curl\boldsymbol{v}} = -\tfrac{1}{3}\textstyle\sum_i \omega_i^2 [\tfrac{1}{2}grad\,(gradf_i(R_i))^2 + \\ +gradf_i(R_i)\cdot\Delta f_i(R_i)] \ , \\ \overline{grad(v^2/2)} = \tfrac{1}{3}\textstyle\sum_i \omega_i^2 grad\,(gradf_i(R_i))^2 \ .\end{aligned} \tag{4.126}$$

By using the spherical symmetry of the function $f_i(R_i)$, these expressions can be cast in the form

$$\overline{(\boldsymbol{v}grad)\boldsymbol{v}} = -\tfrac{2}{3}\sum_i \omega_i^2 f_i'^2 \tfrac{\boldsymbol{R}_i}{R_i^2} \ ,$$

$$-\overline{\boldsymbol{v} \times curl\boldsymbol{v}} = -\tfrac{2}{3}\sum_i \omega_i^2 f_i' \left(f_i'' + \tfrac{1}{R_i} f_i'\right) \tfrac{\boldsymbol{R}_i}{R_i} \ , \tag{4.127}$$

$$\overline{grad(v^2/2)} = \tfrac{2}{3}\sum_i \omega_i^2 f_i' f_i'' \tfrac{\boldsymbol{R}_i}{R_i} \ .$$

We can see that even on average the Navier-Stokes equation is not satisfied, because the first equation (4.127) is not zero, in general; even on average the vorticial liquid develops internal forces which are not equilibrated; it is unstable.

Now, let us assume that the vorticial liquid is sufficiently dense, *i.e.* if we can define a density ρ_v of points $\boldsymbol{r}_i$; further, we assume that the liquid is homogeneous and isotropic, *i.e.* this density is constant and the ω_i and the functions f_i are uniform ($\omega_i = \omega$, $f_i = f$). We assume that this is another distinctive feature of a fully developed turbulence. Then, we may transform the summation over i in equations (4.127) into an integral. By choosing the origin at $\boldsymbol{r} = \boldsymbol{r}_i$ for a fixed $\boldsymbol{r}_i$, and using the notation $\boldsymbol{R} = \boldsymbol{r} - \boldsymbol{r}_i$, we get

$$\overline{(\boldsymbol{v}grad)\boldsymbol{v}} = -\frac{2}{3}\rho_v\omega^2 \int_0^\infty dR \cdot Rf'^2(R) \int do(\boldsymbol{R}/R) \ ; \tag{4.128}$$

the result of integration in this equation is zero, due to the integration over the solid angle o, providing the radial integration is finite. In this case the Navier-Stokes equation is satisfied trivially.

Let us assume that the integral over R is singular at $R = 0$, as another distinctive feature of a fully developed turbulence (the function f is assumed to decrease sufficiently rapid at infinity to have a finite integral in this limit). The integration outside a small region around the i-th point is zero, while the integration over such a small region is indefinite. This singularity implies

$$f(R) \sim (a/R)^n \ , \ n > 0 \tag{4.129}$$

for $R \ll a$, where a is a small characteristic distance. Therefore, we are left with a discrete set of points $\boldsymbol{r}_i$, where the function $f(R_i)$ and the velocity ($v \sim 1/R_i^{n+1}$) are singular. We have now a small region of dimension a, around each point $\boldsymbol{r}_i$, which includes a mass of fluid, say, M, where the inertial term given by equation (4.128) is not defined. Since the positions $\boldsymbol{r}_i$ may change in time, we are left with a classical gas of particle-like vortices (or a solution of vortices in the background fluid). This is an example of emergent physics.[48] Of course, we may have also a mixture of vorticial gases, each characterized by a dimension a and a mass M.

The equation of energy conservation

$$\frac{\partial}{\partial t}\left(\frac{1}{2}v^2\right) + \boldsymbol{v}\cdot(\boldsymbol{v}grad)\boldsymbol{v} = \nu\boldsymbol{v}\Delta\boldsymbol{v} \ , \tag{4.130}$$

averaged over the fluctuating vectors $\boldsymbol{\omega}_i$, is reduced to the viscosity term

$$\nu\overline{\boldsymbol{v}\Delta\boldsymbol{v}} = \frac{2\nu}{3}\sum_i \omega_i^2 f_i^{'}\left(f_i^{''} + 2f_i^{'}/R_i\right)^{'} \ . \tag{4.131}$$

This term should be computed outside the regions with dimension a, where the motion is defined. The non-vanishing value of this term indicates an energy loss, which should be compensated from the outside. The vorticial gas is in quasi equilibrium.

The above considerations are valid for spherical-symmetric functions $f_i(R_i)$; if the vortices have a lower (internal) symmetry, the unit vector $\boldsymbol{R}/R$ in equation (4.128) is replaced by functions which do not have a spherical symmetry, and the inertial term is not vanishing, in general. The vortices are unstable, and, likely, they tend to acquire a spherical symmetry, which ensures a (quasi)-equilibrium.

4.9.4 Vorticial gas

Let us assume a homogeneous, isotropic, fluctuating distribution of (singular) vortices localized at $\boldsymbol{r}_i$ with mass M, as described above. Their density is

$$\rho = M\sum_i \delta(\boldsymbol{r} - \boldsymbol{r}_i) \tag{4.132}$$

[48] P. W. Anderson, "More is different", Science **177** 393 (1972). Sound is another example.

and

$$\frac{\partial \rho}{\partial t} = -M \sum_i \boldsymbol{u}_i grad\delta(\boldsymbol{r} - \boldsymbol{r}_i) \ , \tag{4.133}$$

where $\boldsymbol{u}_i = d\boldsymbol{r}_i/dt$ is their velocity. The velocity field of the vortices is

$$\boldsymbol{u} = v \sum_i \boldsymbol{u}_i \delta(\boldsymbol{r} - \boldsymbol{r}_i) \ , \tag{4.134}$$

where v is the small volume over which the δ-function is localized, such that $v = a^3$ and $M = \rho v$. Let us compute

$$\begin{aligned} div(\rho\boldsymbol{u}) &= vMdiv \textstyle\sum_{ij} \delta(\boldsymbol{r} - \boldsymbol{r}_i)\boldsymbol{u}_j \delta(\boldsymbol{r} - \boldsymbol{r}_j) = \\ &= Mdiv \textstyle\sum_i \boldsymbol{u}_i \delta(\boldsymbol{r} - \boldsymbol{r}_i) = \\ &= M \textstyle\sum_i \boldsymbol{u}_i grad\delta(\boldsymbol{r} - \boldsymbol{r}_i) = -\frac{\partial \rho}{\partial t} \ ; \end{aligned} \tag{4.135}$$

we can see that the continuity equation is satisfied. We note that this is the mass conservation of the ensemble.

Now, let us focus on the Navier-Stokes equation

$$\rho \frac{\partial \boldsymbol{u}}{\partial t} + \rho(\boldsymbol{u} grad)\boldsymbol{u} = -gradp + \eta \Delta \boldsymbol{u} \ , \tag{4.136}$$

where p is the pressure and η is the viscosity coefficient, and let us compute each term in this equation for our vorticial fluid. We have

$$\frac{\partial \boldsymbol{u}}{\partial t} = v \textstyle\sum_i \dot{\boldsymbol{u}}_i \delta(\boldsymbol{r} - \boldsymbol{r}_i) - v \textstyle\sum_i \boldsymbol{u}_i (\boldsymbol{u}_i grad)\delta(\boldsymbol{r} - \boldsymbol{r}_i) \ . \tag{4.137}$$

The inertial term is

$$\begin{aligned} (\boldsymbol{u} grad)\boldsymbol{u} &= v^2 \textstyle\sum_{ij} \boldsymbol{u}_j \delta(\boldsymbol{r} - \boldsymbol{r}_i)(\boldsymbol{u}_i grad)\delta(\boldsymbol{r} - \boldsymbol{r}_j) = \\ &= v \textstyle\sum_i \boldsymbol{u}_i (\boldsymbol{u}_i grad)\delta(\boldsymbol{r} - \boldsymbol{r}_i) \ . \end{aligned} \tag{4.138}$$

By comparing equations (4.137) and (4.138), we can see that the inertial (Euler's) term disappears from equation. This is expected since the δ-function equals the variable $\boldsymbol{r}$ to the function $\boldsymbol{r}_i(t)$, which amounts to Lagrange's approach. The term on the left in equation (4.136) becomes

$$\rho \frac{\partial \boldsymbol{u}}{\partial t} + \rho(\boldsymbol{u} grad)\boldsymbol{u} = M \textstyle\sum_i \dot{\boldsymbol{u}}_i \delta(\boldsymbol{r} - \boldsymbol{r}_i) \ , \tag{4.139}$$

and equation (4.136) reads now

$$M \sum_i \dot{\boldsymbol{u}}_i \delta(\boldsymbol{r}-\boldsymbol{r}_i) = -v \sum_i p_i grad \delta(\boldsymbol{r}-\boldsymbol{r}_i) + \\ +\eta v \sum_i \boldsymbol{u}_i \Delta\delta(\boldsymbol{r}-\boldsymbol{r}_i) \ , \tag{4.140}$$

where p_i is the pressure at the position $\boldsymbol{r}_i$. The pressure term in equation (4.140) is a force per unit volume acting upon the vortex placed at $\boldsymbol{r}_i$; it may arise from the pressure exerted by the background fluid particles. This term may be written as

$$\sum_i \boldsymbol{f}_i \delta(\boldsymbol{r}-\boldsymbol{r}_i) \ , \tag{4.141}$$

where $\boldsymbol{f}_i$ is the force acting at $\boldsymbol{r}_i$. The factor $\Delta\delta(\boldsymbol{r}-\boldsymbol{r}_i)$ is of the order $-\frac{1}{a^2}\delta(\boldsymbol{r}-\boldsymbol{r}_i)$, where a is of the order of the dimension of the vortex ($v=a^3$); consequently, we may replace the viscosity terms in equation (4.140) by

$$-\frac{\eta v}{a^2} \sum_i \boldsymbol{u}_i \delta(\boldsymbol{r}-\boldsymbol{r}_i) \ . \tag{4.142}$$

A similar estimation holds also for the ζ-term. The equation of motion (4.140) describes a set of independent particles with mass M, subjected to an external force $\boldsymbol{f}_i$ and a friction force; the equation of motion of each such particle can be written as

$$M\dot{\boldsymbol{u}}_i = \boldsymbol{f}_i - \eta a \boldsymbol{u}_i \ , \tag{4.143}$$

which is Newton's law of motion. The damping coefficient caused by the friction force is very small, such that we can consider the ensemble of vortices as a (quasi) ideal classical gas of independent, identical, pointlike particles. Therefore, a fully developed turbulence may be viewed as the (quasi) thermodynamic equilibrium of such a gas (or a solution of vortices in the background fluid). This is an example of emergent physics.[49] We can define a temperature of turbulence, which is approximately the mean kinetic energy of the translational motion of a vortex. Also, we can estimate a chemical potential from evaluating $\overline{v^2}/2$. We note that the density ρ_v of the vorticial gas and the dimension a of the vortex remain undetermined; these parameters can be estimated from experiment.

[49] P. W. Anderson, *loc. cit.*

4.10 Particular vorticial liquids

We present below three particular types of vorticial liquids, with strong singularities. Two are particular cases of the vorticial liquid described above, with $f_i(R_i) = -\ln R_i$ in two-dimensions and $f_i(R_i) = -1/R_i$ in three dimensions; we call them coulombian vorticial liquids. The other is a dipolar vorticial liquid.

4.10.1 Filamentary liquid

Let us consider a set of rectilinear, parallel filaments, directed along the z-axis, placed at positions $\boldsymbol{r}_i$ in the (x, y)-plane, with vorticities $\boldsymbol{\omega}_i$. This is a two-dimensional vorticial liquid of "cyclons". The velocity field is given by

$$\boldsymbol{v} = -\sum_i \boldsymbol{\omega}_i \times grad \ln R_i = \sum_i \frac{\boldsymbol{R}_i \times \boldsymbol{\omega}_i}{R_i^2} \quad , \tag{4.144}$$

where $\boldsymbol{R}_i = \boldsymbol{r} - \boldsymbol{r}_i$.

Equation (4.144) shows that the velocity is derived from a vector potential $\boldsymbol{A}$, through $\boldsymbol{v} = curl\boldsymbol{A}$. By taking the $curl$ in this equation, we get

$$\Delta \boldsymbol{A} = -2\boldsymbol{\omega} \quad , \tag{4.145}$$

providing $div\boldsymbol{A} = 0$ (and $div\boldsymbol{\omega} = 0$), where $curl\boldsymbol{v} = 2\boldsymbol{\omega}$; $\boldsymbol{\omega}$ is called vorticity. The vorticity distribution corresponding to equation (4.144)

$$\boldsymbol{\omega}(\boldsymbol{r}) = \frac{1}{2} curl\boldsymbol{v} = -\pi \sum_i \boldsymbol{\omega}_i \delta(\boldsymbol{R}_i) \quad , \tag{4.146}$$

gives the vector potential

$$\boldsymbol{A} = \sum_i \boldsymbol{\omega}_i \ln R_i \ . \tag{4.147}$$

According to equation (4.121), the inertial term has the components

$$\boldsymbol{f} = -\boldsymbol{v} \times curl\boldsymbol{v} = -2\pi \sum_{i \neq j} \omega_i \omega_j grad_i \ln R_{ij} \cdot \delta(\boldsymbol{R}_i) \quad , \tag{4.148}$$

where $\boldsymbol{R}_{ij} = \boldsymbol{r}_i - \boldsymbol{r}_j$, and $grade$, where

$$e = \frac{1}{2}v^2 = \sum_{i \neq j} \omega_i \omega_j \frac{\boldsymbol{R}_i \boldsymbol{R}_j}{2R_i^2 R_j^2} \tag{4.149}$$

is a density of kinetic energy (per unit mass). Apart from the force $\boldsymbol{f}$, which acts at the positions of the vortices, there exist internal forces given by $grade$, which make the liquid unstable. We may call these forces Euler's forces. We can see that the filamentary vorticial liquid is unstable. The motion and the statistics of parallel, rectilinear filaments have been extensively investigated.[50] The instability is associated with a negative temperature in an attempt of a statistical theory.[51]

The total force given by equation (4.148)

$$\boldsymbol{F} = \int d\boldsymbol{r} \boldsymbol{f} = -2\pi \sum_{i \neq j} \omega_i \omega_j grad_i \ln R_{ij} \tag{4.150}$$

is zero. We can see that a force

$$\boldsymbol{f}_{ij} = -\boldsymbol{f}_{ji} = -2\pi \omega_i \omega_j grad_i \ln R_{ij} \tag{4.151}$$

acts between any pair (ij) of vortices. This force derives from a potential

$$U_{ij} = 2\pi \omega_i \omega_j \ln R_{ij} \ , \tag{4.152}$$

[50]C. C. Lin, "On the motion of vortices in two-dimensions, I, II", Proc. Nat. Acad. Sci. USA **27** 570 (1941); *On the Motion of Vortices in Two Dimensions*, University of Toronto Press,Toronto (1943); L.Onsager, "Statistical hydrodynamics", Suppl. to Nuovo Cimento **6** 279 (1949); A. J. Chorin, *Vorticity and Turbulence*, Springer, NY (1994); D. Montgomery and G. Joyce, "Statistical mechanics of "negative temperature" states", Phys. Fluids **17** 1139 (1974); also V. I. Berezinsky, "Destruction of long-range order in one-dimensional and two-dimensional systems having a continuous symmetry group. I. Classical systems", Sov. Phys.-JEPT **32** 493 (1971); *ibid*, **34** 610 (1972) (ZhETF **59** 907 (1971); **61** 1144 (1972)); J. M. Kosterlitz and D. J. Thouless, "Ordering, metastability and phase transitions in two-dimensional systems", J. Phys. **C6** 1181 (1973).

[51]L.Onsager, "Statistical hydrodynamics", Suppl. to Nuovo Cimento **6** 279 (1949); G. K. Batchelor, *The Theory of Homogeneous Turbulence*, Cambridge University Press, Cambridge (1953); A. J. Chorin, *Vorticity and Turbulence*, Springer, NY (1994).

such that

$$\boldsymbol{F} = -\sum_{i \neq j} grad_i U_{ij} \tag{4.153}$$

The density of kinetic energy e can be written as

$$e = \frac{1}{2}v^2 = \frac{1}{2}\boldsymbol{v} curl \boldsymbol{A} = -\frac{1}{2}div(\boldsymbol{v} \times \boldsymbol{A}) + \boldsymbol{A\omega} \ . \tag{4.154}$$

The first term in equation (4.154) is singular at $\boldsymbol{r} = \boldsymbol{r}_i$; we integrate this term over the whole space, transform it into surface integrals, both at infinity and over small circles around each filament, and neglect their contributions. The result of such integrations is a self-energy (or a self-force), which may be left aside. We call this procedure a "renormalization".[52] By doing so, we are left with a total kinetic energy

$$\begin{aligned} E = \int d\boldsymbol{r}e = \int d\boldsymbol{r}\boldsymbol{A\omega} = -\pi \textstyle\sum_{i \neq j} \omega_i \omega_j \ln R_{ij} = \\ = -\frac{1}{2} \textstyle\sum_{i \neq j} U_{ij} = -U \ , \end{aligned} \tag{4.155}$$

where U is the total potential energy. We can see that the total energy is conserved, *i.e.* $E + U = const$. Also, the total force $\boldsymbol{F} = 0$, such that the total momentum is conserved. The total angular momentum is $-2 \int d\boldsymbol{r}\boldsymbol{A}$; it is proportional to $\sum_i \boldsymbol{\omega}_i$. The total torque

$$2\pi \sum_{i \neq j} \omega_i \omega_j \frac{\boldsymbol{r}_i \times \boldsymbol{r}_j}{R_{ij}^2} \tag{4.156}$$

is zero. By this "renormalization" procedure the points $\boldsymbol{r}_i$ are completely decoupled from the fluid, and they may have their own motion. This is an example of emergent dynamics.

Since the liquid is unstable, the vortices may disappear and, if the motion is going to be maintained, they reappear. The vortices may be viewed as elementary excitations of the vorticial liquid. Consequently, we may assume that the vectors $\boldsymbol{\omega}_i$ fluctuate randomly in time. The

[52] A. Sommerfeld, *Vorlesungen ueber Theoretische Physik*, Band II (*Mechanik der Deformierbaren Medien*), Akad. Verlagsgesellschaft, Leipzig (1970).

average over fluctuating vorticities can be computed straightforwardly, by using $\overline{\boldsymbol{\omega}_i^2} = \frac{1}{2}\omega_i^2$; it is given by

$$\overline{(\boldsymbol{v}grad)\boldsymbol{v}} = -\sum_i \omega_i^2 \frac{\boldsymbol{R}_i}{2R_i^4} \ . \tag{4.157}$$

We can see that for a sufficiently dense, homogeneous and isotropic liquid we get a gas of vortices, as discussed above.

4.10.2 Coulombian liquid

For $f_i(R_i) = -1/R_i$ in equation (4.123) we get a coulombian vorticial liquid with the velocity field

$$\boldsymbol{v} = -\sum_i \boldsymbol{\omega}_i \times grad(1/R_i) = \sum_i \frac{\boldsymbol{\omega}_i \times \boldsymbol{R}_i}{R_i^3} \tag{4.158}$$

and the vector potential

$$\boldsymbol{A} = \sum_i \frac{\boldsymbol{\omega}_i}{R_i} \ . \tag{4.159}$$

The equation $\Delta \boldsymbol{A} = -2\boldsymbol{\omega}$ is satisfied for

$$\boldsymbol{\omega}(\boldsymbol{r}) = 2\pi \sum_i \boldsymbol{\omega}_i \delta(\boldsymbol{R}_i) \ , \tag{4.160}$$

but this vorticity differs from

$$\frac{1}{2} curl\boldsymbol{v} = 2\pi \sum_i \boldsymbol{\omega}_i \delta(\boldsymbol{R}_i) - \frac{1}{2} grad\, div \sum_i (\boldsymbol{\omega}_i / R_i) \ . \tag{4.161}$$

By applying the "renormalization" procedure we get

$$\boldsymbol{F} = \int d\boldsymbol{r} \boldsymbol{f} = -\int d\boldsymbol{r} \boldsymbol{v} \times curl\boldsymbol{v} = 4\pi \sum_{i \neq j} grad_i \frac{\boldsymbol{\omega}_i \boldsymbol{\omega}_j}{R_{ij}} \tag{4.162}$$

and

$$E = \int d\boldsymbol{r} e = \frac{1}{2} \int d\boldsymbol{r} v^2 = 2\pi \sum_{i \neq j} \frac{\boldsymbol{\omega}_i \boldsymbol{\omega}_j}{R_{ij}} \ , \tag{4.163}$$

where, in both cases, $\boldsymbol{v} = curl\boldsymbol{A}$ is used. The potential from equation (4.152) is now $U_{ij} = -4\pi\frac{\boldsymbol{\omega}_i\boldsymbol{\omega}_j}{R_{ij}}$. The total energy is conserved, the total force is zero, the total angular momentum is proportional to $\sum_i \boldsymbol{\omega}_i$ and the total torque (zero) is

$$4\pi\sum_{i\neq j}\boldsymbol{\omega}_i\boldsymbol{\omega}_j\frac{\boldsymbol{r}_i\times\boldsymbol{r}_j}{R_{ij}^3} \ . \tag{4.164}$$

The average of the inertial term over fluctuating vortices is obtained from equation (4.127)

$$\overline{(\boldsymbol{v}grad)\boldsymbol{v}} = -\frac{2}{3}\sum_i \omega_i^2\frac{\boldsymbol{R}_i}{R_i^6} \ . \tag{4.165}$$

We get again a vorticial gas.

4.10.3 Dipolar liquid

A dipolar liquid is defined by the vorticity

$$\boldsymbol{\omega} = -2\pi\sum_i \boldsymbol{m}_i\times grad\delta(\boldsymbol{r}-\boldsymbol{r}_i) \ , \tag{4.166}$$

where the vectors $\boldsymbol{m}_i$ may depend on the time, at most. We get the vector potential

$$\boldsymbol{A} = \sum_i\frac{\boldsymbol{m}_i\times\boldsymbol{R}_i}{R_i^3} = -\sum_i \boldsymbol{m}_i\times grad(1/R_i) \tag{4.167}$$

and the velocity field

$$\begin{aligned}\boldsymbol{v}(\boldsymbol{r}) &= \textstyle\sum_i\left[-\boldsymbol{m}_i/R_i^3+3\boldsymbol{R}_i(\boldsymbol{m}_i\boldsymbol{R}_i)/R_i^5\right] = \\ &= \textstyle\sum_i grad\left[\boldsymbol{m}_i grad(1/R_i)\right] \ .\end{aligned} \tag{4.168}$$

We recognize in these equations magnetic (dipole) moments $\boldsymbol{m}_i$, a dipolar vector potential $\boldsymbol{A}$ and a magnetic field $\boldsymbol{v}$.
The inertial term has the components

$$\begin{aligned}\boldsymbol{f} &= -\boldsymbol{v}\times curl\boldsymbol{v} = 2\boldsymbol{\omega}\times\boldsymbol{v} = \\ &= 4\pi\textstyle\sum_{i\neq j} grad\left[\boldsymbol{m}_j grad(1/R_j)\right]\times\left[\boldsymbol{m}_i\times grad\delta(\boldsymbol{R}_i)\right]\end{aligned} \tag{4.169}$$

and *grade*, where

$$e = \frac{1}{2}v^2 = \frac{1}{2}\sum_{i\neq j}\left[\frac{\boldsymbol{m}_i\boldsymbol{m}_j}{R_i^3R_j^3} - \frac{3(\boldsymbol{m}_i\boldsymbol{R}_j)(\boldsymbol{m}_j\boldsymbol{R}_j)}{R_i^3R_j^5} - \right.$$

$$-\frac{3(\boldsymbol{m}_j\boldsymbol{R}_i)(\boldsymbol{m}_i\boldsymbol{R}_i)}{R_i^5R_j^3} + \tag{4.170}$$

$$\left. + \frac{9(\boldsymbol{R}_i\boldsymbol{R}_j)(\boldsymbol{m}_i\boldsymbol{R}_i)(\boldsymbol{m}_j\boldsymbol{R}_j)}{R_i^5R_j^5}\right] .$$

We can see that the dipolar liquid is unstable.

Let us check equations (4.83) for the velocity field of a single vortex. By making use of the identity

$$(\boldsymbol{v}grad)\boldsymbol{v} = -\boldsymbol{v}\times curl\boldsymbol{v} + grad(v^2/2) \ , \tag{4.171}$$

we identify a force

$$-\boldsymbol{v}\times curl\boldsymbol{v} = 4\pi\boldsymbol{v}\times[\boldsymbol{m}\times grad\delta(\boldsymbol{r})] \tag{4.172}$$

and a density of kinetic energy

$$\frac{1}{2}v^2 = \frac{m^2}{2r^6} + \frac{3(\boldsymbol{mr})^2}{2r^8} \ , \tag{4.173}$$

which gives rise to another force $grad(v^2/2)$ (Euler's forces). The singular force given by equation (4.172) acts at the location of the vorticity ($\boldsymbol{r}=0$), such that it is a self-force; we may leave it aside. The second equation (4.83) is reduced to

$$grad(v^2/2) = 0 \ , \tag{4.174}$$

which is not satisfied.

Similarly, the third equation (4.83) becomes

$$grad\left[\dot{\boldsymbol{m}}grad(1/r)\right] = -\frac{1}{\rho}gradp - 4\pi\nu grad\left[\boldsymbol{m}grad\delta(\boldsymbol{r})\right] \ ; \tag{4.175}$$

we should apply this equation for $\boldsymbol{r}\neq 0$, where it would give the pressure

$$p = -\rho\dot{\boldsymbol{m}}grad\frac{1}{r} \tag{4.176}$$

which would generate the time derivative of the dipole moment $\boldsymbol{m}$. The fourth equation (4.83) implies that $gradT$ would be proportional to the vectorial product $\boldsymbol{m}\times\boldsymbol{r}$, which is not a valid equation, because $curl(\boldsymbol{m}\times\boldsymbol{r})=2\boldsymbol{m}\neq 0$.

We conclude that the dipolar vortex does not satisfy equations (4.83) (the heat-transfer equation remains an independent equation).

The total force and the total kinetic energy are

$$\begin{gathered}\boldsymbol{F}=\int d\boldsymbol{r}\boldsymbol{f}=2\int d\boldsymbol{r}\boldsymbol{\omega}\times\boldsymbol{v}=-\sum_{i\neq j}grad_iU_{ij}\ ,\\ E=\int d\boldsymbol{r}e=\int d\boldsymbol{r}\boldsymbol{\omega}\boldsymbol{A}=-\tfrac{1}{2}\sum_{i\neq j}U_{ij}\ ,\end{gathered}\tag{4.177}$$

where

$$U_{ij}=-4\pi\boldsymbol{m}_i grad_i\left[\boldsymbol{m}_j grad_i(1/R_{ij})\right]\ .\tag{4.178}$$

By this "renormalization" procedure, the liquid is reduced to a set of interacting particle-like vortices. We can see that the energy is conserved, the total force is zero, the total angular momentum and the total torque are zero. (This makes unlikely that such a vorticial liquid may appear in turbulence).

The time average over vorticities in equation (4.170) gives

$$\overline{e}=\frac{1}{2}\sum_i\left[\frac{m_i^2}{R_i^6}+\frac{3\overline{(\boldsymbol{m}_i\boldsymbol{R}_i)^2}}{R_i^8}\right]=\sum_i\frac{m_i^2}{R_i^6}\ ,\tag{4.179}$$

which gives a force

$$grad\overline{e}=-6\sum_i\frac{m_i^2\boldsymbol{R}_i}{R_i^8}\ .\tag{4.180}$$

For a dense, homogeneous and isotropic liquid the force given by equation (4.180) is zero (for $\boldsymbol{R}_i\neq 0$); also, the force $\boldsymbol{f}$ is zero for $\boldsymbol{R}_i\neq 0$, and the Navier-Stokes equation is satisfied trivially (on average, with zero pressure). We note that $\Delta\boldsymbol{v}$ is zero for $\boldsymbol{R}_i\neq 0$ (equation (4.168)), such that the viscosity contribution is zero. We are left with a set of positions $\boldsymbol{r}_i$, each surrounded by a small region, where the motion is not defined. According to the above discussion, such a structure may be viewed as a (quasi) ideal classical gas of vortices (or a solution of vortices in the background fluid).

Other vorticial liquids can be imagined, like ring, or spherical-shell vortices, with similar results.

4.10.4 Concluding remarks

"At any moment, according to Kolmogorov, there exists a narrow layer between trivial and impossible where mathematical discoveries are made. Therefore, an applied problem is either solved trivially or not solved at all. It is an altogether different story if an applied problem is found to fit (or made to fit!) the new formalism interesting for a mathematician".

The Navier-Stokes equation, as any differential equation, is valid for smooth functions, *i.e.* functions which are continuous and infinitely differentiable. The turbulence appears for large variations of the velocity over small distances. Consequently, the turbulence cannot be described by the Navier-Stokes equation. The smooth solutions of the Navier-Stokes equation become unstable on the initiation of the turbulence. If the turbulence is going to subsists, the instabilities should manifest themselves as fluctuations of the velocity. We may arrive at the Reynolds equations. In the absence of a mean velocity these fluctuations are produced by an external force (for instance an external pressure gradient), transport momentum by collisions (viscosity), which produces heat (also by viscosity). The energy is conserved. A mean velocity in turbulence is coupled to the fluctuations, in the sense that the rate of the work done by the external force per unit time is controlled by the mean velocity and dissipated by viscosity. Since the fluctuations are generated by the external force, the mean velocity is thus determined. An example is the logarithmic profile (the wall law), another is the singular wind (see the Reynolds equation section). In this sense it is said that the energy is transferred between small and large swirls.

A characteristic feature of the turbulence, besides fluctuations, is vortices. Turbulence appears at the contact of the fluid flow with obstacles, over short distances, where the flow forms swirls. These vortices do not satisfy the Navier-Stokes equation, but on averaging over fluctuations, the velocity and the inertial force of a homogeneous and isotropic set of vortices are zero, and the Navier-Stokes equation is satisfied trivially. If such vortices are singular, we are left with indefinite singularities at the positions of the vortices, which behave as a particle-like set of pointlike objects, each with a fluid mass and a small dimension. Therefore, we are left with an almost ideal, and classical, vorticial gas, which is in quasi equilibrium (or a solution of vortices in the background fluid). The Navier-Stokes equation for this gas is reduced to Newton's equation. The turbulent sub-layer in the boundary layer may be a vorticial gas, over which the logarithmic boundary layer is superimposed. Also, the drop of the drag force in turbulent motion (drag crisis) may be due to the occurrence of the vorticial gas.

The turbulence is produced by some external action, which tends to satisfy the Navier-Stokes equation. But such external forces are not able to satisfy completely the equation, or to maintain its fulfilment. In order to get equilibrium and to satisfy the Navier-Stokes equation the turbulent flow develops fluctuations, homogeneity and isotropy, and the equation is satisfied on average. In some cases, where the turbulence is fully developed, the flow exhibits additional singularities in a discrete set of points; then, the flow looks like an ideal classical gas, in quasi thermodynamic equilibrium, for which the Navier-Stokes equation is reduced to Newton's law of motion, which is satisfied (or a solution of vortices in the background fluid).

We know several exact solutions of the Navier-Stokes equation, in realistic, interesting cases, including numerical calculations. Of course, these solutions are smooth, as any solution with a physical meaning. There is not a general method of solving this equation, since it is a nonlinear equation. The situation may look satisfactory, except that there are reasonable indications, and particular proofs, that, under special conditions, these solutions are unstable; it is sufficient a certain small perturbation, and the solution may explode. We have not a rigorous proof of this circumstance, this is the Millennium problem of the Navier-Stokes equation. In one of the above sections we give arguments towards the existence of such instabilities. They appear for large variations of the velocity over small distances. They are different, in principle, from other instabilities, like those produced for instance by thermal convection. The existence of the instabilities may be related to the lack of an extremum principle for the Navier-Stokes equation. Even if we accept the existence of the instabilities, and the Millennium problem would be over, the problem of the Navier-Stokes does not end, because the fluid continues to flow, in a particular way, including such instabilities, such that we need to know such a flow. The particular way of this flow is the turbulent motion. The flow develop irregular, erratic, fluctuating swirls, of various spatial extent, randomly distributed, with intermittencies, often displaying pointlike, or linelike, singularities. These are the turbulent vortices. We can have mathematical representations for such vortices. Such vortices do not satisfy the Navier-Stokes equation; they are unstable; they form an unstable vorticial liquid. On the other hand, for such a fluctuating flow, the problem of a solution of the Navier-Stokes equation is meaningless, because the flow is not smooth; an average over fluctuations is a meaningful way of seeking a solution. However, in general, even on average, the vorticial liquid does not satisfy the Navier-Stokes equation; it continues to remain unstable. But, if the turbulence is homogeneous and isotropic, as a fully developed turbulence is, both the average of the velocity and the average of the non-linear term of the equation are zero, and the equation is satisfied trivially. With this, the problem of the turbulent flow would be solved. But, if the vortices are singular, as often happens in a fully developed turbulence (recall the large variations over small distances), we are left with a discrete set of pointlike (or linelike) singularities, which is another, emergent fluid, with its own Navier-Stokes equation. The problems seem to recur, but the Navier-Stokes equation for the vortices fluid is reduced to Newton's equation for a set of pointlike (or linelike) particles (vortices), with a small friction and, possibly, an external force on behalf of the background fluid. We have a classical ideal gas, or a solution of vortices in the background fluid, in quasi thermal equilibrum. Now, the problem is fully solved. We may consider the equations of the fluid mechanics for the vortices gas, viewed as a continuous medium, at a higher scale; this is an emergent physics.

4.11 Statistical approach

The Navier-Stokes equation

$$\frac{\partial v_i}{\partial t} = -\frac{1}{\rho}\partial_i p - v_j \partial_j v_i + \nu \Delta v_i \tag{4.181}$$

for an incompressible fluid with the continuity equation $\partial_i v_i = 0$ can also be written as

$$\frac{\partial v_i}{\partial t} = -\frac{1}{\rho}\partial_i p - \partial_j (v_i v_j) + \nu \Delta v_i \ . \tag{4.182}$$

We make the decomposition $v_i = \overline{v}_i + v_i^{'}$, where $\overline{v}_i$ is a time average (so, $\overline{v}_i$ does not depend on the time) and $v_i^{'}$ is a fluctuating velocity ($\overline{v_i^{'}} = 0$); we get

$$\begin{array}{c}\frac{\partial v_i}{\partial t} = \frac{\partial \overline{v}_i}{\partial t} + \frac{\partial v_i^{'}}{\partial t} = \frac{\partial v_i^{'}}{\partial t} = -\frac{1}{\rho}\partial_i \overline{p} - \frac{1}{\rho}\partial_i p^{'} - \\ -\partial_j(\overline{v}_i\overline{v}_j + v_i^{'}\overline{v}_j + \overline{v}_i v_j^{'} + v_i^{'}v_j^{'}) + \nu\Delta\overline{v}_i + \nu\Delta v_i^{'} \ ,\end{array} \tag{4.183}$$

where the pressure is also decomposed into its mean value $\overline{p}$ and its fluctuating part $p^{'}$. Taking here the time average we get

$$0 = -\frac{1}{\rho}\partial_i \overline{p} - \partial_j(\overline{v}_i\overline{v}_j + \overline{v_i^{'}v_j^{'}}) + \nu\Delta\overline{v}_i \ ; \tag{4.184}$$

and the continuity equations

$$\partial_i \overline{v}_i = 0 \ , \ \partial_i v_i^{'} = 0 \ , \ \partial_i \overline{v_i^{'}} = 0 \ . \tag{4.185}$$

Equation (4.184) represents the Reynolds equations. It is currently thought that the Reynolds stress tensor $\overline{v_i^{'}v_j^{'}}$ provides an additional force to the mean flow, but there is no apriori reason to connect this tensor to the mean velocity $\overline{v}_i$. Equation (4.184) shows that the mean flow is decoupled from the fluctuating flow, *i.e.* equation (4.184) should be read in fact

$$0 = -\frac{1}{\rho}\partial_i \overline{p} - \partial_j(\overline{v}_i\overline{v}_j) + \nu\Delta\overline{v}_i \ , \ \partial_j(\overline{v_i^{'}v_j^{'}}) = 0 \ ; \tag{4.186}$$

in general, the fluctuating flow does not satisfy a Navier-Stokes equation (second equation (4.186)).

Nevertheless, we keep the Reynolds stress in equation (4.183), multiply this equation by v_i, sum over i, take the time average and use the continuity equations (4.185), to get the energy conservation equation

$$\begin{gathered} 0=\overline{\frac{\partial}{\partial t}\left(\frac{1}{2}v_i^2\right)}=\overline{v}_i\left[-\frac{1}{\rho}\partial_i\overline{p}-\partial_j(\overline{v}_i\overline{v}_j+\overline{v_i'v_j'})+\nu\Delta\overline{v}_i\right]- \\ -\overline{v_i'v_j'}\partial_j\overline{v}_i-\frac{1}{2}\overline{v}_j\partial_j\left(\overline{v_i'^2}\right)+\nu\overline{v_i'\Delta v_i'}\ , \end{gathered} \tag{4.187}$$

where we have dropped out third-order terms involving products of three fluctuating velocities and their derivatives, as well as the contribution of the fluctuating part of the pressure; we assume that the velocity fluctuations are much smaller than the mean flow velocities. We note that $\overline{v}_i\partial_j\overline{v_i'v_j'}$ and $\overline{v_i'v_j'}\partial_j\overline{v}_i$ can be written as $\partial_j\left(\overline{v}_i\overline{v_i'v_j'}\right)$, such that equation (4.187) becomes

$$\begin{gathered} 0=\overline{\frac{\partial}{\partial t}\left(\frac{1}{2}v_i^2\right)}=\overline{v}_i\left[-\frac{1}{\rho}\partial_i\overline{p}-\partial_j(\overline{v}_i\overline{v}_j)+\nu\Delta\overline{v}_i\right]- \\ -\partial_j(\overline{v}_i\overline{v_i'v_j'})-\frac{1}{2}\overline{v}_j\partial_j\left(\overline{v_i'^2}\right)+\nu\overline{v_i'\Delta v_i'}\ . \end{gathered} \tag{4.188}$$

This equation shows that the density of the mean-flow momentum flux (the term in bracket) is transported with the mean-flow velocity, which transports also the fluctuation energy density (the term $\sim\overline{v_i'^2}$); the fluctuations transport their viscosity momentum density ($\sim\overline{v_i'\Delta v_i'}$). The term $\partial_j(\overline{v}_i\overline{v_i'v_j'})$ shows that a fluctuating momentum flux density comes, or goes out, through an enclosing surface, with the mean-flow velocity. Thre is no evidence in this equation of a coupling between the mean flow and the fluctuating flow. The only connection between the two is that the mean flow transports fluctuations, as expected.

The ν-terms in equation (4.187) can be written as

$$\begin{gathered} \nu\left(\overline{v}_i\Delta\overline{v}_i+\overline{v_i'\Delta v_i'}\right)= \\ =\frac{1}{2}\nu\Delta\left(\overline{v}_i^2+\overline{v_i'^2}\right)-\nu\left(\partial_j\overline{v}_i\right)^2-\nu\overline{\left(\partial_j v_i'\right)^2}\ . \end{gathered} \tag{4.189}$$

The first term on the right in this equation is related to the transport of the energy flux density caused by the internal friction. The last term is different: it represents the heat source generated by viscosity. This heat production is partly conducted (transported) thermally and partly it increases the entropy; therefore, we have

$$T\frac{ds}{dt}=T\left(\frac{\partial s}{\partial t}+v_i\partial_i s\right)=-\frac{1}{\rho}div\boldsymbol{q}+\nu\left(\partial_j\overline{v}_i\right)^2+\nu\overline{\left(\partial_j v_i^{'}\right)^2}\ , \quad (4.190)$$

or

$$T\left(\frac{\partial s}{\partial t}+v_i\partial_i s\right)=\frac{\kappa}{\rho}\Delta T+\nu\left(\partial_j\overline{v}_i\right)^2+\nu\overline{\left(\partial_j v_i^{'}\right)^2}\ , \quad (4.191)$$

where s is the entropy per unit mass, T is the temperature, $\boldsymbol{q}=-\kappa gradT$ is the heat flux density and κ (uniform) is the thermal conductivity. We note that the total change in entropy

$$\begin{gathered}\frac{d}{dt}\int dVs=\int dV\frac{\partial s}{\partial t}=\frac{\kappa}{\rho}\int d\mathbf{r}\frac{(gradT)^2}{T^2}+\\ +\nu\int dV\left[\left(\partial_j\overline{v}_i\right)^2+\overline{\left(\partial_j v_i^{'}\right)^2}\right]\end{gathered} \quad (4.192)$$

is positive (κ and ν are positive). In equation (4.191) the entropy and the temperature should be decomposed into their mean and fluctuating values and the average should be taken:

$$\begin{gathered}\overline{T}\left(\overline{v}_i\partial_i\overline{s}+\overline{v_i^{'}\partial_i s^{'}}\right)+\overline{T^{'}\frac{\partial s^{'}}{\partial t}}+\overline{v}_i\overline{T^{'}\partial_i s^{'}}+\overline{T^{'}v_i^{'}}\partial_i\overline{s}=\\ =\frac{\kappa}{\rho}\Delta\overline{T}+\nu\left(\partial_j\overline{v}_i\right)^2+\nu\overline{\left(\partial_j v_i^{'}\right)^2}\ ,\end{gathered} \quad (4.193)$$

whence we can see that the change in entropy implies fluctuations; in view of our assumptions of small fluctuations we may neglect their contributions to equation (4.193), which becomes

$$\overline{T}\overline{v}_i\partial_i\overline{s}=\frac{\kappa}{\rho}\Delta\overline{T}+\nu\left(\partial_j\overline{v}_i\right)^2\ ; \quad (4.194)$$

here we may use $Td\overline{s}=c_p dT$, where c_p is the specific heat at constant pressure, in order to get the temperature distribution. We can see that the heat-transfer equation is reduced to the mean flow, as expected.

Usually, the fluctuating term $v_i^{'}v_j^{'}$ is kept in the Navier-Stokes equations (4.183) and (4.184); by multiplying by $\overline{v}_i$, summing over i and taking the average, we get the first part of equation (4.187),

$$0 = \overline{\frac{\partial}{\partial t}\left(\frac{1}{2}\overline{v}_i^2\right)} = \overline{v}_i\left[-\frac{1}{\rho}\partial_i\overline{p} - \partial_j(\overline{v}_i\overline{v}_j + \overline{v_i^{'}v_j^{'}}) + \nu\Delta\overline{v}_i\right] \quad , \qquad (4.195)$$

which is in fact equation (4.184). By multiplying by $v_i^{'}$, summing over i and taking the average, we get the second part of equation (4.187),

$$0 = \overline{\frac{\partial}{\partial t}\left(\frac{1}{2}v_i^{'2}\right)} = -\overline{v_i^{'}v_j^{'}}\partial_j\overline{v}_i - \frac{1}{2}\overline{v}_j\partial_j\left(\overline{v_i^{'2}}\right) + \nu\overline{v_i^{'}\Delta v_i^{'}} \quad , \qquad (4.196)$$

By taking a spatial average in equation (4.196) the middle term on the right disappears, while the ν-term gives a constant ($\overline{v}_j$ may be transferred under the derivative ∂_j). Thus, equation (4.196) provides a source term for the mean flow in equation (4.184). However, it is worth noting that a spatial average in equation (4.184) removes the Reynolds stress. It is thought that such a spatial average is relevant only for fluctuations, not for the mean flow; both the Navier-Stokes equation and the energy conservation equation for the mean flow are not spatially averaged. The Reynolds stress is viewed as a source, or a sink, for the mean flow. The existence of such a coupling is due to the fact that the terms $\overline{v}_i\partial_j\overline{v_i^{'}v_j^{'}}$ and $\overline{v_i^{'}v_j^{'}}\partial_j\overline{v}_i$ are viewed as not being related through $\partial_j\left(\overline{v}_i\overline{v_i^{'}v_j^{'}}\right)$; the latter is spatially averaged, the former is not. We note that equation (4.74) includes second-order effects of the fluctuations, which are neglected in the mean heat-transfer equation (4.71). It is easy to see that such a decoupling procedure may allow for singular solutions of the mean flow, as for the singular wind described above. The above statistical procedure may hold for a coupling between the mean flow and the fluctuating flow.

It is worth noting that keeping the Reynolds stress tensor $\overline{v_i^{'}v_j^{'}}$ in the Navier-Stokes equation amounts to introducing unknown quantities. We cannot find these quantities from energy conservation, since the energy conservation brings nothing new to the Navier-Stokes equation. The only way of giving a meaning to the Reynolds stress tensor is by using model assumptions.

In the above sections we described a homogeneous and isotropic turbulence with fluctuating, singular velocities. We showed that this turbulence is equivalent to a solution of singularities in the background fluid. A mixture of such solutes would imply a phase equilibrium. A dissociation of large singularites in smaller ones, or the recombination of small singularities in larger ones looks like a statistical equilibrium of chemical reactions. Also, a coupling between small singularities to a larger one may exist, as in the singular wind.

4.12 Fluctuating velocities

The statistical approach described above is based on the additional hypothesis of a coupling between the mean flow and the fluctuating flow. This leads to two distinct equations, one the averaged Navier-Stokes equation (4.184) and another the energy-conservation equation (4.196). Such a hypothesis may be valid for a moderate turbulence, where the mean velocity is large and the fluctuations are small, includ ing model assumptions. This situation is encountered in the singular wind, where the flow takes place along a plane boundary. In this case the mean flow may exhibit singularities.

The turbulent instabilities ocurring in a fully developed turbulence exhibit large variations of the velocity over small distances. In this case we may assume that the velocity is split into a mean-flow velocity $\boldsymbol{v}_0$ and a fluctuating part $\boldsymbol{v}$, where the mean-flow velocity $\boldsymbol{v}_0$ may have a slight time variation, while the fluctuating velocity $\boldsymbol{v}$ is a rapidly varying velocity. By using this decomposition for an incompressible fluid (density $\rho = const$), the Navier-Stokes equation reads

$$\begin{gathered}\frac{\partial \boldsymbol{v}_0}{\partial t} + \frac{\partial \boldsymbol{v}}{\partial t} + (\boldsymbol{v}_0 grad)\,\boldsymbol{v}_0 + (\boldsymbol{v}_0 grad)\,\boldsymbol{v} + (\boldsymbol{v} grad)\,\boldsymbol{v}_0 + \\ + (\boldsymbol{v} grad)\,\boldsymbol{v} = -\frac{1}{\rho} grad p_0 - \frac{1}{\rho} grad p + \nu\Delta\boldsymbol{v}_0 + \nu\Delta\boldsymbol{v} \ ,\end{gathered} \tag{4.197}$$

where p_0 is the pressure corresponding to the main flow, p is the fluctuating part of the pressure and ν is the kinematical viscosity. In this equation we have three distinct types of time variations, such that it should be viewed as three equations

$$\frac{\partial \boldsymbol{v}_0}{\partial t} + (\boldsymbol{v}_0 grad)\, \boldsymbol{v}_0 = -\frac{1}{\rho} grad p_0 + \nu \Delta \boldsymbol{v}_0 \ ,$$

$$\frac{\partial \boldsymbol{v}}{\partial t} + (\boldsymbol{v}_0 grad)\, \boldsymbol{v} + (\boldsymbol{v} grad)\, \boldsymbol{v}_0 = -\frac{1}{\rho} grad p + \nu \Delta \boldsymbol{v} \ , \tag{4.198}$$

$$(\boldsymbol{v} grad)\, \boldsymbol{v} = 0 \ ;$$

similarly, the continuity equation should be split into

$$div \boldsymbol{v}_0 = 0 \ , \ div \boldsymbol{v} = 0 \ . \tag{4.199}$$

The first equation (4.198) is an independent equation, which should be solved for the main flow velocity $\boldsymbol{v}_0$. Since the main part of the velocity is taken by the fluctuating velocity, we may neglect the quadratic term in this equation. The energy conservation and the heat transfer for this equation are given by

$$\frac{\partial}{\partial t}(v_0^2/2) = -\partial_i \left[v_{0i} \left(p_0/\rho + v_0^2/2 \right) - \nu \partial_i (v_0^2/2) \right] -$$

$$-\nu (\partial_i v_{0j})^2 \ , \tag{4.200}$$

$$T_0 \frac{ds_0}{dt} = \chi \Delta T_0 + \nu (\partial_i v_{0j})^2 \ ,$$

where T_0 is the temperature of the main flow, s_0 is the entropy per unit mass of the main flow and χ is the thermometric conductivity.

Having solved the mean-flow equation we can pass to solve the second equation (4.198) for the fluctuating velocity $\boldsymbol{v}$, with $\boldsymbol{v}_0$ as a parameter. This equation has its own energy-conservation and heat-transfer equations. We note that the temperature and the entropy of the mean flow are different from the temperature and the entropy of the fluctuating part of the flow, which means that the two components of the flow (the mean flow and the fluctuating flow) are not in thermal equilbrium. Indeed, if we multiply the first equation (4.198) by $\boldsymbol{v}$ and the second equation (4.198) by $\boldsymbol{v}_0$, we get cross-terms of the form $T_0 \frac{ds}{dt} + T \frac{ds_0}{dt}$ in the heat-transfer equation, where T and s are the temperature and the entropy of the fluctuating flow. This indicates a heat exchange between the two components of the flow.

We are left with the third equation (4.198), which, in general, is not satisfied. We conclude that the fully developed turbulence does not satisfy the Navier-Stokes equation. The quadratic term of the third equation (4.198) is equivalent to a rapidly varying internal force (Euler's force), which cannot be compensated by any physical external force. The fully developed turbulence is unstable.

Under these conditions it is reasonable to be interested in time averaged quantities. Then, the fluctuating part of the flow is reduced to

$$\overline{(\boldsymbol{v}grad)\,\boldsymbol{v}} = 0 \ . \tag{4.201}$$

Since $div\boldsymbol{v} = 0$, the components of the velocity $\boldsymbol{v}$ are not independent. In general, equation (4.201) is not satisfied, which means that the turbulent motion is unstable even on average.

Since a fully developed turbulence originates in large variations of the velocity across small distances, it is reasonable to associate these variations with a discrete distribution of positions $\boldsymbol{r}_i$,which we call centres of turbulence. Further on, we assume that this is a homogeneous and isotropic distribution, such that we may write the velocity field as

$$\boldsymbol{v} = \sum_i \boldsymbol{v}_i(t, R_i) \ , \tag{4.202}$$

where $\boldsymbol{R}_i = \boldsymbol{r} - \boldsymbol{r}_i$. If $curl\boldsymbol{v}_i \neq 0$, this velocity field represents a vorticial liquid, which is unstable. We assume that the fluctuating velocities are independent at distinct positions, *i.e.* $\overline{\boldsymbol{v}_i} = 0$ and $\overline{\boldsymbol{v}_i\boldsymbol{v}_j} \sim \delta_{ij}$. Equation (4.201) becomes

$$\overline{(\boldsymbol{v}grad)\,\boldsymbol{v}} = \sum_i \overline{(\boldsymbol{v}_i\boldsymbol{R}_i/R_i)\,(d\boldsymbol{v}_i/dR_i)} \ . \tag{4.203}$$

The conditions of homogeneity and isotropy imply that $\boldsymbol{v}_i = \boldsymbol{u}$, *i.e.* the velocities $\boldsymbol{v}_i$ are all equal to the same function $\boldsymbol{u}$. For a sufficiently dense set of positions $\boldsymbol{r}_i$ we can define a density ρ_v of such points, which is a constant. Then, equation (4.203) can be transformed into an integral,

$$\overline{(\boldsymbol{v}grad)\,\boldsymbol{v}} = \rho_v \int dR \cdot R^2 \int do\overline{(\boldsymbol{u}\boldsymbol{R}/R)\,(d\boldsymbol{u}/dR)} \ , \tag{4.204}$$

which is similar to equation (4.128). If the radial integral is finite, the result of integration in the above equation is zero, due to the integration over the solid angle o, such that the term $\overline{(\boldsymbol{v}grad)\,\boldsymbol{v}}$ is zero. In this case we can say that the Navier-Stokes equation is trivially satisfied on average, being reduced to the mean-flow equation (first equation (4.198)).

If the radial integral in equation (4.204) is singular for $\boldsymbol{R} = 0$, as it may often happen for vortices, we are left with a discrete set of singularities, extending over a small characteristic distance a, where the singularity is

$$\boldsymbol{u} \sim (a/R)^n \ , \ n > 1 \tag{4.205}$$

(compare with equation (4.129)). In each of these regions there exists a mass M of fluid, which can be carried by the background fluid and, at the same time, they may have their own motion.

The averaged energy-conservation equation derived from the second equation (4.198)

$$\frac{1}{2}v_0 grad\overline{v^2} + \partial_j\left(\overline{v_i v_j} v_{0i}\right) - \frac{1}{2}\nu\Delta\overline{v^2} = -\nu\overline{(\partial_j v_i)^2} \tag{4.206}$$

shows that the fluctuating motion produces heat which is partly transported by the mean flow (the *div*-terms integrated over a volume are irrelevant). Therefore, we are left with a quasi ideal gas of singularities, or a solution of singularities in the background fluid.

4.13 Qudratic term

Let us write the Navier-Stokes equation as

$$\frac{\partial \boldsymbol{v}}{\partial t} + (\boldsymbol{v}grad)\boldsymbol{v} = -gradw + \nu\Delta\boldsymbol{v} \tag{4.207}$$

for an incompressible fluid, where $w = p/\rho$ is the enthalpy, p is an external pressure, ρ is the density and ν is the kinematic viscosity; we neglect the small variations of temperature. Let us assume that w and $\boldsymbol{v}$ vary slowly in space, and replace this equation by

$$\frac{\partial \boldsymbol{v}}{\partial t} + (\boldsymbol{u}grad)\boldsymbol{v} = -gradw + \nu\Delta\boldsymbol{v} \ , \tag{4.208}$$

where $\boldsymbol{u}$ is a mean, uniform velocity. This equation can be solved by a Fourier transformation,

$$\begin{aligned}\boldsymbol{v} &= \frac{1}{(2\pi)^4}\int d\omega \int d\boldsymbol{k}\frac{\boldsymbol{k}w}{\omega-\boldsymbol{k}\boldsymbol{u}+i\nu k^2}e^{-i\omega t}e^{i\boldsymbol{k}\boldsymbol{r}} = \\ &= \frac{-i}{(2\pi)^3}\int d\boldsymbol{k}\boldsymbol{k}we^{i\boldsymbol{k}(\boldsymbol{r}-\boldsymbol{u}t)}e^{-\nu k^2 t} \ ,\end{aligned} \tag{4.209}$$

where the pole is placed in the lower ω-plane, according to the causality principle. We introduce the notation $\boldsymbol{x} = \boldsymbol{r} - \boldsymbol{u}t$, and rewrite the above equation as

$$\begin{aligned}\boldsymbol{v} &= -\frac{1}{(2\pi)^3}\frac{\partial}{\partial \boldsymbol{x}}\int d\boldsymbol{k}we^{i\boldsymbol{k}\boldsymbol{x}-\nu k^2 t} = \\ &= \frac{i}{(2\pi)^2}\frac{\partial}{\partial \boldsymbol{x}}\int_{-\infty}^{+\infty} dk\frac{wk}{x}e^{ikx-\nu k^2 t} \simeq \\ &\simeq \frac{w}{(4\pi\nu t)^{3/2}}\frac{\partial}{\partial \boldsymbol{x}}e^{-\frac{x^2}{4\nu t}} \ ,\end{aligned} \tag{4.210}$$

or

$$\boldsymbol{v} \simeq \frac{2w\boldsymbol{x}}{(4\pi\nu t)^{5/2}}e^{-\frac{x^2}{4\nu t}} \ . \tag{4.211}$$

In this equation we replace $\boldsymbol{x}$ by $\boldsymbol{r} - \boldsymbol{v}t$, for consistency, and get

$$\boldsymbol{v} \simeq \frac{\boldsymbol{r}}{t + \frac{(4\pi\nu t)^{5/2}}{2w}e^{\frac{|\boldsymbol{r}-\boldsymbol{v}t|^2}{4\nu t}}} \ . \tag{4.212}$$

This equation has the solution $\boldsymbol{v} \simeq \boldsymbol{r}/t$ in the limit $\nu \rightarrow 0$. In this limit $\frac{\partial \boldsymbol{v}}{\partial t} = -\boldsymbol{r}/t^2$, $(\boldsymbol{v}grad)\boldsymbol{v} = \boldsymbol{r}/t^2$, $\Delta\boldsymbol{v} = 0$ and $gradw = 0$. The Fourier transform $\boldsymbol{v}(\boldsymbol{k}) = -\frac{i}{(2\pi)^3 t}grad\delta(\boldsymbol{k})$. This is a quasi-free motion in the Lagrange description. The fluid moves uniformly as a whole.

5 Navier-Stokes and Turbulence

In fairly general conditions we give explicit (smooth) solutions for a potential fow. We show that, rigorously speaking, the equations of the fluid mechanics have not rotational solutions. Within the usual approximations of an incompressible fluid and an isentropic flow, the remaining Navier-Stokes equation has approximate vorticial (rotational) solutions, generated by viscosity. In general, the vortices are unstable, and a discrete distribution of vorticial solutions is not in mechanical equilibrium; it forms an unstable vorticial liquid. On the other hand, these solutions may exhibit turbulent, fluctuating instabilities for large variations of the velocity over short distances. We represent a fully developed turbulence as a homogeneous, isotropic and highly-fluctuating distribution of singular centres of turbulence. A regular mean flow can be included. In that case the Navier-Stokes equation exhibits three time scales. The equations of the mean flow can be disentangled from the equations of the fluctuating part, which is reduced to a vanishing inertial term. This latter equation is not satisfied after averaging out the temporal fluctuations. However, for a homogeneous and isotropic distribution of non-singular turbulence centres the equation for the inertial term is satisfied trivially, *i.e.* the average of the fluctuating velocity and the average of the fluctuating inertial term are both zero. If the velocity is singular at the turbulence centres, we are left with a quasi-ideal classical gas of singularities, or a solution of singularities in quasi thermal equilibrium in the background fluid. This is an example of an emergent dynamics. We give three examples of vorticial liquids.

5.1 Introduction

In fairly general conditions we give explicit (smooth) solutions of the potential flow.

However, as it is well known, the fluids may develop turbulence. In its extreme manifestation the turbulent flow displays very irregular, disordered velocities, fluctuating in time at each point in space. This is known as a fully developed turbulence. By using such fluctuating velocities, besides a steady mean velocity, the Navier-Stokes equation becomes an infinite hierarchy of equations for velocity mean correlation functions, known as Reynolds's equations,[1] which need closure assumptions. According to the experimental observations, it was realized that such irregular movements of the fluid exhibit distributions of swirls (eddies, vortices), of various magnitude and vorticities; it is likely that the large eddies transfer energy to the small eddies, which dissipate it.[2] Statistical concepts like correlations, homogeneity and isotropy have been introduced in the theory of turbulence,[3] and dimensional analysis and similarity arguments allowed the derivation of the energy spectrum of the turbulent eddies.[4]

Meanwhile, the relation of this statistical turbulence with the Navier-Stokes equation remained unclear.[5] Could the Navier-Stokes equation describe a turbulent motion? To what extent and in what sense? Has

[1] O. Reynolds, "On the dynamical theory of the incompressible viscous fluids and the determination of the criterion", Phil. Trans. Roy. Soc. London **A186** 123 (1895).

[2] L. F. Richardson, "The supply of energy from and to atmospheric eddies", Proc. Roy. Soc. London **A97** 354 (1920); *Weather Prediction by Numerical Process*, Cambridge University Press, Cambridge (1922); L. Prandtl, "The generation of vortices in fluids of small viscosity", J. Roy. Aeronaut. Soc. **31** 720 (1927).

[3] G. I. Taylor, "Statistical theory of turbulence", Proc. Roy. Soc. London **A151** 421; II. 444 (1935); G. K. Batchelor, *The Theory of Homogeneous Turbulence*, Cambridge University Press, Cambridge (1953).

[4] A. N. Kolmogorov, "The local structure of turbulence in incompressible viscous fluid for very large Reynolds numbers", Dokl. Akad. Nauk SSSR **30** 299 (1941) (in Russian) (Proc. Roy. Soc. London **A434** 9 (1991)); "Dissipation of energy in locally isotropic turbulence", Dokl. Akad. Nauk SSSR **32** 19 (1941) (in Russian) (Proc. Roy. Soc. Lond. **A434** 15 (1991)); A. M. Obukhov, "On the distribution of energy in the spectrum of turbulent flow", Dokl. Akad. Nauk SSSR **32** 22 (1941) (in Russian).

[5] Z.-S. She, E. Jackson and S. A. Orszag, "Structure and dynamics of homogeneous turbulence: models and simulations", Proc. Roy. Soc. London **A434**

the Navier-Stokes equation smooth and stable solutions? What is the appropriate representation of a turbulent field of velocities?[6]

The dynamics of the vorticity has enjoyed much interest.[7] The results depend on model assumptions. Dynamical-system concepts and statistical models have been invoked in studies of turbulence, with chaotic behaviour, intermittency and coherent structures.[8] In particular, by analogy with the quantum turbulence, "Turbulent flows may be regarded as an intricate collection of mutually-interacting vortices", and "Vortex filaments may thus be seen as the fundamental structure of turbulence,... ".[9]

The difficulties exhibited by the Navier-Stokes equation are related to the viscosity, which governs the vorticity, and the inertial term, which is quadratic in velocity. We show herein that the viscosity term in the Navier-Stokes equation may produce vorticity, provided the fluid is incompressible and the flow is isentropic. Although such an

---

101 (1991); J. Jimenez, A. A. Wray, P. G. Saffman and R. S. Rogalo, "The structure of intense vorticity in isotropic turbulence", J. Fluid Mech. **255** 65 (1993); C. R. Doering and J. D. Gibbon, *Applied Analysis of the Navier-Stokes Equations*, Cambridge University Press, Cambridge (1995); J. D. Gibbon, D. A. Donzis, A. Gupta, R. M. Kerr, R. Pandit and D. Vincenzi, "Regimes of nonlinear depletion and regularity in the 3D Navier-Stokes equations", Nonlinearity **27** 2605 (2014).

[6] J. Leray, "Sur le mouvement d'un liquide viscueux emplissant l'espace", Acta Mathematica **63** 193 (1934).

[7] P. G. Saffman, *Vortex Dynamics*, Cambridge University Press, Cambridge (1992); J.-Z. Wu, H.-Y. Ma and M.-D. Zhou, *Vorticity and Vortex Dynamics*, Springer, NY (2006).

[8] J. I. Polanco, N. P. Mueller and G. Krstulovic, "Vortex clustering, polarization and circulation intermittency in classical and quantum turbulence", Nature Commun. **12**:7090 (2021) (1-11); N. P. Mueller, J. I. Polanco and G. Krstulovic, "Intermittency of velocity circulation in quantum turbulence", Phys. Rev. **X11** 011053 (2021); G. B. Apolinario, L. Moriconi, R. M. Pereira and V. J. Valadao, "Vortex gas modelling of turbulent circulation statistics", Phys. Rev. **E102** 041102(R) (2020); A. A. Migdal, "Loop equation and area law in turbulence", Int. J. Mod. Phys. **A09** 1197 (1994); U. Frisch, "Fully developed turbulence and intermittency", in *Turbulence and Predictability in Geophysical Fluid Dynamics and Climate Dynamics*, eds. M. Ghil, R. Benzi and G. Parisi, Proc. Int. School of Physics Enrico Fermi, Varenna, North-Holland (1985).

[9] J. I. Polanco, N. P. Mueller and G. Krstulovic, "Vortex clustering, polarization and circulation intermittency in classical and quantum turbulence", Nature Commun. **12**:7090 (2021) (1-11).

approximate treatment may look reasonable, we can see that, rigorously speaking, the fluids cannot exhibit vorticity. Moreover, we give arguments that the vortices are unstable.

Further, we show herein that large variations of the velocity over short distances lead to highly fluctuating, swirling instabilities, controlled by viscosity. This is characteristic for the phenomenon of a fully developed turbulence. In this case, the inertial term acquires a major role in describing the flow. We represent a fully developed turbulence as a superposition of fluctuating velocities, associated to a discrete set of turbulence centres. A mean flow may be included. In general, the Navier-Stokes equation, averaged over fluctuations, is not satisfied. On the other hand, a homogeneous and isotropic distribution of (non-singular) turbulence centres leads to vanishing averages of velocity and inertial term, such that the Navier-Stokes equation is satisfied trivially. If the turbulence centres are singular, we are left with a gas of singularities (or a solution of singularities in the background fluid), which is in quasi thermal equilibrium. The corresponding Navier-Stokes equation for the fluid of singularities is reduced to Newton's equation of motion, with a small friction.

We illustrate the above descripton with three examples of vorticial liquids (filamentary liquid, coulombian and dipolar liquid).

5.2 Potential flow. Incompressible fluid

Let us consider a potential flow of an incompressible fluid. The velocity $\boldsymbol{v} = grad\Phi$ is given by the gradient of a potential Φ, which satisfies the Laplace equation

$$\Delta\Phi = 0 \tag{5.1}$$

(incompressiblity condition $div\boldsymbol{v} = 0$). The viscosity term $\sim \Delta\boldsymbol{v}$ is zero, such that we are left with Euler's equation

$$\frac{\partial \boldsymbol{v}}{\partial t} + (\boldsymbol{v}grad)\boldsymbol{v} = -\frac{1}{\rho}gradp \ , \tag{5.2}$$

where ρ is the density and p denotes the pressure. By using the well-known identity

$$(\boldsymbol{v}grad)\boldsymbol{v} = -\boldsymbol{v} \times curl\boldsymbol{v} + grad(v^2/2) \ , \tag{5.3}$$

equation (5.2) becomes

$$\frac{\partial \boldsymbol{v}}{\partial t} + grad(v^2/2 + p/\rho) = 0 \ , \tag{5.4}$$

where $curl\boldsymbol{v} = 0$. As it is well known, this equation leads to

$$\frac{\partial \Phi}{\partial t} + \frac{1}{2}(grad\Phi)^2 + \frac{p}{\rho} = 0 \ . \tag{5.5}$$

In this equation p should be viewed as the variation of the pressure with respect to equilibrium. We assume that p does not depend on Φ and the time.[10]

In equations (5.1) and (5.5) the variables may be separated. Let $g(\boldsymbol{r})$ be a solution of equation (5.1) (satisfying the boundary conditions); the potential can be written as $\Phi = f(t)g(\boldsymbol{r})$, where the function $f(t)$ satisfies equation (5.5),

$$\frac{df}{dt} + \frac{1}{2g}f^2(gradg)^2 + \frac{p}{\rho g} = 0 \ . \tag{5.6}$$

The acceptable solution of this equation (for $f(0) = 0$) is

$$f(t) = \frac{\sqrt{2 \mid p \mid /\rho}}{\mid gradg \mid} \tanh \frac{\sqrt{\mid p \mid /2\rho} \mid gradg \mid}{g} t \tag{5.7}$$

for $p < 0$. It may happen that the boundary conditions for the equation $\Delta\Phi = 0$ depend on time, thus providing the time derivative $\dot{\Phi}$; in that case equation (5.5) gives the pressure.

[10] Phenomena like gravity waves (as well as the sound) are left aside in this treatment.

5.3 Potential flow. Compressible fluid

Let us write down the equations of the fluid mechanics

$$\begin{aligned}
&\frac{\partial\rho}{\partial t}+\rho div\boldsymbol{v}+\boldsymbol{v}grad\rho=0\ ,\\
&\rho\frac{\partial\boldsymbol{v}}{\partial t}+\rho(\boldsymbol{v}grad)\boldsymbol{v}=-gradp-\rho grad\varphi+\\
&+\eta\Delta\boldsymbol{v}+\left(\tfrac{1}{3}\eta+\zeta\right)grad\,div\boldsymbol{v}\ ,\\
&\rho T\left(\frac{\partial s}{\partial t}+\boldsymbol{v}grads\right)=\kappa\Delta T+\sigma'_{ij}\partial_j v_i\ ,
\end{aligned}\tag{5.8}$$

where p is the internal pressure, φ is an external potential, η and ζ are the viscosity coefficients, T is the temperature, s is the entropy per unit masss, κ is the thermoconductivity and

$$\sigma'_{ij}=\eta\left(\partial_i v_j+\partial_j v_i-\frac{2}{3}\delta_{ij}div\boldsymbol{v}\right)+\zeta\delta_{ij}div\boldsymbol{v}\ . \tag{5.9}$$

is the viscosity tensor. In the Navier-Stokes equation (the second equation (5.8)) the forces which determine the velocity are $gradp$ and $\rho grad\varphi$, where $p=p(\rho,T)$ is a function of density and temperature. For all the usual flows the relative variations $\delta\rho/\rho_0$ of the density, $\delta T/T_0$ of the temperature, $\delta p/p_0$ of the pressure, $\delta s/s_0$ of the entropy as well as the variation $\delta\varphi/\varphi_0$ of the external potential are small, in comparison with their equilibrium values, labelled by the suffix 0, when the fluid is at rest. Consequently, we may view the velocity $\boldsymbol{v}$ as a first-order quantity, and linearize the above equations as

$$\begin{aligned}
&\frac{\partial\rho}{\partial t}+\rho_0 div\boldsymbol{v}=0\ ,\\
&\rho_0\frac{\partial\boldsymbol{v}}{\partial t}=-gradp-\rho_0 grad\varphi+\\
&+\eta\Delta\boldsymbol{v}+\left(\tfrac{1}{3}\eta+\zeta\right)grad\,div\boldsymbol{v}\ ,\\
&\rho_0 T_0\frac{\partial s}{\partial t}=\kappa\Delta T\ .
\end{aligned}\tag{5.10}$$

We note that within this approximation there is no heat source, and the first-order equation of energy conservation is reduced to an identity.

The density and entropy variations can be written as

$$\begin{aligned} \delta\rho &= \tfrac{\rho_0}{K}\delta p - \beta\rho_0\delta T \ , \\ \delta s &= -\tfrac{\beta}{\rho_0}\delta p + \tfrac{c_p}{T_0}\delta T \ , \end{aligned} \tag{5.11}$$

where K is the isothermal modulus of compressibility ($1/K = -\frac{1}{V}(\partial V/\partial p)_T$, $V = 1/\rho$), $\beta = \frac{1}{V}(\partial V/\partial T)_p$ is the dilatation coefficient and c_p is the specific heat per unit mass at constant pressure, all at equilibrium (see the Ideal Fluids chapter). In deriving equations (5.11) the Gibbs free energy $d\Phi = Vdp - sdT$ is used.

Part of the temperature variation in equation (5.11) is compensated by pressure variation, as in an adiabatic process; we denote this contribution by δT_1. The remaining part, denoted by δT, corresponds to the conducted heat. Therefore, we write

$$\begin{aligned} \delta\rho &= \tfrac{\rho_0}{K}\delta p - \beta\rho_0\delta T_1 - \beta\rho_0\delta T \ , \\ \delta s &= -\tfrac{\beta}{\rho_0}\delta p + \tfrac{c_p}{T_0}\delta T_1 + \tfrac{c_p}{T_0}\delta T = \tfrac{c_p}{T_0}\delta T \ , \end{aligned} \tag{5.12}$$

whence

$$\frac{\beta}{\rho_0}\delta p = \frac{c_p}{T_0}\delta T_1 \tag{5.13}$$

and

$$\delta\rho = \frac{\rho_0}{K}\left(1 - \frac{\beta^2 T_0 K}{\rho_0 c_p}\right)\delta p - \beta\rho_0\delta T \ . \tag{5.14}$$

In this equation we use the thermodynamic relation

$$\frac{\beta^2 T_0 K}{\rho_0} = c_p - c_v \ , \tag{5.15}$$

where c_v is the specific heat per unit mass at constant volume (see Appendix). Therefore, equation (5.14) becomes

$$\delta\rho = \frac{\rho_0 c_v}{K c_p}\delta p - \beta\rho_0\delta T \ , \tag{5.16}$$

or

$$\delta p = \frac{K c_p}{\rho_0 c_v}\delta\rho + \frac{\beta K c_p}{c_v}\delta T \ . \tag{5.17}$$

Now we use two other thermodynamic relations

$$\frac{c_p}{c_v}K = K_{ad} \ , \ \beta K = \alpha \ , \tag{5.18}$$

where K_{ad} is the adiabatic modulus of compressibility and $\alpha = (\partial p/\partial T)_V$ is the thermal pressure coefficient (see Appendix and the Ideal Fluids chapter). Finally, we get

$$\delta p = \frac{K_{ad}}{\rho_0}\delta\rho + \frac{c_p}{c_v}\alpha\delta T \ , \tag{5.19}$$

which is used in the Navier-Stokes equation. Equations (5.10) become

$$\begin{gathered}\frac{\partial \rho}{\partial t} + \rho_0 div\boldsymbol{v} = 0 \ , \\ \rho_0\frac{\partial \boldsymbol{v}}{\partial t} = -\frac{K_{ad}}{\rho_0}grad\rho - \rho_0 grad\varphi - \frac{c_p}{c_v}\alpha gradT + \\ +\eta\Delta\boldsymbol{v} + \left(\frac{1}{3}\eta + \zeta\right) grad\, div\boldsymbol{v} \ , \\ \rho_0 c_p\frac{\partial T}{\partial t} = \kappa\Delta T \ ,\end{gathered} \tag{5.20}$$

where the second equation (5.12) is used. We can see that the temperature equation (5.20) is independent; it describes the transport of an external temperature, which may provide a source for the velocity in the Navier-Stokes equation. We may leave aside this external temperature.

Let us seek a potential-flow solution of the above equations, where the velocity is derived from a potential Φ, by

$$\boldsymbol{v} = grad\Phi \ . \tag{5.21}$$

We notice that $curl\boldsymbol{v} = 0$ and $curl\, curl\boldsymbol{v} = 0$, *i.e.* $\Delta\boldsymbol{v} = grad\, div\boldsymbol{v}$. Therefore, the Navier-Stokes equation can be written as

$$\begin{gathered}\rho_0\frac{\partial \boldsymbol{v}}{\partial t} = -\frac{K_{ad}}{\rho_0}grad\rho - \rho_0 grad\varphi + \\ + \left(\frac{4}{3}\eta + \zeta\right) grad\, div\boldsymbol{v} \ .\end{gathered} \tag{5.22}$$

By using equation (5.21) in this equation and the continuity equation (first equation (5.20)), we obtain

$$\begin{aligned} &\frac{\partial \rho}{\partial t} + \rho_0 \Delta\Phi = 0 \ , \\ &\frac{\partial \Phi}{\partial t} + \frac{K_{ad}}{\rho_0^2}\rho + \varphi - \frac{1}{\rho_0}\left(\frac{4}{3}\eta + \zeta\right)\Delta\Phi = 0 \ , \end{aligned} \tag{5.23}$$

up to a function of time, where ρ and φ should be viewed as their corresponding variations. By an additional time differentiation we obtain

$$\frac{\partial^2 \Phi}{\partial t^2} - \frac{K_{ad}}{\rho_0}\Delta\Phi + \dot{\varphi} - \frac{1}{\rho_0}\left(\frac{4}{3}\eta + \zeta\right)\Delta\dot{\Phi} = 0 \ . \tag{5.24}$$

This equation provides the potential Φ, therefore the velocity $\boldsymbol{v}$ through equation (5.21) and the density ρ through the first equation (5.23). Equation (5.24) is the wave equation with friction (the term $\sim \Delta\dot{\Phi}$) and sources ($-\dot{\varphi}$). The ratio K_{ad}/ρ_0 is the square of the sound velocity $c = \sqrt{K_{ad}/\rho_0}$. The elementary solutions $e^{-i\omega t}e^{i\boldsymbol{kr}}$ of this (homogeneous) equation are damped plane waves

$$e^{\mp ickt}e^{i\boldsymbol{kr}}e^{-\frac{\sigma k^2}{c}t} \ , \tag{5.25}$$

for $\sigma k \ll c$, where $\sigma = (4\eta/3 + \zeta)/2\rho_0$. The relaxation time is much longer than the wave period. A wave propagating along the x-direction is proportional to $e^{-\gamma x}$, with the attenuation coefficient $\gamma = \sigma k^2/c = \sigma\omega^2/c^3$. This coincides with the result given in the Sound chapter (without the κ-contribution).

5.4 Vorticity

Euler's equation for an ideal fluid can be written as

$$\frac{d\boldsymbol{v}}{dt} = -grad w \ , \tag{5.26}$$

where w is the enthalpy ($dw = \frac{1}{\rho}dp$); the pressure p is a function of density ρ. By taking the *curl*, we get

$$curl\frac{d\boldsymbol{v}}{dt} = 0 \ . \tag{5.27}$$

On the other hand,

$$\frac{d\boldsymbol{v}}{dt} = \frac{\partial \boldsymbol{v}}{\partial t} + (\boldsymbol{v}grad)\boldsymbol{v} = \frac{\partial \boldsymbol{v}}{\partial t} + \left(\frac{\partial \boldsymbol{v}}{\partial t}\right)_f \quad , \tag{5.28}$$

where the suffix f indicates that the derivative is taken along the flow. Equation (5.27) becomes

$$\frac{\partial}{\partial t} curl\boldsymbol{v} + \left(\frac{\partial}{\partial t} curl\boldsymbol{v}\right)_f = 0 \ ; \tag{5.29}$$

since the two variations of the $curl\boldsymbol{v}$ are independent, we get

$$curl\boldsymbol{v} = 0 \quad , \tag{5.30}$$

i.e. the vorticity $curl\boldsymbol{v}$ is conserved along the flow. Therefore, we cannot create, or destroy, vorticity $curl\boldsymbol{v}$ in the flow of an ideal fluid. Equation (5.30) is valid in the absence of special external force, which do not derive from a gradient. This is Helmholtz's circulation law. As it is well known, an ideal fluid supports only an irrotational (potential) flow, where the velocity is derived from a scalar potential ($\boldsymbol{v} = grad\Phi$). By knowing the equation of state of the fluid, Euler's equation and the continuity equation are fully determined.

For a real, viscid, fluid the Navier-Stokes equation is

$$\rho\frac{d\boldsymbol{v}}{dt} = -gradp + \eta\Delta\boldsymbol{v} + \left(\frac{1}{3}\eta + \zeta\right) grad\, div\boldsymbol{v} \quad , \tag{5.31}$$

where η, ζ are the viscosity coefficients. By taking the $curl$, we get

$$\begin{aligned} curl\left(\rho\tfrac{d\boldsymbol{v}}{dt}\right) &= grad\rho \times \tfrac{d\boldsymbol{v}}{dt} + \rho curl\tfrac{d\boldsymbol{v}}{dt} = \\ &= \eta\Delta curl\boldsymbol{v} \ ; \end{aligned} \tag{5.32}$$

we can see that the viscosity η can generate vorticity ($curl\boldsymbol{v} \neq 0$). In general, the velocity $\boldsymbol{v}$ rotates about the vorticity $curl\boldsymbol{v}$. This is a vortex.

Equation (5.32) is in conflict with the continuity equation

$$\frac{d\rho}{dt} + \rho div\boldsymbol{v} = 0 \ . \tag{5.33}$$

Indeed, if $curl\boldsymbol{v} \neq 0$, the velocity should be derived from a *curl* (not from a *grad*!), *i.e.* we should have $\boldsymbol{v} = curl\boldsymbol{A}$, where $\boldsymbol{A}$ is a vector potential. Consequently, the fluid should be incompressible ($div\boldsymbol{v} = div\, curl\boldsymbol{A} = 0$). The density should be constant, both in time and space (along the flow). This indicates that in a compressible fluid we cannot have vortices. Usually, the variations of the density are small, such that they may be neglected for the present purpose.

Therefore, we may limit to an incompressible fluid ($div\boldsymbol{v} = 0$), for which equation (5.32) becomes

$$curl\frac{d\boldsymbol{v}}{dt} = \nu\Delta curl\boldsymbol{v} \ , \tag{5.34}$$

or

$$\frac{\partial}{\partial t}curl\boldsymbol{v} - curl\,(\boldsymbol{v} \times curl\boldsymbol{v}) = \nu\Delta curl\boldsymbol{v} \ , \tag{5.35}$$

where $\nu = \eta/\rho$ is the kinematical viscosity and we have used the identity $(\boldsymbol{v}grad)\boldsymbol{v} = -\boldsymbol{v} \times curl\boldsymbol{v} + grad(v^2/2)$. This is the equation of vorticity; it can also be written as

$$\frac{\partial}{\partial t}curl\boldsymbol{v} + curl\left[(\boldsymbol{v}grad)\boldsymbol{v}\right] = \nu\Delta curl\boldsymbol{v} \ . \tag{5.36}$$

In principle, this equation gives the velocity. From the Navier-Stokes equation we get the pressure. If we write the Navier-Stokes equation as

$$\frac{\partial}{\partial t}curl\boldsymbol{A} + (\boldsymbol{v}grad)\boldsymbol{v} = -\frac{1}{\rho}gradp + \nu\Delta curl\boldsymbol{A} \ , \tag{5.37}$$

we can see that

$$div\left[(\boldsymbol{v}grad)\boldsymbol{v} + \tfrac{1}{\rho}gradp\right] = 0 \ , \tag{5.38}$$

and, of course,

$$div\left(\frac{\partial\boldsymbol{v}}{\partial t} - \nu\Delta\boldsymbol{v}\right) = 0 \ , \tag{5.39}$$

which is an identity ($div\boldsymbol{v} = 0$, $\boldsymbol{v} = curl\boldsymbol{A}$). In some cases, particular solutions are provided by

$$\begin{aligned} (\boldsymbol{v}grad)\boldsymbol{v} &= -grad(p/\rho) \ , \\ \tfrac{\partial\boldsymbol{v}}{\partial t} &= \nu\Delta\boldsymbol{v} \ . \end{aligned} \tag{5.40}$$

We can see that the Navier-Stokes equation is split into (the derivatives of) a diffusion (heat) equation and an equilibrium equation; the first equation (5.40) indicates an equilibrium between the pressure force $-grad(p/\rho)$ and Euler's force $(\boldsymbol{v}grad)\boldsymbol{v}$. The diffusion equation (5.40) holds also for the vorticity, because the above equations are valid for a non-vanishing vorticity. It is easy to see that these equations generalize the equations for the Couette flow.

However, we have also the heat-transfer equation. For an incompressible fluid it reads

$$\rho c_p \frac{dT}{dt} = \kappa \Delta T + \frac{1}{2}\eta \left(\partial_i v_j + \partial_j v_i\right)^2 \ , \tag{5.41}$$

or

$$\frac{dT}{dt} = \chi \Delta T + \frac{1}{2}\frac{\nu}{c_p} \left(\partial_i v_j + \partial_j v_i\right)^2 \ , \tag{5.42}$$

where c_p is the specific heat per unit mass at constant pressure, κ is the thermoconductivity, and $\chi = \kappa/\rho c_p$ is the thermometric conductivity. For an incompressible fluid this equation can be transformed into an equation for the derivatives of the pressure,

$$\frac{dp}{dt} = \chi \Delta p + \frac{1}{2}\frac{\alpha\nu}{c_p} \left(\partial_i v_j + \partial_j v_i\right)^2 \ , \tag{5.43}$$

where $\alpha = (\partial p/\partial T)_v$ is the thermal pressure coefficient. In general, equation (5.43) is not compatible with the Navier-Stokes equation (or equation (5.38)). Therefore, rigorously speaking, we cannot have vorticity in an incompressible fluid either. Usually, the coefficient ν/c_p is very small (of the order $10^{-24} - 10^{-25} g \cdot cm^2/s$), such that, for small gradients of velocity, we have a low rate of entropy production (though, a factor of the order $10^{23} K/erg$ should be taken into account). Under these conditions, we may assume that the flow is isentropic and the heat-transfer equation may be neglected.

In order to get an idea of how large the variations of the density and the temperature can be, we may estimate a change δp in pressure from $\delta p \simeq \rho v^2$. A velocity $v = 100 km/h$, which is fairly large, produces a change $\delta p \simeq 10^4 dyn/cm^2$ in air ($\rho = 10^{-3} g/cm^3$), whose normal pressure is $10^6 dyn/cm^2$; therefore, $\delta p/p \simeq 10^{-2}$. Such a velocity ($\simeq 3 \times 10^3 cm/s$) is close to the mean thermal velocity $\simeq 10^4 cm/s$

(for normal air), and close to the sound velocity in normal air $c \simeq 3.5 \times 10^4 cm/s$. For this velocity we still expect local thermal equilibrium. The change in density is given by $\delta p = K(\delta\rho/\rho)$, where $K = -V(\partial p/\partial V)$ is the (say, isothermal) modulus of compression. For air $K \simeq 10^6 dyn/cm^2$, for water $K \simeq 10^{10} dyn/cm^2$, such that we get $\delta\rho/\rho \simeq 10^{-2}$, 10^{-6}. The change in temperature is obtained from $\delta p = \alpha T(\delta T/T)$, where $\alpha = (\partial p/\partial T)_V$ is the thermal pressure coefficient (at constant volume V). For water $\alpha \simeq 10^{22}/cm^3$, for gases it is much higher; for normal temperature $T = 300K$ we get $\delta T/T \simeq 10^{-4}$, or much lower. Consequently, we may expect an almost ideal, incompressible flow. We note that, although we neglect the viscosity in the heat-transfer equation, we keep it in the Navier-Stokes equation.

Therefore, within these approximations (incompressibility and constant entropy), we are left with equation (5.36) and the Navier-Stokes equation for a vorticial flow. In general, an external pressure which satisfies equation (5.38) (or the Navier-Stokes equation) is very special, such that, if they exist, the vortices might be, in fact, unstable. They develop an Euler's force, which is difficult to be compensated by an external force.

We note that under the conditions stated above, the viscosity may generate (unstable) vortices. In the next section we show that the viscosity may generate another type of instabilities.

For small variations of the velocity we may neglect the inertial term in the vorticity equation (5.36), which becomes

$$\frac{\partial}{\partial t} curl\boldsymbol{v} = \nu \Delta curl\boldsymbol{v} \ . \tag{5.44}$$

By making use of $\boldsymbol{v} = curl\boldsymbol{A}$, we get

$$(\Delta - grad\,div)\left(\frac{\partial \boldsymbol{A}}{\partial t} - \nu\Delta\boldsymbol{A}\right) = 0 \ . \tag{5.45}$$

A solution of this equation is provided by

$$\frac{\partial \boldsymbol{A}}{\partial t} - \nu\Delta\boldsymbol{A} = 0 \ , \tag{5.46}$$

which leads to

$$\boldsymbol{A} = \boldsymbol{A}_0 e^{-\lambda\nu t} e^{\pm i\sqrt{\lambda} r}/r \ , \tag{5.47}$$

where $\boldsymbol{A_0}$ and λ are two constants. The velocity acquires the form

$$\boldsymbol{v} = -\boldsymbol{A}_0 \times grad\left(e^{-\lambda\nu t}e^{\pm i\sqrt{\lambda}r}/r\right) \ , \tag{5.48}$$

and the pressure is uniform within this approximation. We note that, although the spatial dependence of the solution does not depend on viscosity, it is generated by the viscosity term $\nu\Delta\boldsymbol{v}$.

5.5 Instabilities

The equation of energy conservation for an incompressible fluid is

$$\begin{gathered}\frac{\partial}{\partial t}\left(\frac{1}{2}\rho v^2\right) + div\left[\boldsymbol{v}\left(\frac{1}{2}\rho v^2 + p\right) - \frac{1}{2}\eta grad(v^2)\right] + \\ +\eta\left(\partial_j v_i\right)^2 = 0 \ ;\end{gathered} \tag{5.49}$$

it is obtained by multiplying by $\boldsymbol{v}$ the Navier Stokes equation (5.31) for an incompressible fluid ($div\boldsymbol{v} = 0$). The div-term represents a transport of energy and mechanical work of the pressure, and an energy flux associated with collisions (viscosity); the term $\eta\left(\partial_j v_i\right)^2$ represents the heat produced by viscosity. We integrate this equation over a volume V enclosed by a surface S,

$$\begin{gathered}\frac{\partial}{\partial t}\int dV\left(\frac{1}{2}\rho v^2\right) + \oint dS\left[v_n\left(\frac{1}{2}\rho v^2 + p\right) - \frac{1}{2}\eta\partial_n(v^2)\right] + \\ +\eta\int dV\left(\partial_j v_i\right)^2 = 0 \ ,\end{gathered} \tag{5.50}$$

where v_n is the velocity component normal to the surface and ∂_n is the derivative along the normal to the surface.

We compare the orders of magnitude of the surface terms and the η-volume term, and get ratios of the form $\frac{Sl}{V}R$, $\frac{Sl}{V}(p/\rho v^2)R$, $\frac{Sl}{V}$, where l is the distance over which the velocity varies and $R = vl/\nu$ is the Reynolds number. For moderate Reynolds numbers and $Sl/V \ll 1$ we can neglect the surface contributions in comparison with the heat term. By writing

$$\boldsymbol{v} = f(t)\boldsymbol{u}(\boldsymbol{r}) \ , \tag{5.51}$$

the above equation becomes

$$\frac{\partial}{\partial t} f^2 \cdot \int dV \left(\frac{1}{2}\rho u^2\right) + \eta f^2 \int dV \left(\partial_j u_i\right)^2 = 0 \ . \tag{5.52}$$

We can see that the time dependence of the velocity is a damped exponential. The flow is stable, as a consequence of the dissipated heat. The η-term in equation (5.52) gives, in fact, the increase of entropy. The total energy is conserved, while the entropy increases and the mechanical energy decreases.

Let us assume that the integration domains are sufficiently small, such that Sl/V is of the order of unity; the velocity varies over a distance l inside the domains, but we assume that it suffers a large discontinuity across the surface, over a small distance $\delta \ll l$. Then, it is easy to see that the dominant term in equation (5.50) is the collision term, such that equation (5.50) becomes

$$\frac{\partial}{\partial t} f^2 \cdot \int dV \left(\frac{1}{2}\rho u^2\right) - \frac{1}{2}\eta f^2 \oint dS \partial_n (u^2) = 0 \ . \tag{5.53}$$

We can see that for a positive normal derivative the flow is unstable. The viscosity is insufficient to disipate the energy as heat, and the energy is transferred by molecular collisions (viscosity) through surfaces of discontinuities. The process occurs in small domains, with large discontinuities of velocity across their surface, and the instabilities imply returning, swirling and fluctuating, velocities. This is the turbulence phenomenon. We note that the instabilities are governed by viscosity, which gives also vorticity (when the entropy production is neglected). Moreover, we note that the inertial term does not appear in instabilities, though it plays an important role in turbulence.

The above arguments can be extended to compressible fluids, including the variations of the temperature. Indeed, the energy conservation in this case reads

$$\frac{\partial}{\partial t}\left(\frac{1}{2}\rho v^2 + \rho\varepsilon\right) = -\partial_j \left[\rho v_j \left(\frac{1}{2}v^2 + w\right) - v_i \sigma'_{ij} - \kappa \partial_j T\right] \ , \tag{5.54}$$

where ε is the internal energy per unit mass, w is the enthalpy per unit mass and

$$\sigma'_{ij} = \eta \left(\partial_i v_j + \partial_j v_i - \frac{2}{3}\delta_{ij} div \boldsymbol{v}\right) + \zeta \delta_{ij} div \boldsymbol{v} \tag{5.55}$$

is the viscosity tensor. For a smooth flow and a sufficiently large volume the surface term in equation (5.54) can be neglected, and the energy is conserved, as it is well known. However, in the surface integral we have terms of the form

$$\oint dS \left[\eta v_i \left(\partial_i v_n + \partial_n v_i\right) - \left(\frac{2}{3}\eta - \zeta\right) v_n div\boldsymbol{v} + \kappa \partial_n T\right] \quad , \tag{5.56}$$

which imply normal derivatives to the surface, both of velocity and temperature. By collecting these contributions, we get

$$\oint dS \left[\eta \partial_n (v^2/2) + \left(\frac{1}{3}\eta + \zeta\right) \partial_n (v_n^2/2) + \kappa \partial_n T\right] \quad . \tag{5.57}$$

We can see that for large normal derivatives across the surface, both for velocity and temperature, these terms may lead to instabilities.

5.6 Turbulence

As it is well known, for a moderate turbulence, *i.e.* for slowly varying fluctuations, we may decompose the velocity field into a mean velocity and a fluctuating part, and limit ourselves to the time averaged Navier-Stokes equation. This way we get the Reynolds equations, for which the mean energy is coupled to the fluctuating energy, via model assumptions. A fully developed turbulence exhibits highly-varying fluctuations, such that we need to consider the time-dependent Navier-Stokes equation.

The turbulent instabilities ocurring in a fully developed turbulence exhibit large variations of the velocity over small distances. In this case we may assume that the velocity is split into a mean-flow velocity $\boldsymbol{v}_0$ and a fluctuating part $\boldsymbol{v}$, where the mean-flow velocity $\boldsymbol{v}_0$ may have a slight time variation, while the fluctuating velocity $\boldsymbol{v}$ is a rapidly varying velocity. By using this decomposition for an incompressible fluid, the Navier-Stokes equation reads

$$\begin{gathered} \frac{\partial \boldsymbol{v}_0}{\partial t} + \frac{\partial \boldsymbol{v}}{\partial t} + \left(\boldsymbol{v}_0 grad\right) \boldsymbol{v}_0 + \left(\boldsymbol{v}_0 grad\right) \boldsymbol{v} + \left(\boldsymbol{v} grad\right) \boldsymbol{v}_0 + \\ + \left(\boldsymbol{v} grad\right) \boldsymbol{v} = -\frac{1}{\rho} grad p_0 - \frac{1}{\rho} grad p + \nu \Delta \boldsymbol{v}_0 + \nu \Delta \boldsymbol{v} \quad , \end{gathered} \tag{5.58}$$

where p_0 is the pressure corresponding to the main flow and p is the fluctuating part of the pressure. In this equation we have three distinct types of time variations, such that it should be viewed as three equations

$$\frac{\partial \boldsymbol{v}_0}{\partial t} + (\boldsymbol{v}_0 grad)\,\boldsymbol{v}_0 = -\frac{1}{\rho} grad p_0 + \nu\Delta\boldsymbol{v}_0 \ ,$$

$$\frac{\partial \boldsymbol{v}}{\partial t} + (\boldsymbol{v}_0 grad)\,\boldsymbol{v} + (\boldsymbol{v} grad)\,\boldsymbol{v}_0 = -\frac{1}{\rho} grad p + \nu\Delta\boldsymbol{v} \ , \tag{5.59}$$

$$(\boldsymbol{v} grad)\,\boldsymbol{v} = 0 \ ;$$

similarly, the continuity equation should be split into

$$div\boldsymbol{v}_0 = 0 \ , \ div\boldsymbol{v} = 0 \ . \tag{5.60}$$

The first equation (5.59) is an independent equation, which gives the main flow velocity $\boldsymbol{v}_0$. Since the main part of the velocity is taken by the fluctuating velocity, we may neglect the quadratic term in this equation. The energy conservation and the heat transfer for this equation are given by

$$\frac{\partial}{\partial t}(v_0^2/2) = -\partial_i \left[v_{0i}\left(p_0/\rho + v_0^2/2\right) - \nu\partial_i(v_0^2/2)\right] -$$

$$-\nu(\partial_i v_{0j})^2 \ , \tag{5.61}$$

$$T_0\frac{ds_0}{dt} = \chi\Delta T_0 + \nu(\partial_i v_{0j})^2 \ ,$$

where T_0 is the temperature of the main flow, s_0 is the entropy per unit mass of the main flow and χ is the thermometric conductivity.

Having solved the mean-flow equation we can pass to solve the second equation (5.59) for the fluctuating velocity $\boldsymbol{v}$, with $\boldsymbol{v}_0$ as a parameter. This equation has its own energy-conservation and heat-transfer equations. We note that the temperature and the entropy of the mean flow are different from the temperature and the entropy of the fluctuating part of the flow, which means that the two components of the flow (the mean flow and the fluctuating flow) are not in thermal equilibrium. Indeed, if we multiply the first equation (5.59) by $\boldsymbol{v}$ and the second equation (5.59) by $\boldsymbol{v}_0$, we get cross-terms of the form $T_0\frac{ds}{dt} + T\frac{ds_0}{dt}$ in the heat-transfer equation, where T and s are the temperature and the entropy of the fluctuating flow. This indicates a heat exchange between the two components of the flow.

We are left with the third equation (5.59), which, in general, is not satisfied. We conclude that the fully developed turbulence does not satisfy the Navier-Stokes equation. The quadratic term of the third equation (5.59) is equivalent to a rapidly varying internal force (Euler's force), which cannot be compensated by any physical external force. The fully developed turbulence is unstable.

Under these conditions it is reasonable to be interested in time averaged quantities. Then, the second equation (5.59) is zero and the fluctuating part of the flow is reduced to

$$\overline{(\boldsymbol{v} grad)\, \boldsymbol{v}} = 0 \ . \tag{5.62}$$

Since $div\boldsymbol{v} = 0$, the components of the velocity $\boldsymbol{v}$ are not independent. In general, equation (5.62) is not satisfied, which means that the turbulent motion is unstable even on average. The above decomposition differentiates between the mean flow and the fluctuating flow, which, after time averaging, is reduced to equation (5.62). We note that a similar decomposition is valid for a compressible fluid, as long as the velocity, density and entropy fluctuations are independent of one another.

Since a fully developed turbulence originates in large variations of the velocity across small distances, it is reasonable to associate these variations with a discrete distribution of positions $\boldsymbol{r}_i$,which we call centres of turbulence. Further on, we assume that this is a homogeneous and isotropic distribution, such that we may write the velocity field as

$$\boldsymbol{v} = \sum_i \boldsymbol{v}_i(t, R_i) \ , \tag{5.63}$$

where $\boldsymbol{R}_i = \boldsymbol{r} - \boldsymbol{r}_i$. If $curl\boldsymbol{v}_i \neq 0$, this velocity field represents a vorticial liquid, which is unstable. We assume that the fluctuating velocities are independent at distinct positions, *i.e.* $\overline{\boldsymbol{v}_i} = 0$ and $\overline{\boldsymbol{v}_i \boldsymbol{v}_j} \sim \delta_{ij}$. Equation (5.62) becomes

$$\overline{(\boldsymbol{v} grad)\, \boldsymbol{v}} = \sum_i \overline{(\boldsymbol{v}_i \boldsymbol{R}_i / R_i)\, (d\boldsymbol{v}_i / dR_i)} \ . \tag{5.64}$$

The conditions of homogeneity and isotropy imply that the velocities $\boldsymbol{v}_i$ in this equation can be replaced by the same function $\boldsymbol{u}$. For a sufficiently dense set of positions $\boldsymbol{r}_i$ we can define a density ρ_v of such points, which is a constant. Then, equation (5.64) can be transformed into the integral

$$\overline{(\boldsymbol{v}grad)\,\boldsymbol{v}} = \rho_v \int dR \cdot R^2 \int d\overline{o(\boldsymbol{uR}/R)\,(d\boldsymbol{u}/dR)} \ . \qquad (5.65)$$

If the radial integral is finite, the result of integration in the above equation is zero, due to the integration over the solid angle o, such that the term $\overline{(\boldsymbol{v}grad)\,\boldsymbol{v}}$ is zero. In this case we can say that the Navier-Stokes equation is trivially satisfied on average, being reduced to the mean-flow equation (first equation (5.59)).

If the radial integral in equation (5.65) is singular for $\boldsymbol{R} = 0$, as it may often happen for vortices, we are left with a discrete set of singularities, extending over a small characteristic distance a, where the singularity is

$$\boldsymbol{u} \sim (a/R)^n \ , \ n > 1 \ . \qquad (5.66)$$

In each of these regions there exists a mass M of fluid, which can be carried by the background fluid and, at the same time, they may have their own motion.

The averaged energy-conservation equation derived from the second equation (5.59),

$$\frac{1}{2}\boldsymbol{v}_0 grad\overline{v^2} + \partial_j\left(\overline{v_i v_j} v_{0i}\right) - \frac{1}{2}\nu\Delta\overline{v^2} = -\nu\overline{(\partial_j v_i)^2} \ , \qquad (5.67)$$

shows that the fluctuating motion produces heat which is partly transported by the mean flow (the div-terms integrated over a volume are irrelevant). This dissipated heat should be compensated from the outside. Therefore, we are left with a quasi ideal classical gas of singular vortices, or a solution of vortices in the background fluid, in thermal quasi equilibrium. This is an example of emergent dynamics.[11]

[11] P. W. Anderson, "More is different", Science **177** 393 (1972).

5.7 Gas of singularities

Let us assume a homogeneous, isotropic, fluctuating distribution of singular centres of turbulence localized at $\boldsymbol{r}_i$ with mass M, as described above. Their density is

$$\rho = M \sum_i \delta(\boldsymbol{r} - \boldsymbol{r}_i) \tag{5.68}$$

and

$$\frac{\partial \rho}{\partial t} = -M \sum_i \boldsymbol{u}_i grad \delta(\boldsymbol{r} - \boldsymbol{r}_i) \ , \tag{5.69}$$

where $\boldsymbol{u}_i = d\boldsymbol{r}_i/dt$ is their velocity. The velocity field of the singularities is

$$\boldsymbol{u} = v \sum_i \boldsymbol{u}_i \delta(\boldsymbol{r} - \boldsymbol{r}_i) \ , \tag{5.70}$$

where v is the small volume over which the δ-function is localized, such that $v = a^3$ and $M = \rho v$. Let us compute

$$\begin{aligned} div(\rho \boldsymbol{u}) &= vM div \textstyle\sum_{ij} \delta(\boldsymbol{r} - \boldsymbol{r}_i) \boldsymbol{u}_j \delta(\boldsymbol{r} - \boldsymbol{r}_j) = \\ &= M div \textstyle\sum_i \boldsymbol{u}_i \delta(\boldsymbol{r} - \boldsymbol{r}_i) = \\ &= M \textstyle\sum_i \boldsymbol{u}_i grad \delta(\boldsymbol{r} - \boldsymbol{r}_i) = -\frac{\partial \rho}{\partial t} \ ; \end{aligned} \tag{5.71}$$

we can see that the continuity equation is satisfied.

Now, let us focus on the Navier-Stokes equation

$$\rho \frac{\partial \boldsymbol{u}}{\partial t} + \rho (\boldsymbol{u} grad) \boldsymbol{u} = -grad p + \eta \Delta \boldsymbol{u} \ , \tag{5.72}$$

and let us compute each term in this equation for our fluid of singularities. We have

$$\frac{\partial \boldsymbol{u}}{\partial t} = v \textstyle\sum_i \dot{\boldsymbol{u}}_i \delta(\boldsymbol{r} - \boldsymbol{r}_i) - v \sum_i \boldsymbol{u}_i (\boldsymbol{u}_i grad) \delta(\boldsymbol{r} - \boldsymbol{r}_i) \ . \tag{5.73}$$

The inertial term is

$$\begin{aligned} (\boldsymbol{u} grad) \boldsymbol{u} &= v^2 \textstyle\sum_{ij} \boldsymbol{u}_j \delta(\boldsymbol{r} - \boldsymbol{r}_i) (\boldsymbol{u}_i grad) \delta(\boldsymbol{r} - \boldsymbol{r}_j) = \\ &= v \textstyle\sum_i \boldsymbol{u}_i (\boldsymbol{u}_i grad) \delta(\boldsymbol{r} - \boldsymbol{r}_i) \ . \end{aligned} \tag{5.74}$$

By comparing equations (5.73) and (5.74), we can see that the inertial (Euler's) term disappears from equation. This is expected, since the δ-function equals the variable $\boldsymbol{r}$ to the function $\boldsymbol{r}_i(t)$, which amounts to Lagrange's approach. The term on the left in equation (5.72) becomes

$$\rho\frac{\partial \boldsymbol{u}}{\partial t} + \rho(\boldsymbol{u}grad)\boldsymbol{u} = M\sum_i \dot{\boldsymbol{u}}_i\delta(\boldsymbol{r}-\boldsymbol{r}_i) \ , \tag{5.75}$$

and equation (5.72) reads now

$$\begin{aligned} M\sum_i \dot{\boldsymbol{u}}_i\delta(\boldsymbol{r}-\boldsymbol{r}_i) = -v\sum_i p_i grad\delta(\boldsymbol{r}-\boldsymbol{r}_i)+ \\ +\eta v\sum_i \boldsymbol{u}_i\Delta\delta(\boldsymbol{r}-\boldsymbol{r}_i) \ , \end{aligned} \tag{5.76}$$

where p_i is the pressure at the position $\boldsymbol{r}_i$. The pressure term in equation (5.76) is a force per unit volume acting upon the vortex placed at $\boldsymbol{r}_i$; it may arise from the pressure exerted by the background fluid particles. This term may be written as

$$\sum_i \boldsymbol{f}_i\delta(\boldsymbol{r}-\boldsymbol{r}_i) \ , \tag{5.77}$$

where $\boldsymbol{f}_i$ is the force acting at $\boldsymbol{r}_i$. The factor $\Delta\delta(\boldsymbol{r}-\boldsymbol{r}_i)$ is of the order $-\frac{1}{a^2}\delta(\boldsymbol{r}-\boldsymbol{r}_i)$, where a is of the order of the dimension of the vortex ($v = a^3$); consequently, we may replace the viscosity terms in equation (5.76) by

$$-\frac{\eta v}{a^2}\sum_i \boldsymbol{u}_i\delta(\boldsymbol{r}-\boldsymbol{r}_i) \ . \tag{5.78}$$

A similar contribution brings the ζ-term. The equation of motion (5.76) describes a set of independent particles with mass M, subjected to an external force $\boldsymbol{f}_i$ and a friction force; the equation of motion of each such particle can be written as

$$M\dot{\boldsymbol{u}}_i = \boldsymbol{f}_i - \eta a\boldsymbol{u}_i \ , \tag{5.79}$$

which is Newton's law of motion. The damping coefficient caused by the friction force is very small, such that we can consider the ensemble of singularities as a (quasi) ideal classical gas of independent, identical, pointlike particles. Therefore, a (singular) fully developed turbulence may be viewed as the (quasi) thermodynamic equilibrium of such a gas (or a solution of singularities in the background fluid). We can define a temperature of turbulence, which is approximately the mean kinetic energy of the translational motion of a singularity. Also, we can estimate a chemical potential by evaluating $\overline{v^2}/2$. We note that the density ρ_v and the dimension a of the singularities remain undetermined; these parameters can be estimated from experiment.

Also, it is worth noting that we may consider the equations of the fluid mechanics for this new gas of singularities, viewed as a continuous medium, at a higher scale.

5.8 Vorticial liquids

5.8.1 Vortex

For an incompressible fluid with an isentropic flow we consider a velocity field given by

$$\boldsymbol{v} = \boldsymbol{\omega} \times gradf(r) \ , \tag{5.80}$$

where $\boldsymbol{\omega}$ is a vector which may depend only on the time and the function $f(r)$ is smooth everywhere, except, possibly, at the origin $\boldsymbol{r} \neq 0$, and vanishing rapidly at infinity. The velocity $\boldsymbol{v}$ rotates about $\boldsymbol{\omega}$; such a velocity field defines a vortex.

We can check that $div\boldsymbol{v} = 0$ and

$$\boldsymbol{v} = -curl\,[\boldsymbol{\omega} f(r)] \ , \tag{5.81}$$

such that we can define a vector potential $\boldsymbol{A} = -\boldsymbol{\omega} f(r)$ ($\boldsymbol{v} = curl\boldsymbol{A}$). We note that $div\boldsymbol{A} \neq 0$ and the vorticity differs from $\Delta\boldsymbol{A}$, in general,

$$curl\boldsymbol{v} = \Delta\,[\boldsymbol{\omega} f(r)] - grad\,div\,[\boldsymbol{\omega} f(r)] \neq -\Delta\boldsymbol{A} \ . \tag{5.82}$$

If the velocity given by equation (5.81) satisfies the second equation (5.40), we should have $\boldsymbol{\omega} \sim e^{-\nu\lambda t}$ and $\Delta f + \lambda f = 0$, *i.e.* $f \sim e^{\pm i\sqrt{\lambda}r}/r$, where λ is, in general, complex. In two dimensions f is a Bessel function. The Navier-Stokes equation gives the pressure.

The most common example of a velocity field given by equation (5.80) is the filamentary vortex (the "cyclon"), with $\boldsymbol{\omega} = const$, ($\lambda = 0$),

$f(r) = -\ln r$ and the two-dimensional position vector $\boldsymbol{r}$ perpendicular to $\boldsymbol{\omega}$. In this case $\boldsymbol{v} = \boldsymbol{r} \times \boldsymbol{\omega}/r^2$, $\boldsymbol{A} = \boldsymbol{\omega} \ln r$ and $curl\boldsymbol{v} = -2\pi\boldsymbol{\omega}\delta(\boldsymbol{r})$. The velocity is singular at $\boldsymbol{r} = 0$. For the filamentary vortex the Navier-Stokes equation can be satisfied for $p = -\rho\omega^2/2r^2$, *i.e.* for an external potential $\varphi = p/\rho = -\omega^2/2r^2$. This can be provided by the gravitational field for a fluid with a free surface.

For convenience, we give the expression of the various terms in the Navier-Stokes equation for the vortex given by equation (5.80),

$$\begin{aligned} \frac{\partial \boldsymbol{v}}{\partial t} &= (\dot{\boldsymbol{\omega}} \times \boldsymbol{r})\frac{f'}{r} , \\ (\boldsymbol{v}grad)\boldsymbol{v} &= \left[\boldsymbol{\omega}(\boldsymbol{\omega r}) - \omega^2\boldsymbol{r}\right]\frac{f'^2}{r^2} , \\ \Delta\boldsymbol{v} &= (\boldsymbol{\omega} \times \boldsymbol{r})\frac{(f'' + 2f'/r)'}{r} , \end{aligned} \tag{5.83}$$

where the primes denote the derivatives of the function f; the second equation (5.83) is derived by using the identity

$$(\boldsymbol{v}grad)\boldsymbol{v} = -\boldsymbol{v} \times curl\boldsymbol{v} + grad(v^2/2) \ . \tag{5.84}$$

For a filamentary vortex the above expressions become

$$\begin{aligned} (\boldsymbol{v}grad)\boldsymbol{v} &= -\omega^2\boldsymbol{r}\frac{f'^2}{r^2} , \\ \Delta\boldsymbol{v} &= (\boldsymbol{\omega} \times \boldsymbol{r})\frac{(f'' + f'/r)'}{r} . \end{aligned} \tag{5.85}$$

In general, the vortex given by equation (5.80) does not satisfy the Navier-Stokes equation; it develops internal (Euler's) forces which cannot be compensated; the vortex is unstable. Such vortices are examples of singular velocities.

5.8.2 Vorticial liquid

A set of vectors $\boldsymbol{\omega}_i$ placed at $\boldsymbol{r}_i$ form a vorticial liquid. The velocity field is

$$\boldsymbol{v} = \sum_i \boldsymbol{\omega}_i \times gradf_i(R_i) \ , \tag{5.86}$$

where $\boldsymbol{R}_i = \boldsymbol{r} - \boldsymbol{r}_i$. This velocity field is a superposition of independent vortices. Each i-th vortex develops an internal force, while the inertial term generates an interaction which brings an additional force; this additional force depends on all the other j-th vortices, $j \neq i$. Therefore, the Navier-Stokes equation is not satisfied by the velocity field given by equation (5.86), in the sense that there is no physical external pressure to compensate the Euler force.

We assume randomly fluctuating vectors $\boldsymbol{\omega}_i$, as a distinctive feature of a fully developed vorticial turbulence. It is likely that the fluid develops fluctuations as a reaction to its uncompensated internal forces. Specifically, we assume $\overline{\boldsymbol{\omega}_i} = 0$ and $\overline{\omega_i^\alpha \omega_j^\beta} = \frac{1}{3}\overline{\omega_i^2}\delta_{ij}\delta^{\alpha\beta}$, where α, $\beta = 1, 2, 3$ are the cartesian labels of the components of the vectors $\boldsymbol{\omega}_i$, and the overbar indicates the average over time. Then, the average velocity is zero ($\overline{\boldsymbol{v}} = 0$) and we are left with the inertial term

$$\overline{(\boldsymbol{v}grad)\boldsymbol{v}} = \overline{v_j \partial_j v_i} \ . \tag{5.87}$$

The calculation of this term is straightforward; we get

$$\begin{aligned} \overline{(\boldsymbol{v}grad)\boldsymbol{v}} = \tfrac{1}{3}\textstyle\sum_i \omega_i^2[\tfrac{1}{2}grad\,(gradf_i(R_i))^2 - \\ -gradf_i(R_i)\cdot\Delta f_i(R_i)] \ . \end{aligned} \tag{5.88}$$

The averaged Euler forces are

$$\begin{aligned} -\overline{\boldsymbol{v}\times curl\boldsymbol{v}} = -\tfrac{1}{3}\textstyle\sum_i \omega_i^2[\tfrac{1}{2}grad\,(gradf_i(R_i))^2 + \\ +gradf_i(R_i)\cdot\Delta f_i(R_i)] \ , \\ \overline{grad(v^2/2)} = \tfrac{1}{3}\textstyle\sum_i \omega_i^2 grad\,(gradf_i(R_i))^2 \ . \end{aligned} \tag{5.89}$$

By using the spherical symmetry of the function $f_i(R_i)$, these expressions can be cast in the form

$$\begin{aligned} \overline{(\boldsymbol{v}grad)\boldsymbol{v}} = -\tfrac{2}{3}\textstyle\sum_i \omega_i^2 f_i'^2 \frac{\boldsymbol{R}_i}{R_i^2} \ , \\ -\overline{\boldsymbol{v}\times curl\boldsymbol{v}} = -\tfrac{2}{3}\textstyle\sum_i \omega_i^2 f_i' \left(f_i'' + \frac{1}{R_i}f_i'\right)\frac{\boldsymbol{R}_i}{R_i} \ , \\ \overline{grad(v^2/2)} = \tfrac{2}{3}\textstyle\sum_i \omega_i^2 f_i' f_i'' \frac{\boldsymbol{R}_i}{R_i} \ . \end{aligned} \tag{5.90}$$

We can see that even on average the Navier-Stokes equation is not satisfied, in the sense discussed above. Even on average the vorticial liquid develops internal forces which are not equilibrated; it is unstable. We shall give specific examples of such an instability below.

Now, let us assume that the vorticial liquid is sufficiently dense, *i.e.* if we can define a density ρ_v of points $\boldsymbol{r}_i$; further, we assume that the liquid is homogeneous and isotropic, *i.e.* this density is constant and the ω_i and the functions f_i in the above equations may be replaced by the same functions ω and f. We assume that this is another distinctive feature of a fully developed turbulence. Then, we may transform the summation over i in equations (5.90) into an integral, like in equation (5.65). By choosing the origin at $\boldsymbol{r} = \boldsymbol{r}_i$ for a fixed $\boldsymbol{r}_i$, and using the notation $\boldsymbol{R} = \boldsymbol{r} - \boldsymbol{r}_i$, we get

$$\overline{(\boldsymbol{v}grad)\boldsymbol{v}} = -\frac{2}{3}\rho_v\omega^2\int_0^\infty dR\cdot Rf'^2(R)\int do(\boldsymbol{R}/R)\ ; \qquad (5.91)$$

the result of integration in this equation is zero, due to the integration over the solid angle o, providing the radial integration is finite. In this case the Navier-Stokes equation is satisfied trivially.

Let us assume that the integral over R is singular at $R = 0$, as another distinctive feature of a fully developed turbulence (the function f is assumed to decrease sufficiently rapid at infinity to have a finite integral in this limit). The integration outside a small region around the i-th point is zero, while the integration over such a small region is indefinite. This singularity implies

$$f(R) \sim (a/R)^n\ ,\ n > 0 \qquad (5.92)$$

for $R \ll a$, where a is a small characteristic distance (compare with equation (5.66)). Therefore, we are left with a discrete set of points $\boldsymbol{r}_i$, where the function $f(R_i)$ and the velocity ($v \sim 1/R_i^{n+1}$) are singular. We have now a small region of dimension a, around each point $\boldsymbol{r}_i$, which includes a mass of fluid, say, M, where the inertial term given by equation (5.91) is not defined. Since the positions $\boldsymbol{r}_i$ may change in time, we are left with a classical gas of particle-like vortices (or a solution of vortices in the background fluid), as discussed above. Of course, we may have also a mixture of vorticial gases, each characterized by a dimension a and a mass M.

The equation of energy conservation (equation (5.67))

$$\frac{\partial}{\partial t}\left(\frac{1}{2}v^2\right) + \boldsymbol{v}\cdot(\boldsymbol{v}grad)\boldsymbol{v} = \nu\boldsymbol{v}\Delta\boldsymbol{v} \ , \tag{5.93}$$

averaged over the fluctuating vectors $\boldsymbol{\omega}_i$, is reduced to the viscosity term

$$\nu\overline{\boldsymbol{v}\Delta\boldsymbol{v}} = \frac{2\nu}{3}\sum_i \omega_i^2 f_i^{'}\left(f_i^{''} + 2f_i^{'}/R_i\right)^{'} \ . \tag{5.94}$$

This term should be computed outside the regions with dimension a, where the motion is defined. The non-vanishing value of this term indicates an energy loss, which should be compensated from the outside. The vorticial gas is in quasi equilibrium.

The above considerations are valid for spherical-symmetric functions $f_i(R_i)$; if the vortices have a lower (internal) symmetry, the unit vector $\boldsymbol{R}/R$ in equation (5.91) is replaced by functions which do not have a spherical symmetry, and the inertial term is not vanishing, in general. The vortices are unstable, and, likely, they could tend to acquire a spherical symmetry, which ensures a (quasi)-equilibrium.

5.8.3 Filamentary liquid

Let us consider a set of rectilinear, parallel filaments, directed along the z-axis, placed at positions $\boldsymbol{r}_i$ in the (x, y)-plane, with vorticities $\boldsymbol{\omega}_i$. This is a two-dimensional vorticial liquid of "cyclons". The velocity field is given by

$$\boldsymbol{v} = -\sum_i \boldsymbol{\omega}_i \times grad \ln R_i = \sum_i \frac{\boldsymbol{R}_i \times \boldsymbol{\omega}_i}{R_i^2} \ , \tag{5.95}$$

where $\boldsymbol{R}_i = \boldsymbol{r} - \boldsymbol{r}_i$.

Equation (5.95) shows that the velocity is derived from a vector potential $\boldsymbol{A}$, through $\boldsymbol{v} = curl\boldsymbol{A}$. By taking the *curl* in this equation, we get

$$\Delta\boldsymbol{A} = -2\boldsymbol{\omega} \tag{5.96}$$

(providing $div\boldsymbol{A} = 0$ and $div\boldsymbol{\omega} = 0$), where we introduce the notation $curl\boldsymbol{v} = 2\boldsymbol{\omega}$; $\boldsymbol{\omega}$ is called vorticity. The vorticity distribution corresponding to equation (5.95)

$$\boldsymbol{\omega}(\boldsymbol{r}) = \frac{1}{2} curl \boldsymbol{v} = -\pi \sum_i \boldsymbol{\omega}_i \delta(\boldsymbol{R}_i) \ , \tag{5.97}$$

gives the vector potential

$$\boldsymbol{A} = \sum_i \boldsymbol{\omega}_i \ln R_i \ . \tag{5.98}$$

According to equation (5.84), the inertial term has the components

$$\boldsymbol{f} = -\boldsymbol{v} \times curl \boldsymbol{v} = -2\pi \sum_{i \neq j} \omega_i \omega_j grad_i \ln R_{ij} \cdot \delta(\boldsymbol{R}_i) \ , \tag{5.99}$$

where $\boldsymbol{R}_{ij} = \boldsymbol{r}_i - \boldsymbol{r}_j$, and $grad e$, where

$$e = \frac{1}{2} v^2 = \sum_{i \neq j} \omega_i \omega_j \frac{\boldsymbol{R}_i \boldsymbol{R}_j}{2R_i^2 R_j^2} \tag{5.100}$$

is a density of kinetic energy (per unit mass). Apart from the force $\boldsymbol{f}$, which acts at the positions of the vortices, there exist internal forces given by $grad e$, which make the liquid unstable. The motion and the statistics of parallel, rectilinear filaments have been extensively investigated,[12] the instability being associated with a negative temperature in an attempt of a statistical theory.[13]

[12] C. C. Lin, "On the motion of vortices in two-dimensions, I, II", Proc. Nat. Acad. Sci. USA **27** 570 (1941); *On the Motion of Vortices in Two Dimensions*, University of Toronto Press,Toronto (1943); L.Onsager, "Statistical hydrodynamics", Suppl. to Nuovo Cimento **6** 279 (1949); D. Montgomery and G. Joyce, "Statistical mechanics of "negative temperature" states", Phys. Fluids **17** 1139 (1974); A. J. Chorin, *Vorticity and Turbulence*, Springer, NY (1994); V. I. Berezinsky, "Destruction of long-range order in one-dimensional and two-dimensional systems having a continuous symmetry group. I. Classical systems", Sov. Phys.-JETP **32** 493 (1971) (ZhETF **59** 907 (1971)); "Destruction of long-range order in one-dimensional and two-dimensional systems having a continuous symmetry group. II. Quantum systems", Sov. Phys.-JETP **34** 610 (1972) (ZhETF **61** 1144 (1972)); J. M. Kosterlitz and D. J. Thouless, "Ordering, metastability and phase transitions in two-dimensional systems", J. Phys. **C6** 1181 (1973).

[13] L.Onsager, "Statistical hydrodynamics", Suppl. to Nuovo Cimento **6** 279 (1949); A. J. Chorin, *Vorticity and Turbulence*, Springer, NY (1994).

The total force given by equation (5.99)

$$\boldsymbol{F} = \int d\boldsymbol{r}\boldsymbol{f} = -2\pi \sum_{i \neq j} \omega_i \omega_j grad_i \ln R_{ij} \tag{5.101}$$

is zero. We can see that a force

$$\boldsymbol{f}_{ij} = -\boldsymbol{f}_{ji} = -2\pi\omega_i\omega_j grad_i \ln R_{ij} \tag{5.102}$$

acts between any pair (ij) of vortices. This force derives from a potential

$$U_{ij} = 2\pi\omega_i\omega_j \ln R_{ij} \quad , \tag{5.103}$$

such that

$$\boldsymbol{F} = -\sum_{i \neq j} grad_i U_{ij} \tag{5.104}$$

The density of kinetic energy e can be written as

$$e = \frac{1}{2}v^2 = \frac{1}{2}\boldsymbol{v} curl \boldsymbol{A} = -\frac{1}{2}div(\boldsymbol{v} \times \boldsymbol{A}) + \boldsymbol{A}\boldsymbol{\omega} \ . \tag{5.105}$$

The first term in equation (5.105) is singular at $\boldsymbol{r} = \boldsymbol{r}_i$; we integrate this term over the whole space, transform it into surface integrals, both at infinity and over small circles around each filament, and neglect their contributions. The result of such integrations is a self-energy (or a self-force), which may be left aside. We call this procedure a "renormalization".[14] By doing so, we are left with a total kinetic energy

$$E = \int d\boldsymbol{r}e = \int d\boldsymbol{r}\boldsymbol{A}\boldsymbol{\omega} = -\pi \sum_{i \neq j} \omega_i\omega_j \ln R_{ij} = \\ = -\tfrac{1}{2}\sum_{i \neq j} U_{ij} = -U \quad , \tag{5.106}$$

where U is the total potential energy. We can see that the total energy is conserved, *i.e.* $E + U = const$. Also, the total force $\boldsymbol{F} = 0$, such that the total momentum is conserved. The total angular momentum is $-2\int d\boldsymbol{r}\boldsymbol{A}$; it is proportional to $\sum_i \boldsymbol{\omega}_i$. The total torque

$$2\pi \sum_{i \neq j} \omega_i\omega_j \frac{\boldsymbol{r}_i \times \boldsymbol{r}_j}{R_{ij}^2} \tag{5.107}$$

[14] A. Sommerfeld, *Vorlesungen ueber Theoretische Physik*, Band II (*Mechanik der Deformierbaren Medien*), Akad. Verlagsgesellschaft, Leipzig (1970).

is zero. By this "renormalization" procedure the points $\boldsymbol{r_i}$ are completely decoupled from the fluid, and they may have their own motion. The average over fluctuating vorticities can be computed straightforwardly, by using $\overline{\boldsymbol{\omega}_i^2} = \frac{1}{2}\omega_i^2$; it is given by

$$\overline{(\boldsymbol{v} grad)\boldsymbol{v}} = -\sum_i \omega_i^2 \frac{\boldsymbol{R}_i}{2R_i^4} \ . \tag{5.108}$$

We can see that for a sufficiently dense, homogeneous and isotropic liquid we get a gas of (singular) vortices, as discussed above.

5.8.4 Coulombian liquid

For $f_i(R_i) = -1/R_i$ in equation (5.86) we get a coulombian vorticial liquid with the velocity field

$$\boldsymbol{v} = -\sum_i \boldsymbol{\omega}_i \times grad(1/R_i) = \sum_i \frac{\boldsymbol{\omega}_i \times \boldsymbol{R}_i}{R_i^3} \tag{5.109}$$

and the vector potential

$$\boldsymbol{A} = \sum_i \frac{\boldsymbol{\omega}_i}{R_i} \ . \tag{5.110}$$

The equation $\Delta \boldsymbol{A} = -2\boldsymbol{\omega}$ is satisfied for

$$\boldsymbol{\omega}(\boldsymbol{r}) = 2\pi \sum_i \boldsymbol{\omega}_i \delta(\boldsymbol{R}_i) \ , \tag{5.111}$$

but this vorticity differs from

$$\frac{1}{2} curl \boldsymbol{v} = 2\pi \sum_i \boldsymbol{\omega}_i \delta(\boldsymbol{R}_i) - \frac{1}{2} grad\, div \sum_i (\boldsymbol{\omega}_i / R_i) \ . \tag{5.112}$$

By applying the "renormalization" procedure we get the force

$$\boldsymbol{F} = \int d\boldsymbol{r} \boldsymbol{f} = -\int d\boldsymbol{r} \boldsymbol{v} \times curl \boldsymbol{v} = 4\pi \sum_{i \neq j} grad_i \frac{\boldsymbol{\omega}_i \boldsymbol{\omega}_j}{R_{ij}} \tag{5.113}$$

and the energy

$$E = \int dre = \frac{1}{2}\int drv^2 = 2\pi \sum_{i \neq j} \frac{\omega_i \omega_j}{R_{ij}} \ , \tag{5.114}$$

where, in both cases $\boldsymbol{v} = curl\boldsymbol{A}$ is used. The potential from equation (5.103) is now $U_{ij} = -4\pi\frac{\omega_i\omega_j}{R_{ij}}$. The total energy is conserved, the total force is zero, the total angular momentum is proportional to $\sum_i \boldsymbol{\omega}_i$ and the total torque (zero) is

$$4\pi \sum_{i \neq j} \omega_i \omega_j \frac{\boldsymbol{r}_i \times \boldsymbol{r}_j}{R_{ij}^3} \ . \tag{5.115}$$

The average of the inertial term over fluctuating vortices is obtained from equation (5.90)

$$\overline{(\boldsymbol{v}grad)\boldsymbol{v}} = -\frac{2}{3}\sum_i \omega_i^2 \frac{\boldsymbol{R}_i}{R_i^6} \ , \tag{5.116}$$

such that we may get a gas of singular vortices.

5.8.5 Dipolar liquid

A dipolar liquid is defined by the vorticity

$$\boldsymbol{\omega} = -2\pi \sum_i \boldsymbol{m}_i \times grad\delta(\boldsymbol{r} - \boldsymbol{r}_i) \ , \tag{5.117}$$

where the vectors $\boldsymbol{m}_i$ may depend on the time, at most. We get the vector potential

$$\boldsymbol{A} = \sum_i \frac{\boldsymbol{m}_i \times \boldsymbol{R}_i}{R_i^3} = -\sum_i \boldsymbol{m}_i \times grad(1/R_i) \tag{5.118}$$

and the velocity field

$$\begin{aligned} \boldsymbol{v}(\boldsymbol{r}) &= \textstyle\sum_i \left[-\boldsymbol{m}_i/R_i^3 + 3\boldsymbol{R}_i(\boldsymbol{m}_i\boldsymbol{R}_i)/R_i^5\right] = \\ &= \textstyle\sum_i grad\left[\boldsymbol{m}_i grad(1/R_i)\right] \ . \end{aligned} \tag{5.119}$$

We recognize in these equations magnetic (dipole) moments $\boldsymbol{m}_i$, a dipolar vector potential $\boldsymbol{A}$ and a magnetic field $\boldsymbol{v}$.

The inertial term has the components

$$\boldsymbol{f} = -\boldsymbol{v} \times curl\boldsymbol{v} = 2\boldsymbol{\omega} \times \boldsymbol{v} =$$
$$= 4\pi \sum_{i \neq j} grad\left[\boldsymbol{m}_j grad(1/R_j)\right] \times \left[\boldsymbol{m}_i \times grad\delta(\boldsymbol{R}_i)\right] \tag{5.120}$$

and $grade$, where

$$e = \frac{1}{2}v^2 = \frac{1}{2}\sum_{i \neq j}\left[\frac{\boldsymbol{m}_i\boldsymbol{m}_j}{R_i^3 R_j^3} - \frac{3(\boldsymbol{m}_i\boldsymbol{R}_j)(\boldsymbol{m}_j\boldsymbol{R}_j)}{R_i^3 R_j^5} -\right.$$
$$-\frac{3(\boldsymbol{m}_j\boldsymbol{R}_i)(\boldsymbol{m}_i\boldsymbol{R}_i)}{R_i^5 R_j^3} + \tag{5.121}$$
$$\left.+\frac{9(\boldsymbol{R}_i\boldsymbol{R}_j)(\boldsymbol{m}_i\boldsymbol{R}_i)(\boldsymbol{m}_j\boldsymbol{R}_j)}{R_i^5 R_j^5}\right] .$$

We can see that the dipolar liquid is unstable.

The total force and the total kinetic energy are

$$\boldsymbol{F} = \int d\boldsymbol{r}\boldsymbol{f} = 2\int d\boldsymbol{r}\boldsymbol{\omega} \times \boldsymbol{v} = -\sum_{i \neq j} grad_i U_{ij} \ ,$$
$$E = \int d\boldsymbol{r}e = \int d\boldsymbol{r}\boldsymbol{\omega}\boldsymbol{A} = -\frac{1}{2}\sum_{i \neq j} U_{ij} \ , \tag{5.122}$$

where

$$U_{ij} = -4\pi\boldsymbol{m}_i grad_i\left[\boldsymbol{m}_j grad_i(1/R_{ij})\right] . \tag{5.123}$$

By this "renormalization" procedure, the liquid is reduced to a set of interacting particle-like vortices. We can see that the energy is conserved, the total force is zero, the total angular momentum and the total torque are zero.

The time average over vorticities in equation (5.121) leads to

$$\overline{e} = \frac{1}{2}\sum_i\left[\frac{m_i^2}{R_i^6} + \frac{3\overline{(\boldsymbol{m}_i\boldsymbol{R}_i)^2}}{R_i^8}\right] = \sum_i \frac{m_i^2}{R_i^6} \ , \tag{5.124}$$

which gives a force

$$grad\overline{e} = -6\sum_i \frac{m_i^2\boldsymbol{R}_i}{R_i^8} . \tag{5.125}$$

For a dense, homogeneous and isotropic liquid the force given by equation (5.125) is zero (for $\boldsymbol{R}_i \neq 0$); also, the force $\boldsymbol{f}$ is zero for $\boldsymbol{R}_i \neq 0$, and the Navier-Stokes equation is satisfied trivially (on average, with zero pressure). We note that $\Delta \boldsymbol{v}$ is zero for $\boldsymbol{R}_i \neq 0$ (equation (5.119)), such that the viscosity contribution is zero. We are left with a set of positions $\boldsymbol{r}_i$, each surrounded by a small region, where the motion is not defined. According to the above discussion, such a structure may be viewed as a (quasi) ideal classical gas of vortices (or a solution of vortices in the background fluid).

5.9 Concluding remarks

In fairly general conditions we have given above explicit (smooth) solutions for the potential flow. We have shown that, rigorously speaking, the equations of the fluid mechanics have not rotational solutions. However, usually we may neglect the variations of the density and the temperature, such that, in these conditions, the Navier-Stokes equation may also exhibit (approximate) vorticial solutions, governed by the viscosity. We give arguments that the vortices are unstable. On the other hand, for large variations of the velocity over small distances, the fluid velocity exhibits turbulent, highly fluctuating instabilities, controlled by viscosity. Such a fully developed turbulence occurs as a consequence of the insufficiency of the viscosity to dissipate heat. We represent the fully developed turbulence as a superposition of highly fluctuating velocities, associated to a discrete distribution of turbulence centres, and we are interested in the temporal average of this velocity field. A regular mean flow may be added. It is shown that the Navier-Stokes equation is not satisfied on average. However, for a homogeneous and isotropic distribution of (non-singular) turbulence centres (as another distinctive feature of a fully developed turbulence), the temporal average of both the fluctuating velocity and the inertial term is zero, such that the Navier-Stokes equation is satisfied trivially. If the velocity is singular at the turbulence centres we are left with a quasi ideal classical gas of singularities (or a solution of singularities in the background fluid), in thermal quasi equilibrium, as an example of emergent dynamics. The Navier-Stokes equation for this fluid of singularities is reduced to Newton's law of motion (with a small friction).

At a higher scale, equations of fluid mechanics can be considerd for this gas, as a continuous medium. We have illustrated all the above considerations with three examples of vorticial liquids.

5.10 Summary

A flow with velocity $\boldsymbol{v}$ of a fluid with density ρ is governed by the continuity equation (mass conservation)

$$\frac{\partial \rho}{\partial t} + div(\rho \boldsymbol{v}) = 0 \ , \tag{5.126}$$

the Navier-Stokes equation

$$\rho \frac{d\boldsymbol{v}}{dt} = -gradp + \eta \Delta \boldsymbol{v} + \left(\frac{1}{3}\eta + \zeta \right) grad\, div\boldsymbol{v} \tag{5.127}$$

and the entropy equation

$$\rho T \frac{ds}{dt} = \kappa \Delta T + \sigma'_{ij} \partial_j v_i \ , \tag{5.128}$$

where

$$\sigma'_{ij} = \eta \left(\partial_i v_j + \partial_j v_i - \frac{2}{3}\delta_{ij} div\boldsymbol{v} \right) + \zeta \delta_{ij} div\boldsymbol{v} \tag{5.129}$$

is the viscosity tensor, s is the entropy per unit mass, T is the temperature, η, ζ are the viscosity coefficients and κ is the thermoconductivity. The total derivative is $\frac{d}{dt} = \frac{\partial}{\partial t} + vgrad$. The energy conservation reads

$$\frac{\partial}{\partial t}\left(\frac{1}{2}\rho v^2 + \rho \varepsilon \right) = -\partial_j \left[\rho v_j \left(\frac{1}{2}v^2 + w \right) - v_i \sigma'_{ij} - \kappa \partial_j T \right] \ , \tag{5.130}$$

where ε is the internal energy per unit mass and w is the entalpy per unit mass ($d(\varepsilon + p/\rho) = dw = dp/\rho + Tds$).

p in the Navier-Stokes equation is the internal pressure; an external force $-\rho grad\varphi$ may be added, as well as an external force which does not derive from a potential. Additional interactions with boundaries may be considered (like in capilarity phenomena).

All the variations of the quantities in the above equations should be viewed as small variations with respect to a global mechanical equilibrium and a local thermodynamic equilibrium. In particular, the velocity should be viewed as much smaller than the sound velocity, which gives the limit of thermal velocity. Large variations, like the shock waves, or relativistic local equilibrium are special subjects. Under these conditions, the pressure p and the entropy s are functions of ρ and T, such that the above equations are determined. Usually, part of the entropy is vanishing in adiabatic processes, and we are left with $Tds = c_p dT$, where c_p is the specific heat per unit mass at constant pressure.

It is worth noting that the viscosity tensor in the entropy equation gives an increase of entropy, *i.e.* an increase of the internal energy, while the total energy is conserved; therefore, the mechanical energy decreases as a a result of the viscosity action.

Under these conditions the aim is to solve the above equations for the velocity, for given initial and boundary conditions.

The velocity is given by $\boldsymbol{v} = grad\Phi$ or $\boldsymbol{v} = curl\boldsymbol{A}$, where Φ is a scalar potential and $\boldsymbol{A}$ is a vector potential. In the first case we have a potential (irrotational) flow ($curl\boldsymbol{v} = 0$), in the second case we have a vorticial (rotational) flow (since $curl\boldsymbol{v} \neq 0$).

For all the usual flows the relative variations of density, pressure, temperature are very small, such that we may use an incompressible flow, or we may linearize the equations of the fluid mechanics. In particular, for a vorticial flow the fluid is incompressible, since $div\boldsymbol{v} = div\, curl\boldsymbol{A} = 0$. Under these circumstances we can get explicit (smooth) solutions both for the potential flow and the vorticial flow. In particular, the Navier-Stokes equation leads to a vorticity equation, which gives the velocity, while the pressure is given by the full Navier-Stokes equation. The entropy equation gives another equation for pressure, such that the problem is overdetermined. However, we may neglect the entropy equation in all the usual cases, because the entropy variations are small, such that we are left, in fact, with a rotational flow, which is incompressible and isentropic. It is worth noting that the vorticity is governed by the presence of the viscosity.

In general, the solutions of the Navier-Stokes equation exhibit instabilities in certain particular conditions. Indeed, for large variations of the velocity over short distances the energy cannot be dissipated as heat, and the mechanical energy may increase indefinitely. This is the phenomenon of turbulence, characterized by fluctuating swirlings of velocity.

We can superpose fluctuating velocities over a regular mean flow. It is reasonable to be interested only in the averages over fluctuations. The average of the fluctuating velocities leads to the equation

$$\overline{(\boldsymbol{v}grad)\boldsymbol{v}} = 0 \ , \tag{5.131}$$

which is not satisfied. A turbulent flow does not satisfy the Navier-Stokes equation.

We can associate the fluctuating velocities with a discrete distribution of turbulence centres placed at $\boldsymbol{r}_i$, *i.e.* we may write

$$\boldsymbol{v} = \sum_i \boldsymbol{v}_i(t, R_i) \ , \tag{5.132}$$

where $\boldsymbol{R}_i = \boldsymbol{r} - \boldsymbol{r}_i$. We assume a dependence on R_i, as a for a homogeneous and isotropic turbulence, a distinctive feature of a fully developed turbulence. Then, for a dense, homogeneous and isotropic distribution of turbulence centres equation (5.131) is satisfied, such that we may say that the Navier-Stokes equation is satisfied trivially in this case. Possibly, we may be left with singularities at the turbulence centres, which behave as an almost ideal classical gas (or a solution of singularities in the background fluid).

6 Sound

6.1 Simplified equations

The equations of fluid mechanics are the continuity equation, the heat transfer equation and the Navier-Stokes equation. The continuity equation is

$$\frac{\partial\rho}{\partial t} + div(\rho\boldsymbol{v}) = 0 \ , \tag{6.1}$$

where ρ is the density and $\boldsymbol{v}$ is the velocity of the fluid. The viscosity is represented by a viscosity stress tensor

$$\sigma'_{ij} = \eta\left(\partial_i v_j + \partial_j v_i - \frac{2}{3}\partial_k v_k \delta_{ij}\right) + \zeta\partial_k v_k \delta_{ij} \ , \tag{6.2}$$

where η and ζ are the viscosity coefficients. It gives a force $\partial_j\sigma'_{ij}$ per unit volume, such that the Navier-Stokes equation reads

$$\begin{aligned} \rho\tfrac{dv_i}{dt} &= \rho\tfrac{\partial v_i}{\partial t} + \rho(\boldsymbol{v}grad)v_i = \\ &= -\partial_i p + \partial_j\sigma'_{ij} \ , \end{aligned} \tag{6.3}$$

where p is the pressure, or

$$\begin{aligned} \rho\tfrac{d\boldsymbol{v}}{dt} &= \rho\tfrac{\partial\boldsymbol{v}}{\partial t} + \rho(\boldsymbol{v}grad)\boldsymbol{v} = \\ &= -gradp + \left(\tfrac{1}{3}\eta + \zeta\right) grad\, div\boldsymbol{v} + \eta\Delta\boldsymbol{v} \ . \end{aligned} \tag{6.4}$$

The heat transfer equation is

$$\rho c_p\left(\frac{\partial T}{\partial t} + \boldsymbol{v}gradT\right) = div(\kappa gradT) + \sigma'_{ij}\partial_j v_i \ , \tag{6.5}$$

where T is the temperature, c_p is the specific heat per unit mass and κ is the thermoconductivity. In deriving the heat transfer equation we assume that the flow brings small, slightly varying perturbations to local thermodynamic equilibrium. The equilibrium processes are so quick, that local dilatation/compression processes compensate part of temperature variations. Since

$$\begin{aligned}\sigma'_{ij}\partial_j v_i &= \eta\left(\partial_i v_j + \partial_j v_i - \tfrac{2}{3}\delta_{ij}\partial_k v_k\right)\partial_j v_i + \zeta\delta_{ij}\partial_k v_k \partial_j v_i = \\ &= \tfrac{1}{2}\eta\left(\partial_i v_j + \partial_j v_i - \tfrac{2}{3}\delta_{ij}\partial_k v_k\right)^2 + \zeta(div\boldsymbol{v})^2 \ ,\end{aligned} \tag{6.6}$$

the equation of heat transfer can also be written as

$$\begin{aligned}\rho c_p\left(\tfrac{\partial T}{\partial t} + \boldsymbol{v} gradT\right) &= div(\kappa gradT) + \\ &+ \tfrac{1}{2}\eta\left(\partial_i v_j + \partial_j v_i - \tfrac{2}{3}\delta_{ij}\partial_k v_k\right)^2 + \zeta(div\boldsymbol{v})^2 \ .\end{aligned} \tag{6.7}$$

This is the equation of entropy transfer, since the heat is $Tds = c_p dT$, s being the entropy per unit mass; the occurrence of c_p indicates the fact that we take the entropy variations at equilibrium (constant pressure). Combining the Navier-Stokes equation with the continuity equation, we get the law of momentum conservation

$$\begin{aligned}&\tfrac{\partial}{\partial t}(\rho v_i) + \partial_j \Pi_{ij} = 0 \ , \\ &\Pi_{ij} = p\delta_{ij} + \rho v_i v_j - \sigma'_{ij} \ ,\end{aligned} \tag{6.8}$$

where Π_{ij} is the tensor of the momentum flux density.
Finally, we have the law of energy conservation

$$\frac{\partial}{\partial t}\left(\frac{1}{2}\rho v^2 + \rho\varepsilon\right) = -\partial_j\left[\rho v_j\left(\frac{1}{2}v^2 + w\right) - v_i\sigma'_{ij} - \kappa\partial_j T\right] \ , \tag{6.9}$$

where ε is the internal energy per unit mass and w is the enthalpy per unit mass. The energy ε is a function of specific volume per unit mass (*i.e.*, ρ) and entropy (or T); the enthalpy is a function of pressure and entropy; by using the equation of state for pressure, we get the enthalpy as a function of ρ and T. We can see that the three equations (continuity, Navier-Stokes and heat transfer) contain three unknowns (velocity, density and temperature), such that the problem of solving these equations is determined. The parameters η, ζ, c_p and κ are considered constants (and positive).

When the velocity is large, and it varies abruptly over small distances, the inertial term $(\boldsymbol{v}grad)\boldsymbol{v}$ in the Navier-Stokes equations may bring an important contribution to the energy. This happens especially when viscosity tends to zero, a circumstance which forces abrupt spatial variations of the velocity in the viscosity terms. This situation is characterized by large Reynolds numbers $R = vl/\nu$, where l is the distance over which the velocity varies and $\nu = (\eta, \zeta)/\rho$ stands for the kinematic viscosity. In order to have a stable flow, the energy should be dissipated as heat, by viscosity. For large Reynolds numbers, this function of the viscosity becomes insufficient. The fluid responds by taking the mechanical energy in the collision term $v_i\sigma^{'}_{ij}$ in equation (6.9), which develops discontinuities surfaces. Due to its negative sign, it brings an instability of the flow. We have turbulence in this case, and discontinuous and fluctuating velocity, so the solution of the equations of fluid mechanics do not exist (as shown in the Turbulence. Boundary Layer chapter). Therefore, it is reasonable to limit ourselves to small velocities, with small spatial variations, and a finite viscosity. On the other hand, large velocities, for instance larger than the sound velocity, or the thermal molecular velocity, disrupt the flow, and damage the local thermodynamic equilibrium, which makes the above equations invalid.

Within these limitations we may neglect the inertial term in the Navier-Stokes equations and replace the viscosity terms by $-\gamma\rho\boldsymbol{v}$, where $\gamma \simeq (\eta, \zeta)/\rho l^2$ is a small damping coefficient, with dimension of an inverse time. This approximation amounts to the replacement

$$\partial_j\sigma^{'}_{ij} \rightarrow -\rho\gamma v_i \tag{6.10}$$

and

$$\partial_j(v_i\sigma^{'}_{ij}) \rightarrow -2\rho\gamma v^2 \ , \ \sigma^{'}_{ij}\partial_j v_i \rightarrow 2\rho\gamma v^2 \ . \tag{6.11}$$

It is worth noting that the approximation used here is meaningful for all the terms given above, which can be written symbolically as $d\sigma \sim d^2v$, $d(v\sigma) \sim d^2(v^2)$ and $\sigma dv \sim (dv)^2$; while the term

$vd\sigma \sim vd^2v$, which occurs originally in the law of energy conservation, cannot be put in a form suitable for this approximation. In fact, in order to preserve the identity $d(v\sigma) = vd\sigma + \sigma dv$, we need to admit $vd\sigma = 0$. This is in agreement with the introduction of the entropy in the equation of energy conservation, which is made via an identity, a procedure which justifies the separation of this equation in law of energy conservation and equation of heat transfer. In addition, the procedure of approximation adopted here separates the volume and the surface contributions, in accordance with their physical meaning.
A similar approximation can be done in the heat transfer equation, so we get

$$\begin{aligned} \tfrac{\partial \boldsymbol{v}}{\partial t} + \gamma \boldsymbol{v} &= -\tfrac{1}{\rho} grad p \ , \\ c_p \left(\tfrac{\partial T}{\partial t} + \boldsymbol{v} grad T \right) &= \tfrac{1}{\rho} div(\kappa grad T) + 2\gamma v^2 \ . \end{aligned} \tag{6.12}$$

Within this approximation the laws of momentum and energy conservation become

$$\begin{aligned} \tfrac{\partial}{\partial t}(\rho v_i) + \partial_i p &= -\rho\gamma v_i \ , \\ \tfrac{\partial}{\partial t}\left(\tfrac{1}{2}\rho v^2 + \rho\varepsilon\right) &= -\partial_j \left(\rho v_j w - \kappa \partial_j T\right) - 2\rho\gamma v^2 \ . \end{aligned} \tag{6.13}$$

We can see that the entropy increases by thermoconduction and damping ($\gamma > 0$), while the momentum and the energy decrease by damping.
A further simplification can be made in these equations. The source term in the heat transfer equation is $2\gamma v^2$, which is of second order in powers of v. It generates a corresponding variation of the temperature, so we may neglect the term $\boldsymbol{v} grad T$ in the equation of heat transfer, which becomes

$$c_p \frac{\partial T}{\partial t} = \frac{1}{\rho} div(\kappa grad T) + 2\gamma v^2 \ . \tag{6.14}$$

Let us focus now on the continuity equation. For $\boldsymbol{v} = 0$ the density is constant and uniform, let we denote it by ρ_0. Small velocities generate small changes $\delta\rho$ in density. The linearization of the continuity equation leads to

$$\frac{\partial \delta\rho}{\partial t} + \rho_0 div \boldsymbol{v} = 0 \ , \tag{6.15}$$

whence we can see that

$$\delta\rho = -\rho_0 div\boldsymbol{u} \ , \qquad (6.16)$$

where $\boldsymbol{u}$ is a displacement field, depending on time and position; the velocity is $\boldsymbol{v} = \frac{\partial \boldsymbol{u}}{\partial t}$. The change in density entails a change in volume $\delta V = V_0 div\boldsymbol{u}$, where V_0 is the volume at rest. Consequently, we can write the change in pressure as

$$\delta p = \left(\frac{\partial p}{\partial \rho}\right)_T \delta\rho + \left(\frac{\partial p}{\partial T}\right)_V \delta T = \frac{K}{\rho_0}\delta\rho + \alpha\delta T \ , \qquad (6.17)$$

where $K = -V\left(\frac{\partial p}{\partial V}\right)_T$ (> 0) is the isothermal modulus of compression (bulk modulus) and $\alpha = \left(\frac{\partial p}{\partial T}\right)_V$ is the thermal pressure coefficient ($\alpha = K\beta$, where $\beta = \frac{1}{V}(\partial V/\partial T)_p$ is the thermal expansion coefficient, see Appendix). We consider these parameters (K, α) constants. The Navier-Stokes equation (6.12) becomes

$$\frac{\partial^2 \boldsymbol{u}}{\partial t^2} - \frac{K}{\rho_0} grad\, div\boldsymbol{u} + \gamma\dot{\boldsymbol{u}} = -\frac{\alpha}{\rho_0} gradT - grad\varphi \ , \qquad (6.18)$$

where an external potential φ is introduced (force per unit volume $-\rho grad\varphi$). This is a wave equation with sources ($gradT$ and $grad\varphi$) and damping ($\gamma\dot{\boldsymbol{u}}$). It can be solved by using a potential ψ given by $\boldsymbol{u} = grad\psi$. The wave velocity is $c_{th} = \sqrt{K/\rho_0}$ (entropic sound). Equation (6.18) should be solved together with the heat transfer equation (6.14).

However, the usual processes are much slower than the processes which achieve local thermodynamic equilibrium. In this case part of the temperature variations is taken by density variations, such that the isothermal modulus of compression K in the above equation is changed into the adiabatic modulus K_{ad}. This amounts to include the adiabatic part of the flow. This process is illustrated shortly below.

In general, the heat transfer equation (6.14) may include heat sources. These sources are usually given as an external temperature T_0 in some regions of the fluid, or on its boundaries. The equation is solved by viewing these sources as boundary conditions. In the first approximation the damping term $2\gamma v^2$ may be left aside, such that the heat transfer equation is decoupled from the Navier-Stokes equation; the temperature acts as a driving force for this latter equation. Having solved the Navier-Stokes equation, we may compute the damping coefficient of the temperature. From equation (6.13) the force acting upon the unit area of a surface with normal $\boldsymbol{n}$ is $p\boldsymbol{n}$ (small effects of $\delta\rho$ and δT included). The law of energy conservation in the form given by equation (6.13) is useless in this context. The kinetic energy per unit volume is given by $E_{kin} = \frac{1}{2}\rho_0 v^2$. The variations of energy and enthalpy are $\delta\varepsilon = p_0\delta\rho/\rho_0^2 + c_p\delta T$, $\delta w = \delta p/\rho_0 + c_p\delta T$.

A special treatment is required for an ideal fluid, where the thermoconductivity, the damping and external heat sources are absent. In this case the flow is adiabatic. But the density changes produce entropy, which is absorbed by a change in temperature, such that the δT-term in equation (6.18) is present. For an adiabatic process we have

$$\delta s = \left(\frac{\partial s}{\partial V}\right)_T \delta V + \left(\frac{\partial s}{\partial T}\right)_V \delta T = \alpha\delta V + \frac{c_v}{T}\delta T = 0 \ , \qquad (6.19)$$

where we use $(\partial S/\partial V)_T = -\partial^2 F/\partial V\partial T = (\partial p/\partial T)_V = \alpha$, F and c_v being the free energy and the specific heat per unit mass at constant volume. We insert δT from this equation in equation (6.18), and get a change of the bulk modulus given by

$$K \to K\left(1 + \frac{K\beta^2 T}{\rho_0 c_v}\right) = K_{ad} \ , \qquad (6.20)$$

which is the adiabatic modulus of compression (within our approximation, according to the Ideal Fluids chapter; see Appendix[1]). Equation (6.18) with the adiabatic modulus of compression is the equation of sound. This is an example of emergent dynamics.[2] The motion of the fluid particles with velocity $\boldsymbol{v}$ is transmitted, moves through the fluid, with velocity $c = \sqrt{K_{ad}/\rho}$.

[1] L. Landau and E. Lifshitz, *Course of Theoretical Physics*, vol. 5 (*Statistical Physics*), Elsevier, Oxford (1980).
[2] P. W. Anderson, "More is different", Science **177** 393 (1972).

6.2 Sound

We consider an ideal fluid, without viscosity, thermoconduction and external heating. Equation (6.18) is its flow equation; we write it as

$$\ddot{\boldsymbol{u}} - c^2 grad\, div \boldsymbol{u} = -grad\varphi \ , \tag{6.21}$$

where $c = \sqrt{K_{ad}/\rho}$ is the sound velocity and φ is an external potential. ρ is the constant and uniform density of the fluid, and φ is an external pressure divided by density. For air $c \simeq 350m/s$, for water $c \simeq 1500m/s$.[3] Since the displacement $\boldsymbol{u}$ in the above equation is generated by a gradient, its *curl* is zero, so its *curl curl* $= -\Delta + grad\, div = 0$, and $grad\, div = \Delta$. The sound is a potential flow. We can write the above equation as

$$\ddot{\boldsymbol{u}} - c^2 \Delta \boldsymbol{u} = -grad\varphi \ , \tag{6.22}$$

which is the standard form of the wave equation. Its solution is a superposition of plane waves of the form $e^{i(\boldsymbol{kr}-\omega t)}$, with frequency given by $\omega^2 = c^2 k^2$, where the source $-grad\varphi$ plays the role of a boundary condition. This is achieved by causal solutions, *i.e.* solutions which propagate in the future. The other boundary condition is provided by the behaviour of the solution at infinity, or at a finite boundary, in which case we have acausal vibrations.[4]

Let us consider an elementary solution $\boldsymbol{u} = \boldsymbol{u}_0 e^{i(\boldsymbol{kr}-\omega t)}$; we should not forget that its *curl* is zero, so the amplitude $\boldsymbol{u}_0$ is directed along the wave vector $\boldsymbol{k}$ ($\boldsymbol{k} \times \boldsymbol{u}_0 = 0$). The sound is a longitudinal wave. The amplitude of the velocity is $\boldsymbol{v}_0 = \omega \boldsymbol{u}_0 = ck\boldsymbol{u}_0$. The condition of validity of the sound equation (6.21) requires $v_0 \ll c$, *i.e.* $ku_0 \ll 1$: the sound amplitude should be much smaller than its wavelength $\lambda =$

[3] For an ideal gas with the equation of state $p = \frac{\rho}{M}T$, where M is the molecular mass, $K_{th} = p$. The velocity $c = \sqrt{p/\rho}$ for sound was established by Newton, *Philosophiae Naturalis Principia Mathematica*, Royal Society, London (1687). The necessity of the adiabatic modulus in this equation was introduced by Laplace. For the monoatomic ideal gas $K_{ad}/K_{th} = c_p/c_v$. Noteworthy, the sound velocity is of the order $c \simeq \sqrt{p/\rho} = \sqrt{T/M}$, which is the order of the thermal molecular velocity.

[4] M. Apostol, *Singular Equations of Waves and Vibrations*, Cambridge Scholars Publishing, Newcastle upon Tyne (2023).

$2\pi/k$. The relative change in volume and density is of the order $ku_0 \ll 1$ ($div\boldsymbol{u}$). The change in pressure is of the order $\delta p \simeq c^2\rho k u_0 \ll \rho c^2$, much smaller than the equilibrium pressure (equation (6.17)). This change affects our auditory senses, the human ear can hear sounds with frequency in the range $20 - 2\times 10^4 s^{-1}$ (s^{-1} is also called Hertz, Hz). The change in temperature is of the order $\delta T/T \simeq \frac{\alpha}{C_v}ku_0 \ll \alpha/C_v$, which is an extremely small fraction (equation (6.19)).
By multiplying equation (6.21) by $\dot{u}_i$ we get

$$\frac{\partial}{\partial t}\left[\frac{1}{2}\dot{u}^2 + \frac{1}{2}c^2(div\boldsymbol{u})^2\right] - c^2 div\left(\dot{\boldsymbol{u}}div\boldsymbol{u}\right) = -\dot{\boldsymbol{u}}grad\varphi\ . \qquad (6.23)$$

This equation shows that the work -$\rho\dot{\boldsymbol{u}}grad\varphi$ done by the external force per unit time and unit volume is equal to the change in energy

$$e = \frac{1}{2}\rho\dot{u}^2 + \frac{1}{2}\rho c^2(div\boldsymbol{u})^2 \qquad (6.24)$$

per unit time and unit volume and the energy transported per unit time and unit area

$$\boldsymbol{s} = -\rho c^2\left(\dot{\boldsymbol{u}}div\boldsymbol{u}\right) \qquad (6.25)$$

(energy flux density), such that

$$\frac{\partial e}{\partial t} + div\boldsymbol{s} = -\rho\dot{\boldsymbol{u}}grad\varphi\ . \qquad (6.26)$$

We can see that the energy of the sound consists of the kinetic energy $\frac{1}{2}\rho\dot{u}^2$ and the potential energy $\frac{1}{2}c^2\rho(div\boldsymbol{u})^2$, the latter arising from the compression-dilatation processes. Since $\omega^2 = c^2k^2$, the time average of the mechanical energy of the sound (per unit volume) is

$$\overline{e} = \rho\overline{\dot{u}^2}\ . \qquad (6.27)$$

From equation (6.26) the energy which falls on an area S per unit time, averaged over time and space, is the intensity

$$I = \overline{s} = \overline{\frac{1}{S}\frac{\partial}{\partial t}\int dVe}\ . \qquad (6.28)$$

It is measured in decibells (dB), defined by $10\lg(I/I_0)$, where I_0 is a reference intensity (for instance, $10^{-9}erg/cm^2\cdot s$). We can see that the displacement and the velocity in sound are very small.

If we write equation (6.21) as

$$\rho\frac{\partial v_i}{\partial t} - \rho c^2 div\, grad u_i = -\rho\partial_i\varphi \ , \tag{6.29}$$

we can see that the force exerted by sound on the unit area of a surface with normal $\boldsymbol{n}$ (pressure) is $\boldsymbol{P} = \rho c^2(\boldsymbol{n} grad)\boldsymbol{u}$.

6.3 Reflection and refraction of sound waves

It is convenient to introduce in equation (6.21) the potential ψ through $\boldsymbol{u} = grad\psi$, which transforms this equation in

$$\ddot{\psi} - c^2\Delta\psi = -\varphi \ . \tag{6.30}$$

Let us consider a plane wave $\psi = Ae^{i(\boldsymbol{kr}-\omega t)}$, where $\omega^2 = c^2k^2$, which is a solution of the homogeneous equation (6.30). We use its real part. We have

$$\boldsymbol{u} = i\boldsymbol{k}\psi \ , \ \boldsymbol{v} = \dot{\boldsymbol{u}} = \omega\boldsymbol{k}\psi \tag{6.31}$$

and the pressure exerted by the plane wave

$$\delta p = \rho c^2 div\boldsymbol{u} = -\rho\omega^2\psi \tag{6.32}$$

($-\delta p$ is the pressure exerted on the particles in the wave). Also, from equations (6.24) and (6.25)

$$\overline{e} = \frac{1}{2}\rho\omega^2k^2 \mid A \mid^2 \ , \ \overline{\boldsymbol{s}} = c\overline{e}\frac{\boldsymbol{k}}{k} \ , \tag{6.33}$$

where the temporal averages are taken.

Let us consider a separation surface $x = 0$ between two media denoted by 1 and 2 and an incident plane wave with the wavevector $\boldsymbol{k} = (k_x, k_y, 0)$, given by

$$\psi_1 = A_1 e^{i\omega\left(\frac{x}{c_1}\cos\theta_1 + \frac{y}{c_1}\sin\theta_1 - t\right)} \ . \tag{6.34}$$

The angle θ_1 made by $\boldsymbol{k}_1$ and the normal to the surface is the incidence angle. The boundary conditions act only for $x = 0$, so the y, z, t-dependence is not affected, *i.e.* ω, k_y and k_z are the same as in the incident wave. The directions of propagation of the incident, reflected and refracted waves are coplanar. From the same ω and k_y we get

$$\sin\theta_1 = \sin\theta_1^{'} \ , \ \frac{\sin\theta_1}{c_1} = \frac{\sin\theta_2}{c_2} \ , \tag{6.35}$$

where $\theta_1^{'}$, θ_2 are the angles of reflection and refraction, and $c_{1,2}$ are the sound velocities in the two media. It follows that the angle of reflection is equal to the angle of incidence and the ratio of the $\sin' s$ of the angles of incidence and refraction is equal to the ratio of the velocities. This latter result is known as Snell's law.[5]

The potentials of the reflected and refracted waves can be written as

$$\begin{aligned} \psi_1^{'} &= A_1^{'} e^{i\omega\left(-\frac{x}{c_1}\cos\theta_1 + \frac{y}{c_1}\sin\theta_1 - t\right)} \ , \\ \psi_2 &= A_2 e^{i\omega\left(\frac{x}{c_2}\cos\theta_2 + \frac{y}{c_2}\sin\theta_2 - t\right)} \ . \end{aligned} \tag{6.36}$$

On the surface of separation the normal velocities must be equal and the pressure must be the same, *i.e.*

$$\begin{aligned} \tfrac{\cos\theta_1}{c_1}(A_1 - A_1^{'}) &= \tfrac{\cos\theta_2}{c_2} A_2 \ , \\ \rho_1(A_1 + A_1^{'}) &= \rho_2 A_2 \ . \end{aligned} \tag{6.37}$$

The reflection coefficient R is defined as the ratio of the energy flux densities in the reflected and incident waves, time averaged. By making use of equations (6.25) amd (6.31), we get $R = \mid A_1^{'} \mid^2 / \mid A_1 \mid^2$. Solving equations (6.37) we obtain

$$R = \left(\frac{\rho_2\tan\theta_2 - \rho_1\tan\theta_1}{\rho_2\tan\theta_2 + \rho_1\tan\theta_1}\right)^2 \ , \tag{6.38}$$

where the angles θ_2 and θ_1 are related through equation (6.35). This relationship gives

$$R = \left(\frac{\rho_2 c_2\cos\theta_1 - \rho_1\sqrt{c_1^2 - c_2^2\sin^2\theta_1}}{\rho_2 c_2\cos\theta_1 + \rho_1\sqrt{c_1^2 - c_2^2\sin^2\theta_1}}\right)^2 \tag{6.39}$$

[5] Who rediscovered it in 1621.

The reflection coefficient can be zero for some incidence angle, in which case we have a total refraction. The refraction coefficient is $1 - R$, since the energy flux densities must be conserved (it differs, in general, from the ratio of the amplitude squares). The surface of separation suffers a pressure in the reflection-refraction process, which can be computed with the above formulae.

It is worth noting that the reflection and refraction computed above are valid for plane waves, whose (plane) wavefront has a spatial extension (much larger than the wavelengths). For a superposition of plane waves (solution of the wave equation) the incident wave puts in motion the particles on the separation surface, which, in turn, act as secondary sources of sound waves in both media (according to Huygens principle[6]); the secondary waves produced by these sources are the reflected and refracted waves in this case.[7]

The reflection/refraction described above is valid in the so-called approximation of geometrical acoustics (optics), where the wavefronts are much larger than the wavelength; it is used to say that the wavelength λ goes to zero in this case. The reflection coefficient is valid for $c_1/c_2 \geq \sin\theta_1$; otherwise, the geometrical acoustics is not valid. For the special case $c_1/c_2 = \sin\theta_{10} < 1$ we have a total reflection ($R = 1$). In these cases we need to resort to the solution of the wave equation with boundary conditions on the surface. An incident wave may propagate in this case along the surface, which is a "lateral" wave; and it is reflected with a conical wavefront, so this reflected wave is called a "conical" wave.[8]

6.4 Eikonal

Let $Ae^{i\Phi}$ be a complex function of time t and position $\boldsymbol{r}$, where A is an amplitude and Φ is a phase. Let us assume that both A and Φ vary little in time and space. In a small region in the neighbourhood

[6] Ch. Huygens in 1678 (*Dioptrica*) and *Traite de la Lumiere* (1690).

[7] M. Apostol, *loc. cit.*; B. F. Apostol and L. C. Cune, *A Guide to Practical Seismology*, Cambridge Scholars Publishing, Newcastle upon Tyne (2023).

[8] L. Landau and E. Lifshitz, *Course of Theoretical Physics*, vol. 6 (*Fluid Mechanics*), Elsevier, Oxford (1987) and References therein.

of a fixed position $\boldsymbol{r}_0$ and a fixed moment of time t_0 we can write the expansion

$$\Phi = \Phi_0 + \boldsymbol{r} grad\Phi + t\frac{\partial \Phi}{\partial t} \tag{6.40}$$

and neglect the variations of A. $\boldsymbol{k} = grad\Phi$ can be viewed as a wavevector and, similarly, $\omega = -\partial\Phi/\partial t$ can be viewed as a frequency. Therefore, within these limitations, we get a plane wave, which propagates with velocity $c = \omega/k$ along the wavevector $\boldsymbol{k}$. This velocity depends on t_0 and $\boldsymbol{r}_0$. Our complex function $Ae^{i\Phi}$ may satisfy the wave equation with $\omega^2 = c^2k^2$. For this approximation to be meaningful the region must still be much larger than the wavelength λ and the duration must be much longer than the period $T = 2\pi/\omega$ (we say that the approximation is valid in the limit $\lambda \rightarrow 0$). In addition, the equation

$$\left(\frac{\partial \Phi}{\partial x}\right)^2 + \left(\frac{\partial \Phi}{\partial y}\right)^2 + \left(\frac{\partial \Phi}{\partial z}\right)^2 - \frac{1}{c^2}\left(\frac{\partial \Phi}{\partial t}\right)^2 = 0 \tag{6.41}$$

is satisfied, as derived from $\omega^2 = c^2k^2$. In these conditions we can associate to the wave a ray, directed along $\boldsymbol{k}$. This approximation is known as the geometrical-optics (-acoustics) approximation, or the eikonal approximation, and equation (6.41) is known as the eikonal equation; Φ is the eikonal.

This equation is analogous to the Hamilton-Jacobi equation, where the mechanical action is replaced by phase, so $\boldsymbol{k}$ may be viewed as a momentum and ω may be viewed as energy. The propagation proceeds along the ray with velocity

$$\boldsymbol{v} = \partial\omega/\partial\boldsymbol{k} \; ; \tag{6.42}$$

this is the group velocity of the wave. The superposition in a wave packet is proportional to $e^{i\Delta\boldsymbol{k}(\boldsymbol{r}-\boldsymbol{v}t)}$, such that, by an integration, the wavepacket looks like a localized particle ($\sim \delta(\boldsymbol{r}-\boldsymbol{v}t)$), moving with the group velocity $\boldsymbol{v}$; this localization extends in fact up to corrections of the order of $\delta r \sim 1/(\delta k)^2$ (and corresponding durations), where δk is the range of the monochromatic components in the wave packet.

We have obviously

$$\dot{\boldsymbol{k}} = \frac{\partial^2\Phi}{\partial\boldsymbol{r}\partial t} = -\frac{\partial\omega}{\partial\boldsymbol{r}} \; , \tag{6.43}$$

which is Newton's law of motion for a body with position $\boldsymbol{r}$, momentum $\boldsymbol{k}$ and energy ω. Actually, the position is $\boldsymbol{r}_0$, and $\boldsymbol{k}$ and ω are functions of $\boldsymbol{r}_0$ and t_0. According to the above equations we may view ω as a function of t, $\boldsymbol{r}$ (*i.e.*, t_0, $\boldsymbol{r}_0$) and $\boldsymbol{k}$.

In steady flow $\omega = ck$ leads to $v = c$, directed along $\boldsymbol{k}$. In an inhomogeneous medium we know the function $c(\boldsymbol{r})$. It is worth noting that equation (6.26) tells that the energy propagates with velocity c, which in an inhomogeneous medium is the group velocity $v = c(\boldsymbol{r})$. Also, we note that the plane wave can be viewed as a free particle over limited regions around any $\boldsymbol{r}_0$ (according to equation (6.43)), with a slight variation of momentum and energy. This picture is valid over limited spatial regions and during limited times. During such a time interval the particle "disappears" from a region and "appears" in the neighbouring region, with a similar behaviour. The particle has a mean freepath of the order of the spatial extension δr and a lifetime of the order of $\tau \simeq \delta r/c$. We call this picture[9] the quasi-particle picture. From equation (6.43) we get $\dot{\boldsymbol{k}} = -k grad c$, where the magnitude $k = \omega/c$. We introduce $\boldsymbol{n} = \boldsymbol{k}/k = \frac{c}{\omega}\boldsymbol{k}$, such that $\boldsymbol{v} = c\boldsymbol{n}$, and compute

$$\begin{aligned}\dot{\boldsymbol{k}} &= \tfrac{\omega}{c}\dot{\boldsymbol{n}} - \tfrac{\omega}{c^2}\boldsymbol{n}(\boldsymbol{v} grad c) = \\ &= \tfrac{\omega}{c}[\dot{\boldsymbol{n}} - \boldsymbol{n}(\boldsymbol{n} grad c)] = -\tfrac{\omega}{c} grad c \ ,\end{aligned} \tag{6.44}$$

whence

$$\dot{\boldsymbol{n}} = -grad c + \boldsymbol{n}(\boldsymbol{n} grad c) \ ; \tag{6.45}$$

the variation along the ray element of length dl is

$$\frac{d\boldsymbol{n}}{dl} = \frac{1}{c}\left[-grad c + \boldsymbol{n}(\boldsymbol{n} grad c)\right] \ . \tag{6.46}$$

This equation gives the path of the ray. Since $d\boldsymbol{n}/dl$ tends to be maximal, $\boldsymbol{n}$ bends towards regions where c is smaller. For $c = const$ the path is a straight line, which corresponds to a maximal derivative $d\boldsymbol{n}/dl$. Therefore, the propagation is made with a minimal path length and duration. This is Fermat's principle.[10]

[9] Which we owe to Landau, around 1957.
[10] P. de Fermat in 1662.

Indeed, the ray equation (6.46) can be derived from the extremum of the integral

$$\int \boldsymbol{k} d\boldsymbol{l} = \int k dl \ , \tag{6.47}$$

taken between two fixed points. We have

$$\int \delta k dl + k \delta dl = \int (gradk \cdot \delta \boldsymbol{r} dl + k \boldsymbol{n} d\delta \boldsymbol{r}) = 0 \ ; \tag{6.48}$$

hence

$$\frac{d(k\boldsymbol{n})}{dl} = gradk \ . \tag{6.49}$$

On the other hand,

$$\frac{d(k\boldsymbol{n})}{dl} = \frac{dk}{dl}\boldsymbol{n} + k\frac{d\boldsymbol{n}}{dl} = (\boldsymbol{n} gradk)\boldsymbol{n} + k\frac{d\boldsymbol{n}}{dl} \ , \tag{6.50}$$

such that, from equations (6.49) and (6.50),

$$k\frac{d\boldsymbol{n}}{dl} = gradk - (\boldsymbol{n} gradk)\boldsymbol{n} \ , \tag{6.51}$$

and equation (6.46) follows by using $k = \omega/c$. We can see that the integral in equation (6.47) is $\int \frac{dl}{c}$ along the ray, so the propagation time is an extremum.

From equation (6.33) we have $\overline{\boldsymbol{s}} = c\overline{\varepsilon}\boldsymbol{n}$. In steady flow $div\overline{\boldsymbol{s}} = 0$. The distribution of energy is given by the equation

$$div\, c\,(\overline{\varepsilon} grad\Phi / \mid grad\Phi \mid) = 0 \ , \tag{6.52}$$

where the eikonal is given by equation (6.41).

6.5 Moving frame

Let us consider the wave equation

$$\frac{\partial^2 \psi'}{\partial t'^2} - c^2 \Delta' \psi' = 0 \tag{6.53}$$

in a reference frame K'. An elementary solution is the plane wave $\psi' \sim e^{i(\boldsymbol{k}'\boldsymbol{r}' - \omega' t')}$. Let us assume that the frame K' moves with the constant and uniform velocity $\boldsymbol{V}$ with respect to a frame K. The transformations of the coordinates are

$$\boldsymbol{r} = \boldsymbol{r}' + \boldsymbol{V}t' \ , \ t = t' \ , \tag{6.54}$$

such that

$$\frac{\partial}{\partial \boldsymbol{r}'} = \frac{\partial}{\partial \boldsymbol{r}} \ , \ \frac{\partial}{\partial t'} = \frac{\partial}{\partial t} + \boldsymbol{V}\frac{\partial}{\partial \boldsymbol{r}} \ ; \tag{6.55}$$

equation (6.53) becomes

$$\left(\frac{\partial}{\partial t} + \boldsymbol{V} grad\right)^2 \psi - c^2 \Delta \psi = 0 \ , \tag{6.56}$$

where ψ is the transformed function ψ'. We can see that

$$(\omega - \boldsymbol{V}\boldsymbol{k})^2 = c^2 k^2 = \omega'^2 \ . \tag{6.57}$$

The frequency is changed according to

$$\omega' = ck \rightarrow \omega - \boldsymbol{V}\boldsymbol{k} = ck \ , \tag{6.58}$$

while the wavevector remains the same. If ψ' is the potential of the sound (longitudinal wave), ψ describes that sound with a changed frequency. This is the Doppler effect,[11] as resulting from Galilei's transformations on changing the reference frame. Indeed, the transformations (6.54) applied to the plane wave give $e^{i[\boldsymbol{k}'\boldsymbol{r} - (\omega' + \boldsymbol{V}\boldsymbol{k})t]}$. The wave equation is changed by the transport term $\boldsymbol{V} grad$ in the time derivative. We note that the group velocity is changed into $c\boldsymbol{k}/k + \boldsymbol{V}$. The standard wave equation (6.53) is obtained from the equations of fluid mechanics by certain approximations, which lead to the simplified equations (6.14), (6.15) and (6.18). In particular, the Navier-Stokes equation is simplified to

$$\rho_0 \frac{\partial \boldsymbol{v}'}{\partial t'} = -grad\delta p \ , \tag{6.59}$$

where $\delta p = -\frac{K}{\rho_0}\delta\rho + \alpha\delta T$, for constant entropy. In the frame K the continuity equation becomes

$$\frac{\partial \delta\rho}{\partial t} + \rho_0 div\boldsymbol{v} + \boldsymbol{V} grad\delta\rho = 0 \ , \tag{6.60}$$

[11] Ch. Doppler, "Ueber das farbige Licht der Doppelsterne und einiger anderer Gestirne des Himmels", Abhandl. Koenigl. Boem. Gesellschaft **2** 465 (1842).

within the same approximation. It can also be written as

$$\frac{d\delta\rho}{dt}+\rho_0 div\boldsymbol{v}=0\ , \tag{6.61}$$

which is satisfied by $\delta\rho=-\rho_0 div\boldsymbol{v}$, provided we compute the velocity by

$$\boldsymbol{v}=\left(\frac{\partial}{\partial t}+\boldsymbol{V}grad\right)\boldsymbol{u}=\frac{d\boldsymbol{u}}{dt}\ . \tag{6.62}$$

Also, the Navier-Stokes equation should be written in the frame K as

$$\rho_0\left(\frac{\partial}{\partial t}+\boldsymbol{V}grad\right)\boldsymbol{v}=-grad\delta p\ , \tag{6.63}$$

which becomes

$$\rho_0\frac{d^2\boldsymbol{u}}{dt^2}=\rho_0\left(\frac{\partial}{\partial t}+\boldsymbol{V}grad\right)^2\boldsymbol{u}=-grad\delta p\ , \tag{6.64}$$

or

$$\left(\frac{\partial}{\partial t}+\boldsymbol{V}grad\right)^2\boldsymbol{u}-c^2 grad\,div\boldsymbol{u}=0\ . \tag{6.65}$$

By using $\boldsymbol{u}=grad\psi$, this equation is precisely equation (6.56), as expected; it includes the transport term $\boldsymbol{V}grad$ in the time derivative. It is worth noting that the standard wave equation is invariant to Lorentz transformations with the speed of light replaced by the speed of sound.[12]

Let us multiply equation (6.63) by $\boldsymbol{v}$:

$$\frac{\partial}{\partial t}\left(\frac{1}{2}\rho_0 v^2\right)+\boldsymbol{V}grad\left(\frac{1}{2}\rho_0 v^2\right)=-\boldsymbol{v}grad\delta p\ . \tag{6.66}$$

We can see that the kinetic energy is changed by a contribution arising from $\boldsymbol{V}grad$. For a plane wave this change is given by

$$E_{kin}=E_{kin}^{'}+\frac{\boldsymbol{V}\boldsymbol{k}}{\omega'}E_{kin}^{'}\ , \tag{6.67}$$

[12] M. Apostol, *Singular Equations of Waves and Vibrations*, Cambridige Scholars Publishing, Newcastle upon Tyne (2023).

or by making use of equation (6.57),

$$E_{kin} = \frac{\omega}{\omega - \boldsymbol{Vk}} E'_{kin} \ . \tag{6.68}$$

On the other hand $E'_{pot} = E'_{kin}$, from equation (6.59); the term $grad\delta p$ gives the potential energy (equation (6.23)). If we do the same manipulations in equation (6.64), with d/dt instead of $\partial/\partial t$, we get $E_{pot} = E_{kin}$. This is due to the transport term in $\boldsymbol{v}$, occurring in $\delta\rho = -\rho_0 div\boldsymbol{u}$. We get finally

$$E = \frac{\omega}{\omega - \boldsymbol{Vk}} E' \tag{6.69}$$

for the energy E of the sound in the moving medium and E' the energy of the sound in the medium at rest. This relationship can also be written as $E/\omega = E'/\omega'$.

As regards the Doppler equation $\omega = ck + \boldsymbol{Vk}$ (equation (6.58)) we note that the wavevector k can be large and ω small, such that the geometrical acoustics applies, providing V exceeds the sound velocity (supersonic flows). The rays propagate within the cone with angle α, $\cos\alpha = \pm c/V$; V/c is called the Mach number.[13] Also, we note that the geometrical acoustics (optics) may be applied to a spatial distribution of transport velocites $\boldsymbol{V}(\boldsymbol{r})$, providing $V \ll c$ and varies slowly over much larger distances than the wavelength.

6.6 Waves

One of the most usual manner of generating a sound is a short pressure impulse in a small volume. The wave equation for the potential ψ can then be written as

$$\ddot{\psi} - c^2\Delta\psi = \varphi_0 v\delta(\boldsymbol{r})T\delta(t) \ , \tag{6.70}$$

with a singular source placed in the small volume v and during a short time interval T. This is equation (6.30), where the potential

[13] E. Mach and P. Salcher, "Photographische Fixierung der durch Projectile in der Luft eingeleiteten Vorgaenge", Sitzungsber. Kaiserl. Akad. Wiss. Wien **95** 764 (1887).

$\varphi_0 = p_0/\rho$ is a pressure p_0 divided by density ρ. Obviously, the equation has spherical symmetry, so, for $\boldsymbol{r} \neq 0$, we have

$$\ddot{\psi} - \frac{c^2}{r^2}\frac{d}{dr}\left(r^2\frac{d\psi}{dr}\right) = 0 \ , \tag{6.71}$$

or, by using $\psi = \chi/r$,

$$\ddot{\chi} - c^2\chi'' = 0 \ ; \tag{6.72}$$

hence $\chi(t - r/c)$, where we choose the solution propagating in the future (retarded solution); and $\psi = \frac{\chi(t-r/c)}{r}$. For $r \to 0$, equation (6.70) can be written as

$$\Delta\psi = -\frac{\varphi_0 vT}{c^2}\delta(\boldsymbol{r})\delta(t) \ . \tag{6.73}$$

whose solution is[14]

$$\psi = \frac{\varphi_0 vT}{4\pi c^2 r}\delta(t) \ . \tag{6.74}$$

By comparing this solution with $\frac{\chi(t)}{r}$, we get $\chi(t) = \frac{\varphi_0 vT}{4\pi c^2}\delta(t)$ and

$$\psi = \frac{\varphi_0 vT}{4\pi c^2}\frac{\delta(t - r/c)}{r} \ . \tag{6.75}$$

This is called a spherical wave. From the derivation given above we can see that the source is treated as a boundary condition, by taking $r \to 0$ for any t.

In general, since $G = \frac{\delta(t-r/c)}{4\pi c^2 r}$ is solution of

$$\ddot{G} - c^2\Delta G = \delta(\boldsymbol{r})\delta(t) \ , \tag{6.76}$$

the solution of the equation

$$\ddot{\psi} - c^2\Delta\chi = S(r,t) \tag{6.77}$$

[14]Equation (6.73) is known as Poisson equation (with a singular source; S. D. Poisson, "Memoire sur la theorie du magnetisme en mouvement", Mem. Acad. Roy. Sci. Inst. France **6** 441 (1823)); the corresponding homogeneous equation is known as the Laplace equation (P.-S. de Laplace, *Theorie du Mouvement et de la Figure elliptique des Planetes*, Pierres, Paris (1784).

is

$$\psi = \int d\boldsymbol{r}'dt'G(t-t',\boldsymbol{r}-\boldsymbol{r}')S(\boldsymbol{r}',t') = \\ = \frac{1}{4\pi c^2}\int d\boldsymbol{r}'\frac{S(\boldsymbol{r}',t-|\boldsymbol{r}-\boldsymbol{r}'|)}{|\boldsymbol{r}-\boldsymbol{r}'|} \tag{6.78}$$

for a general source S. This is known as Kirchhoff's retarded solution.[15] G is called the Green function.[16]

Let us assume a pressure $p_0 = 10^5 dyn/cm^2$ ($1Kgf$ per $10cm^2$), in volume $v = 10cm^3$ in air ($c = 350m/s$, $\rho = 10^{-3}g/cm^3$); then, $\psi \simeq 10^{-1}/r(cm^2)$ and the displacement is $u \simeq 10^{-1}/r^2(cm)$. We can see that the displacement is very small; the velocity is given by $u/T \simeq 10^{-1}/Tr^2(cm/s)$, where T is the small duration of the shock.

The spherical wave given by equation (6.75), and written as

$$\psi = \frac{AT\delta(t-r/c)}{r}\ ,\ A = \frac{\varphi_0 v}{4\pi c^2}\ , \tag{6.79}$$

is a special wave. It is a singular function with the value

$$\psi = \frac{A}{r} \tag{6.80}$$

for a distance of the order $l = cT$ around $r = ct$, a duration T in this point, and zero outside. Its derivatives can be estimated by using the parameters T and l. For instance, the displacement is $u = \frac{A}{lr}$ in this point, directed along $\boldsymbol{r}$ (longitudinal wave); the energy density $e = \rho\dot{u}^2$ (equation (6.27)) is $e = \rho A^2/T^2l^2r^2$, the energy flux density $\boldsymbol{s} = -\rho c^2(\dot{\boldsymbol{u}}div\boldsymbol{u})$ (equation (6.25)) is $s = \rho c^2A^2/Tl^3r^2$, directed along $\boldsymbol{r}$; the continuity equation $\frac{\partial e}{\partial t}+div\boldsymbol{s} = 0$ is satisfied, because $l = cT$. The external force $\varphi_0 v\delta(\boldsymbol{r})T\delta(t)$ in equation (6.70) performs a mechanical work $\rho\dot{u}\varphi_0/d$ per unit time and unit volume, which can be written as $4\pi\rho c^2A^2/Tvd^3$, where d is the dimension of the "focus", where the external force is active. This mechanical work is transformed into energy $4\pi\rho c^2A^2/vd^3$, *i.e.* in energy e. From this equality we get $4\pi c^2/v = 1/T^2d$, or $v = 4\pi d^3$, with $d = cT$. Indeed, the pressure p_0 acts over the surface $4\pi d^2$ of the focus, so we should write the volume

[15] G. Kirchhoff, *Vorlesungen ueber Mathematische Physik*, Teubner, Leipzig (1883, 1891, 1894).

[16] G. Green, *An Essay on the Application of Mathematical Analysis to the Theories of Electricity and Magnetism*, Wheelhouse, Nottingham (1828).

v in equation (6.70) as $v = 4\pi d^2 \cdot d$. The energy e is transported through space with velocity c, as the energy flux density s, according to the continuity equation.

The solution given above is obtained by treating the source as a boundary condition. The other boundary condition is for $r \to \infty$. First we note that the wave goes like $1/r$, which ensures a finite energy flux through a surface with area $4\pi r^2$ at infinity. This is known as Sommerfeld radiation condition.[17] Next, we see that the wave depends on $t - r/c$, *i.e.* the wave propagates, travels. Such a dependence is specific to waves (which sometimes are also called travelling waves). At infinity the wave preserves this behaviour but we should recall that it has always a decaying factor $e^{-\gamma t}$, with $\gamma > 0$, which means also $e^{-\gamma r/c}$, which ensures the vanishing of the wave at infinity. Indeed, the wave equation (6.70) or (6.76) can be solved by Fourier transform, *e.g.*,

$$G = -\frac{1}{(2\pi)^4}\int d\omega d\boldsymbol{k}\frac{e^{-i\omega t+i\boldsymbol{kr}}}{\omega^2 - k^2} , \tag{6.81}$$

where the poles should be placed in the lower ω-half-plane, *i.e.* $\omega = \pm ck - i\varepsilon$, $\varepsilon \to 0^+$. This condition is necessary for having a zero wave in the past ($t < 0$), which is the causality principle. The integration in equation (6.81) gives $G = \frac{\delta(t-r/c)}{4\pi c^2 r}$, as above, with an additional factor $e^{-\varepsilon t} = e^{-\varepsilon r/c}$.

Also, if we perform only the $\boldsymbol{k}$-integration in equation (6.81) we get

$$G(\omega, \boldsymbol{r}) = \frac{1}{4\pi c^2 r}e^{i\frac{\omega}{c}r} , \tag{6.82}$$

a function which satisfies the Helmholtz equation

$$\Delta G + \frac{\omega^2}{c^2}G = -\frac{1}{c^2}\delta(\boldsymbol{r}), \tag{6.83}$$

derived from equation (6.76);[18] we can check by direct calculation this solution of equation (6.83), including the boundary condition $\int dV\Delta G = \oint d\boldsymbol{f} gradG = -1/c^2$, for an infinitesimal volume around the origin enclosed by a surface f.

[17] A. Sommerfeld, *Vorlesungen ueber Theoretische Physik*, Band VI (*Partielle Differentialgleichugen der Physik*), Akad. Verlagsgesellschaft, Leipzig (1966).

[18] H. Helmholtz, *Vorlesungen ueber die Elektromagnetische Theorie des Lichts*, Voss, Hamburg und Leipzig (1897).

6.7 Poisson's formula

In solving the Navier-Stokes equation (6.12) for waves we need to know the initial distribution of velocity and the initial distribution of pressure. Indeed, the velocity is given by $\boldsymbol{v} = \dot{\boldsymbol{u}} = grad\dot{\psi}$ and the pressure is given by $\delta p = -K_{ad} div\boldsymbol{u} = -K_{ad}\Delta\psi$; from these equations we can have ψ and $\dot{\psi}$ at the initial time, which means precisely the two conditions needed for solving the wave equation. We note that the solution of the wave equation $\ddot{\psi} - c^2\Delta\psi = 0$ depends on $t \pm r/c$, so ψ and $\dot{\psi}$ at the initial time are necessary and sufficient conditions. Let us consider two solutions ψ and φ of this equation. It is easy to see that

$$\begin{aligned} \frac{dI}{dt} = \frac{d}{dt}\int dV(\varphi\dot{\psi} - \psi\dot{\varphi}) = c^2 \int dV(\varphi\Delta\psi - \psi\Delta\varphi) = \\ = c^2 \oint d\boldsymbol{f}(\varphi grad\psi - \psi grad\varphi) = 0 \ , \end{aligned} \tag{6.84}$$

where the integration is performed over the whole space and the functions vanish at infinity. Therefore,

$$I = \int dV(\varphi\dot{\psi} - \psi\dot{\varphi}) = const \ . \tag{6.85}$$

Let us now consider the solution

$$\psi = \frac{1}{r}\delta[r - c(t_0 - t)] \ , \tag{6.86}$$

where r is the distance from the origin O. We have

$$\begin{aligned} \int dV\psi = 4\pi c(t_0 - t) \ , \\ \int dV\dot{\psi} = -4\pi c \ . \end{aligned} \tag{6.87}$$

Then, we have

$$I(t_0) = -4\pi c\varphi(\boldsymbol{r}, t_0) \tag{6.88}$$

and

$$I_0 = -\frac{\partial}{\partial t_0}\int dV\varphi_0\psi_0 - \int dV\dot{\varphi}_0\psi_0 \ , \tag{6.89}$$

where the suffix 0 indicates $t = 0$. We can compute the integrals in this equation, and get

$$I_0 = -\frac{\partial}{\partial t_0}\left(ct_0 \int do\varphi_{0,r=ct_0}\right) - ct_0 \int do\dot{\varphi}_{0,r=ct_0} \ , \tag{6.90}$$

where o is the solid angle. Now, $I(t_0) = I_0$, such that we get

$$\varphi(\boldsymbol{r}, t) = \frac{1}{4\pi}\left[\frac{\partial}{\partial t}\left(t \int do\varphi_{0,r=ct}\right) + t \int do\dot{\varphi}_{0,r=ct}\right] \ , \tag{6.91}$$

where the suffix 0 is omitted. This is Poisson's formula, which gives the solution in terms of the initial distribution and its initial time derivative.[19] We can see that the wave propagates from the initial region with velocity c, with a forward front and a backward front.

6.8 Explosions

The force density acting in a point $\boldsymbol{r}$ in fluid can be represented as $\boldsymbol{f}(t)w(\boldsymbol{r})$, where $\boldsymbol{f}$ is the force, t denotes the time and $w(\boldsymbol{r})$ is the spatial distribution of this force. This force produces a longitudinal displacement $\delta\boldsymbol{r}$, directed along the force, such that the net force acting in a point is

$$\boldsymbol{f}w(\boldsymbol{r} + \delta\boldsymbol{r}) - \boldsymbol{f}w(\boldsymbol{r}) = \boldsymbol{f}(\delta\boldsymbol{r}gradw) = f\delta\boldsymbol{r}gradw \ . \tag{6.92}$$

The product $f\delta r = M$ is a mechanical work; it is twice the energy transmitted to the fluid particle, taking $w = \delta(\boldsymbol{r})$. Let us suppose that the force acts a very short time T, *i.e.* let us take $M(t) = -MT\delta(t)$. For $M > 0$ such a force may be viewed as the source of an explosion in fluid; for $M < 0$ it corresponds to an implosion.[20] The corresponding sound equation reads

$$\rho\ddot{\boldsymbol{u}} - \rho c^2 grad\, div\boldsymbol{u} = -MT\delta(t)grad\delta(\boldsymbol{r}) \ , \tag{6.93}$$

[19] S. D. Poisson, "Memoire sur la theorie du son", J. de l'Ecole Polytechnique **7** 319 (1808).

[20] B. F. Apostol, *Seismology*, Nova, NY (2020), and References therein.

where $\boldsymbol{u}$ is the displacement. The solution is given by $\boldsymbol{u} = grad\psi$,

$$\rho\ddot{\psi} - \rho c^2 \Delta\psi = -MT\delta(t)\delta(\boldsymbol{r}) \ , \tag{6.94}$$

where

$$\psi = -\frac{MT}{4\pi\rho c^2}\frac{\delta(t-r/c)}{r} \ . \tag{6.95}$$

Making use of $\boldsymbol{u} = grad\psi$, we get the displacement in the near-field region ($\sim 1/r^2$) and in the far-field region

$$\boldsymbol{u} = \frac{MT}{4\pi\rho c^3}\frac{\delta'(t-r/c)}{r^2}\boldsymbol{r} \ , \tag{6.96}$$

valid for $r \gg cT = l$, where l is of the order of the spatial extension of the function $\delta(\boldsymbol{r})$ (the focus of the explosion). If we have a general source $gradS(t, \boldsymbol{r})$, the solution is obtained by using the Green function given by equation (6.95).
We can estimate the magnitude of the source M, as well as the parameter T, from measurements of the displacement in the far field regions. To this end we introduce the maximum of the displacement

$$v = \frac{M}{4\pi\rho c^3 T r} \tag{6.97}$$

from equation (6.96) and write the energy

$$E = \rho\dot{u}^2 \times 4\pi r^2 cT = \frac{4\pi\rho c r^2}{T}v^2 \ . \tag{6.98}$$

Since $E = M/2$, we get

$$\begin{aligned} cT = l = \sqrt{2rv} \ , \\ M = 2\pi\rho c^2(2rv)^{3/2} = 2\pi\rho c^2 l^3 \ ; \end{aligned} \tag{6.99}$$

the volume of the focal region is computed by $V = 4\pi(l/2)^2 l = \pi l^3$.

6.9 Waves in two dimensions

The solution of the equation

$$\ddot{G} - c^2\Delta G = \delta(\boldsymbol{R})\delta(t) \tag{6.100}$$

is

$$G = \frac{1}{4\pi c^2} \frac{\delta(t - R/c)}{R} \; . \tag{6.101}$$

If we write $\boldsymbol{R} = (\boldsymbol{r}, z)$ and integrate the equation with respect to z, we get

$$\ddot{G}_2 - \Delta_2 G_2 = \delta(\boldsymbol{r})\delta(t) \; , \tag{6.102}$$

where

$$G_2 = \int dz G \tag{6.103}$$

and Δ_2 is the laplacian in two dimensions. We can view this equation as the wave equation in two dimensions, or a wave equation in three dimensions with a singular source uniformly distributed along the z-coordinate, so that its solution does not depend on z. The integration in equation (6.103) gives immediately

$$G_2 = \frac{1}{2\pi c} \frac{1}{\sqrt{c^2 t^2 - r^2}} \; , \; ct > r \tag{6.104}$$

and zero for $ct < r$. This is called a cylindrical wave. We can see that it has a tail, in contrast with three dimensions.

The Fourier transform of G_2 with respect to time is

$$G_2(\omega, \boldsymbol{r}) = \frac{1}{4\pi c^2} \int dx \frac{e^{i\frac{\omega}{c}\sqrt{x^2+r^2}}}{\sqrt{x^2 + r^2}} = \frac{i}{4c^2} H_0^{(1)}(\omega r/c) \; ; \tag{6.105}$$

the function $H_0^{(1)}$ for $\omega/c > 0$ is the Hankel function (of the first kind and zeroth order). An approximate estimation of the integral in equation (6.105) gives the asymptotic behaviour[21]

$$H_0^{(1)}(\lambda r) = \begin{cases} \frac{2i}{\pi} \ln(\lambda r) \; , \; \lambda r \ll 1 \; , \\ \sqrt{\frac{2}{\pi \lambda r}} e^{i(\lambda r - \pi/40} \; , \; \lambda r \gg 1 \; . \end{cases} \tag{6.106}$$

The Green function G_2 can be used for the solution with a general source (equation (6.78)).

A Poisson's formula can also be obtained in two dimensions from equation (6.91), by noting that the area element $c^2 t^2 do$ of a sphere with

[21] M. Apostol, *Singular Equations of Waves and Vibrations*, Cambridge Scholars Publishing, Newcastle upon Tyne (2023).

radius ct has the projection $df\sqrt{c^2t^2-\rho^2}/ct$, where ρ is the distance from the source to the observation point (centre of the sphere); hence we can have do, such that equation (6.91) becomes

$$\varphi(\boldsymbol{r},t)=\frac{1}{2\pi c}\frac{\partial}{\partial t}\int\frac{\varphi_0 d\boldsymbol{r}'}{\sqrt{c^2t^2-(\boldsymbol{r}-\boldsymbol{r}')^2}}+ \\ +\frac{1}{2\pi c}\int\frac{\dot{\varphi}_0 d\boldsymbol{r}'}{\sqrt{c^2t^2-(\boldsymbol{r}-\boldsymbol{r}')^2}} \ , \tag{6.107}$$

where the integration is performed over the circle with radius ct; the factor 2 occurs because the projection of the area corresponds to two area elements on the sphere. This wave has a forward front, but not a backward one.

6.10 Vibrations

In the infinite space the wave equation has solutions which depend on $distance - c\cdot time$, where c is called sometimes phase velocity. These solutions satisfy the causality principle, in the sense that they exist only in the future. They are usually called waves, especially when they oscillate. The waves travel, propagate, with a small damping at infinity. The source of the waves is a boundary condition; the waves may satisfy boundary conditions on some points, lines, surfaces.

A completely different kind of solutions are admitted by the wave equation in a finite region of space. In these solutions the time dependence is separated by the space dependence. The time is indefinite, possibly with a small damping, while boundary conditions define these solutions. They are called vibrations. They exist both in the future and in the past. For vibrations it is convenient to use the time Fourier transform of the wave equation, which becomes the Helmholtz equation

$$\Delta\psi+k^2\psi=S \ , \tag{6.108}$$

where $k=\omega/c$ and S denotes the source.[22] The Green-function equa-

[22]The vibrations of a fluid sphere (and of a solid sphere) are treated in B. F. Apostol, *Seismology*, Nova, NY (2020) and B. F Apostol and L C. Cune, *A Guide to Practical Seismology*, Cambridge Scholars Publishing, Newcastle upon Tyne (2023).

tion reads

$$\Delta G + k^2 G = \delta(\boldsymbol{R}) \ , \tag{6.109}$$

whose solution is the spherical wave

$$G = -\frac{\cos kR}{4\pi R} \ ; \tag{6.110}$$

it is related to spherical harmonics and spherical Bessel functions. By integration over one or two coordinates, we get the Green functions in two and one dimensions.[23]

Let us suppose a recipient filled with fluid with an attached tube with length l and small cross-section S. The mass of fluid in the tube is $\rho l S$ and the force acting upon this mass is $S\delta p$, where δp is the pressure difference. Therefore, we have the equation of motion $\rho l S\dot{v} = S\delta p$, where v is the velocity of the fluid which flows through the tube. On the other hand, for small displacements, $\dot{\delta p} = c^2\dot{\delta\rho}$, and $\dot{\delta\rho} \simeq -\rho Sv/V$, where V is the volume of the vessel; therefore, we get the equation of motion $l\ddot{v} = -c^2 Sv/V$, which tells that we have a vibration with frequency $\omega = c\sqrt{S/lV}$.[24]

6.11 Turbulence as source of sound

The simplified equations (6.12) to (6.18) have been derived with the simplified Navier-Stokes equation

$$\frac{\partial \boldsymbol{v}}{\partial t} = -\frac{1}{\rho_0} grad\delta p \tag{6.111}$$

(omitting the damping term), where the change in pressure has been identified as

$$\delta p = c^2\delta\rho = -c^2\rho_0 div\boldsymbol{u} \ . \tag{6.112}$$

Let us keep formally in this equation the inertial term $\boldsymbol{T} = (\boldsymbol{v}grad)\boldsymbol{v}$ and write equation (6.111) as

$$\frac{\partial^2 \boldsymbol{u}}{\partial t^2} + \frac{1}{\rho_0} grad\delta p = -T \ . \tag{6.113}$$

---

[23] M. Apostol, *loc. cit.*

[24] L. Landau and E. Lifshitz, *Course of Theoretical Physics*, vol. 6 (*Fluid Mechanics*), Elsevier, Oxford (1987) and References therein.

Let us take the *div* in this equation,

$$\frac{1}{c^2}\frac{\partial^2 \delta p}{\partial t^2} - \Delta \delta p = \rho_0 div\boldsymbol{T} \ . \tag{6.114}$$

This is a wave equation for pressure, with source $\rho_0 div\boldsymbol{T}$.

We may imagine a region in space where we have turbulence. The turbulent velocities enter the term T; this term depends on position and time. Consequently, it can indeed be viewed as a source in the wave equation (6.114), which describes sound produced by turbulence.

It is worth noting that by taking the *div* in equation (6.113) we get, in general,

$$div\left(\frac{1}{\rho}grad\delta p\right) - \frac{1}{\rho c^2}\frac{\partial^2 \delta p}{\partial t^2} = 0 \ , \tag{6.115}$$

where ρ may vary, as in an inhomogeneous medium (and the source is omitted).

6.12 Long narrow tube

In equation (6.114) we need $\Delta\delta p = div\, grad\delta p$. Let us consider a long narrow tube with a slightly varying cross-section S along the x-direction. We have

$$div\ gradp = \frac{1}{\delta x}\frac{1}{\delta x}\frac{\delta F}{S} = \frac{1}{S}\frac{\partial}{\partial x}\left(S\frac{\partial p}{\partial x}\right) \ , \tag{6.116}$$

where F is the force acting upon the fluid. The sound equation reads

$$\frac{1}{c^2}\frac{\partial^2 \delta p}{\partial t^2} - \frac{1}{S}\frac{\partial}{\partial x}\left(S\frac{\partial p}{\partial x}\right) = 0 \ . \tag{6.117}$$

The source of this equation is localized at the end of the tube; it has the form $CS\delta l\delta(\boldsymbol{r})$, where δl is an undetermined length and C is a coefficient. This coefficient can be obtained from

$$C = \frac{1}{c^2}\ddot{\delta p} = -\rho\frac{\delta\ddot{V}}{V} \ , \tag{6.118}$$

where δV is the variation of the volume V. Obviously, we have $\delta\dot{V} = Sv_0$, where v_0 is the velocity at the end of the tube. Therefore $C = -\rho\frac{S\ddot{v}_0}{V} = -\rho\ddot{v}_0/l$, where l is the length of the tube. Open holes along the tube cause a variation of v_0.

The intensity of the sound emitted by the tube is $I = \rho c v_0^2 S$. If we know the function $v_0(t)$, we can compute the pressure of the sound from the wave equation with source, the displacement $\boldsymbol{u}$, the energy and intensity; equating the two intensities we can get an estimate of the parameter δl.

6.13 Emission of sound

The simplified equations of fluid mechanics can take a different form, by starting with the continuity equation and the Navier-Stokes equation

$$\begin{gathered} \delta\dot{\phi} + \rho div\boldsymbol{v} = 0 \ , \\ \rho\frac{\partial \boldsymbol{v}}{\partial t} = -grad\delta p \ . \end{gathered} \tag{6.119}$$

here we write $\delta p = K_{ad}\delta\rho/\rho$, where the entropy conservation is assumed, so $\delta p = c^2\delta\rho$. Now we take the time derivative in the second equation above and use the first equation,

$$\ddot{\boldsymbol{v}} - c^2 grad\, div\boldsymbol{v} = 0 \ , \tag{6.120}$$

which, obviously, is the time derivative of the homogeneous equation (6.21) for displacement. We introduce the velocity potential Φ through $\boldsymbol{v} = grad\Phi$ and get the wave equation

$$\ddot{\Phi} - c^2\Delta\Phi = 0 \ . \tag{6.121}$$

This equation is suitable for boundary conditions, where, for instance, the normal velocity $v_n = \frac{\partial\Phi}{\partial n} = 0$. We recall that the simplified equations are valid for small velocities (much smaller than the sound velocity), with small spatial variations, such that the inertial term may be neglected and, also, the viscosity is sufficiently small to be neglected. In these conditions the displacement u is much smaller than the wavelength; indeed, from $v \ll c$ we get $u \ll c/\omega = \lambda$, where u is the displacement and λ is the wavelength. Also, the flow is potential, and the bodies past which the fluid flows should be much larger than the displacement, otherwise the flow may not be potential (low Reynolds numbers).

The above equation is useful for oscillating bodies in fluid, where they can be viewed as source of sound. Also, we may view sound falling upon bodies at rest, which leads to sound scattering. We distinguish two cases in such situations. First, let the wavelength λ be much smaller than the dimension l of the body ($\lambda \ll l$). Then, the emitted energy flux density is $c\rho u_n^2$, where u_n is the normal velocity at the surface of the body; therefore the emitted intensity is

$$I = \rho c \oint \overline{u_n^2} df \ , \tag{6.122}$$

where the time average is taken and the integration is over the surface of the body.

In the opposite limit $\lambda \gg l$ the temporal term in equation (6.121) can be neglected near the body, so we get the Laplace equation

$$\Delta\Phi = 0 \tag{6.123}$$

for the flow potential. It is worth noting that this is the equation of potential flow for an incompressible fluid ($div\boldsymbol{v} = 0$); sound appears at large distances. The solution of this equation depends on the shape of the body. However, at large distances, but still small compared with λ, we may use

$$\Phi = -\frac{a}{r} + \boldsymbol{A} grad(1/r) \ , \tag{6.124}$$

where a and $\boldsymbol{A}$ are constants. It is easy to see that the first term corresponds to an oscillating volume V of the body, *i.e.* $\frac{a}{r^2} 4\pi r^2 = \dot{V}$ and $a = \frac{\dot{V}}{4\pi}$; this is valid for $l \ll r \ll \lambda$. On the other hand, in the far-field region (wave zone) the solution Φ should have the form $\frac{f(t-r/c)}{r}$, where f is some function. It follows that the potential far away from the body is

$$\Phi = -\frac{\dot{V}(t - r/c)}{4\pi r} \tag{6.125}$$

and the velociy is

$$\boldsymbol{v} = \frac{\ddot{V}(t - r/c)}{4\pi c r}\frac{\boldsymbol{r}}{r} . \tag{6.126}$$

The emission is isotropic, and the sound mean intensity is given by

$$I = \rho c \oint \overline{v^2} df = \frac{\rho}{4\pi c}\overline{\ddot{V}^2} . \tag{6.127}$$

By similar arguments the second term in equation (6.124) can be written as

$$\Phi = div\left[\boldsymbol{A}(t - r/c)/r\right] , \tag{6.128}$$

which gives

$$\Phi = -\dot{\boldsymbol{A}}(t - r/c)\boldsymbol{n}/cr , \tag{6.129}$$

where $n = \boldsymbol{r}/r$, and

$$\boldsymbol{v} = \frac{\boldsymbol{n}(\boldsymbol{n}\ddot{\boldsymbol{A}})}{c^2 r} . \tag{6.130}$$

This is called a dipole emission of sound. The intensity is

$$I = \frac{4\pi\rho}{3c^3}\overline{\ddot{\boldsymbol{A}}^2} . \tag{6.131}$$

The formulae are applicable for viscosity ν satisfying the condition $\sqrt{\nu/\omega} \ll l$ ($\sqrt{\nu/\omega}$ is the thickness of the adherent layer).
Similar formulae can be obtained for a cylinder, moving perpendicular to its axis.[25]

6.14 Scattering of sound

We consider a plane wave incident on a body. The dimension of the body should be larger than the displacement, in order to have a potential flow; also the wavelength should be much larger than the displacement, and the velocity should be much smaller than the sound velocity, with slight spatial variations, in order to use the simplified equations. Once arrived at the body the incident wave ceases its

[25] L. Landau and E. Lifshitz, *Course of Theoretical Physics*, vol. 6 (*Fluid Mechanics*), Elsevier, Oxford (1987) and References therein.

existence; the fluid on the surface of the body is put in motion, it becomes source of sound, which is emitted as a scattered wave.[26] Obviously, this situation is similar to the emission of sound, only that now it is the fluid which moves, not the body. In addition we assume that the body is sufficiently heavy and incompressible, as to view it at rest.

We assume a much larger wavelength λ than the dimension l of the body ($\lambda \gg l$). Then, we may use equations (6.125) and (6.129). The change in volume in equation (6.125) is given by $\delta V/V = \delta u/\delta r = ku = v/c$, with usual notations. Equation (6.125) becomes

$$\Phi^{(1)} = -\frac{V_0\dot{v}(t - r/c)}{4\pi cr} , \tag{6.132}$$

where V_0 is the volume of the body and v is the velocity of the incident wave. The scattered velocity is

$$\boldsymbol{v}_{sc}^{(1)} = \frac{V_0\boldsymbol{n}\ddot{v}(t - r/c)}{4\pi c^2 r} , \tag{6.133}$$

$\boldsymbol{n}$ being the unit vector in the direction of scattering.

In the Ideal Fluids chapter we derived the dipole potential $4\pi\rho A_i = m_{ij}u_j + \rho V_0 u_i$ for a body moving with velocity $\boldsymbol{u}$; now, we must replace the velocity $\boldsymbol{u}$ by the velocity $-\boldsymbol{v}$ of the fluid; we get

$$A_i = -\frac{m_{ij}}{4\pi\rho}v_j - \frac{V_0}{4\pi}v_i \ , \ \Phi^{(2)} = -\frac{\dot{\boldsymbol{A}}(t - r/c)\boldsymbol{r}}{cr^2} , \tag{6.134}$$

and

$$\boldsymbol{v}_{sc}^{(2)} = \frac{\boldsymbol{n}(\boldsymbol{n}\ddot{\boldsymbol{A}})}{c^2 r} . \tag{6.135}$$

The mean energy flux per unit time through the solid angle do is $c\rho\overline{v_{sc}^2}r^2 do$; by integrating over the solid angle we get the scattered intensity

$$I_{sc} = \frac{\rho V_0^2\overline{\ddot{v}^2}}{4\pi c^3} + \frac{4\pi\rho\overline{\ddot{A}^2}}{3c^3} \tag{6.136}$$

[26] See B. F. Apostol, "Scattering of longitudinal waves (sound) by defects in fluids. Rough surface", Centr. Eur. J. Phys. **11** 1036 (2013).

(the mixed term gives zero on integration). The cross-section $d\sigma$ is the ratio of the mean energy scattered in a solid angle (per unit time) to the mean incident intensity $\rho c\overline{v^2}$,

$$d\sigma = \frac{\overline{v_{sc}^2}}{\overline{v^2}} r^2 do \; ; \tag{6.137}$$

the total cross-section is obtained by integratig over the solid angle,

$$\sigma = \frac{V_0^2}{4\pi c^4} \frac{\overline{\ddot{v}^2}}{\overline{v^2}} + \frac{4\pi}{3c^4} \frac{\overline{\ddot{A}^2}}{\overline{v^2}} \; . \tag{6.138}$$

For a sphere with radius R the dipole potential is $\boldsymbol{A} = -\frac{1}{2}R^3\boldsymbol{v}$; the velocity of the incident wave at any point is $v = v_0 \cos\omega t$. We get the cross-section

$$d\sigma = \frac{\omega^4 R^6}{9c^4}\left(1 - \frac{3}{2}\cos\theta\right)^2 do \tag{6.139}$$

and $\sigma = (7\pi/9)(R^3\omega^2/c^2)^2$, where θ is the angle made by the incident wave with the scattering direction. The scattering is maximal for $\theta = \pi$. The force acting upon the sphere is $\overline{\varepsilon}_0 c\sigma/c$ on behalf of the incident wave, where $\overline{\varepsilon}_0$ is the mean energy density in the wave; similarly, the scattered wave produces a force $\overline{\varepsilon}_0 d\sigma$, which should be projected on the direction of the incident wave; therefore, the net force is

$$\overline{\varepsilon}_0 \int (1 - \cos\theta) d\sigma \; . \tag{6.140}$$

6.15 Absorption of sound

At the beginning of this chapter we have shown that the damping coefficient of the flow is $\gamma \simeq (\eta, \zeta)/\rho l^2$ (velocity $\sim e^{-\gamma t}$). For sound l is the wavelength.

The flow of fluids is a non-equilibrium thermodynamic process, with local equilibrium. The thermoconductivity and the viscosity dissipates energy, tending to destroy the flow and to bring the fluid in equilibrium. The mechanical energy in a non-equilibrium process is the released heat, $\delta E_{mech} = (\partial E/\partial S)\delta S = T\delta S$, where E is the energy, T is the temperature and S is the entropy; it is the maximum work done in a reversible process. Therefore, the rate of energy dissipation is $\dot{E}_{mech} = -T\dot{S}$. In the Viscous Fluids chapter we have shown that

$$\dot{S} = \int dV \frac{\kappa (gradT)^2}{T^2} +$$

$$+ \int dV \frac{\eta}{2T} \left(\partial_i v_j + \partial_j v_i - \frac{2}{3}\delta_{ij}\partial_k v_k\right)^2 + \tag{6.141}$$

$$+ \int dV \frac{\zeta}{T} (div\boldsymbol{v})^2 \ .$$

Since T varies slightly, we can write approximately

$$\dot{E}_{mech} = -\frac{\kappa}{T} \int dV (gradT)^2 +$$

$$-\frac{\eta}{2} \int dV \left(\partial_i v_j + \partial_j v_i - \frac{2}{3}\delta_{ij}\partial_k v_k\right)^2 - \tag{6.142}$$

$$-\zeta \int dV (div\boldsymbol{v})^2 \ .$$

Let us consider a plane wave with velocity $v = v_x = v_0 \cos(kx - \omega t)$. The temporal mean of the last two terms in equation (6.142) gives

$$-\frac{1}{2}\left(\frac{4\eta}{3} + \zeta\right) v_0^2 k^2 V_0 \ , \tag{6.143}$$

where V_0 is the volume of the fluid.

Now we would need the temperature variation in sound, for computing the thermoconductivity contribution. The sound is an adiabatic process, so the change of temperature is compensated by the change of the volume, or pressure. The variation of the temperature is given precisely by the variation of the entropy which we wish to compute. All these mean that the κ-contribution should be left aside. However, we follow the current procedure and start with the adiabatic transform

$$\delta s = \left(\frac{\partial s}{\partial V}\right)_T \delta V + \left(\frac{\partial s}{\partial T}\right)_V \delta T = \alpha \delta V + \frac{c_v}{T}\delta T = 0 \ , \tag{6.144}$$

where $\alpha = (\partial p/\partial T)_V$ is the thermal pressure coefficient and c_v is the specific heat per unit mass at constant volume. δV is the variation of volume per unit mass. We get

$$\delta T = -\frac{\alpha T}{c_v}\delta V = -\frac{K\beta T}{c_v}\delta V \ , \tag{6.145}$$

where $K = -V(\partial p/\partial V)_T$ is the isothermal modulus of compression and $\beta = \frac{1}{V}(\partial V/\partial T)_V$ is the thermal expansion coefficient; indeed, $\alpha = K\beta$ (see Appendix). On the other hand, $K/c_v = K_{ad}/c_p$ (see Appendix), such that

$$\delta T = -\frac{K_{ad}\beta T}{c_p}\delta V = -\frac{c^2\beta T}{c_p}\rho\delta V \ . \tag{6.146}$$

The volume variation in sound is given by

$$\delta V = Vku = -V\frac{v}{c} \ , \tag{6.147}$$

where u is the displacement, v is the velocity and k is the wavevector; we get

$$\delta T = \frac{\beta cT}{c_p}\rho V \cdot v = \frac{\beta cT}{c_p}v \ , \tag{6.148}$$

since $\rho V = 1$. Now we can compute the thermoconductivity contribution to equation (6.142); it is

$$-\frac{\kappa T}{2}\left(\frac{\beta c}{c_p}\right)^2 v_0^2 k^2 V_0 \ . \tag{6.149}$$

Here we use the formulae

$$c_p - c_v = T\beta^2(\partial p/\partial\rho)_T = T\beta^2(c_v/c_p)(\partial p/\partial\rho)_s = T\beta^2 c^2\frac{c_v}{c_p} \tag{6.150}$$

(see Appendix), such that equation (6.149) becomes

$$-\frac{\kappa}{2}\left(1/c_v - 1/c_p\right) v_0^2 k^2 V_0 \ . \tag{6.151}$$

Finally, we get

$$\dot{E}_{mech} = -\frac{1}{2}v_0^2 k^2 V_0\left[\frac{4\eta}{3} + \zeta + \kappa\left(1/c_v - 1/c_p\right)\right] \ . \tag{6.152}$$

The energy of the sound wave is $E_{mech} = \rho\overline{v^2}V_0 = \frac{1}{2}\rho v_0^2 V_0$, such that

$$\dot{E}_{mech}/E_{mech} = -\frac{k^2}{\rho}\left[\frac{4\eta}{3} + \zeta + \kappa\left(1/c_v - 1/c_p\right)\right] = -2c\gamma \ , \tag{6.153}$$

where the absorption coefficient γ, defined by equation (6.153), is such that $E_{mech} \sim e^{-2\gamma x}$; we can write also $\gamma = a\omega^2$.

According to the above calculations the phase of a plane wave can be written as $(\omega - ica\omega^2)t - kx$, or $(ck - ica\omega^2)t - \omega x/c$, so the dispersion relation is

$$k = \omega/c + ia\omega^2 \; ; \tag{6.154}$$

this relationship corresponds to a pressure wave described by the equation

$$\frac{\partial \delta p}{\partial x} = -\frac{1}{c}\frac{\partial \delta p}{\partial t} + a\frac{\partial^2 \delta p}{\partial t^2} \; . \tag{6.155}$$

By using the transformation $\tau = t - x/c$, this equation becomes

$$\frac{\partial \delta p}{\partial x} = a\frac{\partial^2 \delta p}{\partial \tau^2} \; , \tag{6.156}$$

which is a diffusion equation (thermal conduction). For a finite-duration initial wave, we get at large distances $\delta p \sim \frac{1}{\sqrt{ax}} e^{-\tau^2/4ax}$, *i.e.* a gaussian, with width $\sim \sqrt{x}$ and height $\sim 1/\sqrt{x}$. The energy density is $\sim \delta p^2 \sim 1/x$, the total energy is $\sim 1/\sqrt{x}$.

The sound absorption derived here is valid for small values of the absorption coefficient γ, *i.e.* $\gamma \ll \omega/c$; indeed, we have estimated this coefficient by using an undamped wave. Usually, $\kappa/c_{v,p} \simeq \eta, \zeta$, such that $\gamma \simeq \omega^2\nu/c^3$, where ν is the kinematic viscosity.[27] The condition becomes $\omega\nu \ll c^2$. On the other hand, in gases, $\nu \sim lv_{th} \simeq lc$, where l is the mean freepath and v_{th} is the thermal velocity, such that the condition becomes $\omega l \ll c$, or $l \ll \lambda$, which is indeed the condition for the existence of the sound. In liquids the viscosity is higher, but the sound velocity is also higher.[28] If the viscosity is very large, such that it gives a force comparable to the variations of the internal pressure, the Navier-Stokes equation itself becomes inapplicable.

Near solid walls the sound absorption can be very large, because the variations of the temperature in the sound wave occur over a very

[27] The thermoconductivity may be very large, at high temperatures, by radiative transfer; then, the flow is not adiabatic anymore, and the entropic sound may appear (see L. Landau and E. Lifshitz, *Course of Theoretical Physics*, vol. 6 (*Fluid Mechanics*), Elsevier, Oxford (1987)).

[28] M. Apostol, *Physical Kinetics*, Cambridge Scholars Publishing, Newcastle upon Tyne (2020).

small distance; therefore, the temperature gradients are very large, and, consequently, the thermoconduction is large, so the absorption is high. Similarly, the viscosity gives large gradients of the tangential velocity at solid walls, which leads to an increased attenuation of sound.

A special case of large sound absorption may occur in emulsions (two phases, *e.g.* oil droplets in water), due to the different properties of the two phases, which may lead to high gradients.[29]

6.16 Streaming of sound

Let us assume a sound wave flowing in the x-direction and a solid wall in the xz-plane. On the wall the velocity is zero, and the viscosity diminishes gradually the velocity of the fluid particles in the sound wave on approaching the wall from the y-direction. An acoustic boundary layer appears, and we know (from the Viscous Fluids chapter) that a penetration depth δ ($=\sqrt{2\nu/\omega}$) exists for the oscillatory motion in viscous fluids. Since the velocity is small, the flow in the boundary layer may be considered incompressible. Also, we consider the wavelength much larger than the extension of the wall, and the latter much larger than the penetration depth. On the other hand, the continuity equation requires a normal-to-the-wall velocity, so the flow is two-dimensional and it has vorticity. The velocity v_x varies much along the y-direction, while the component v_y is small, so we may view the flow as being laminar. We can see that the sound propagation takes place mainly away from the wall. This is the streaming of the sound.[30] It is a second-order effect in the amplitude of the main velocity.

The inertial term in the Navier-Stokes equation should be retained

[29] M. A. Isakovich, ZhETF **18** 907 (1948) (in Russian), quoted by L. Landau and E. Lifshitz, *loc. cit.*

[30] Lord Rayleigh, "The form of standing waves on the surface of running water", Proc. London Math. Soc. **1** 69 (1883); "On the circulation of air observed in Kundt's tubes, and on some allied acoustical problems", Phil. Trans. Roy. Soc. **175** 1 (1884).

now. Consequently, we write

$$\frac{\partial v_x}{\partial t} + v_x \frac{\partial v_x}{\partial x} + v_y \frac{\partial v_x}{\partial y} - \nu \frac{\partial^2 v_x}{\partial y^2} = U \frac{\partial U}{\partial x} + \frac{\partial U}{\partial t} \ , \qquad (6.157)$$

where U is the velocity in the main flow; in addition,

$$\frac{\partial v_x}{\partial x} + \frac{\partial v_y}{\partial y} = 0 \ . \qquad (6.158)$$

These are known as Prandtl's equations.[31] The equations are solved perturbatively.[32] An alternative approach is to use the equation for the stream function, with viscosity. It is found that v_x increases exponentially (governed by terms like $e^{-y/\delta}$) from the wall, and exhibits small stationary contributions, together with v_y (stationary vortices).

6.17 Second viscosity

The second viscosity ζ appears in compressible fluids. The compressibility effects appear in processes with a long relaxation time τ (slow processes). First, a parameter ξ changes in such processes according to

$$\dot{\xi} = -(\xi - \xi_0)/\tau \ , \qquad (6.159)$$

so $\xi = \xi_0/(1 - i\omega\tau)$.[33] The change in pressure can be written as $\delta p \simeq \delta K$, where K is the modulus of compression. We write this variation as $\rho(c_\infty^2 - c_0^2)$, where c_∞ is the "sound" velocity in the non-equilibrium state and c_0 is the sound velocity in equilibrium. The total change in pressure should be of the form

$$\delta p \sim \frac{\rho(c_\infty^2 - c_0^2)}{1 - i\omega\tau} \ . \qquad (6.160)$$

[31] L. Prandtl, "Ueber Fluessigkeitsbewegungen bei sehr kleiner Reibung", in Verhandlungen des III. Internationalen Mathematiker Kongresses, Heidelberg (1904), p. 484; see als H. Blasius, "Grenzschichten in Fluessigkeiten mit kleiner Reibung", Z. angew. Math. Phys. **56** 1 (1908).

[32] H. Schlichting, "Berechnung ebener periodischer Grenzschichtstroemungen", Phys. Z. **33** 327 (1932).

[33] This is known as Debye relaxation (P. Debye, *Polar Molecules*, Dover, NY (1929)).

In order to compare this change in pressure to the contribution $\zeta div\boldsymbol{v}$ to the viscosity tensor, we note that a factor of the order $\tau div\boldsymbol{v}$ should ocurr in such relaxation processes. Therefore, we have

$$\zeta = \frac{\tau\rho(c_\infty^2 - c_0^2)}{1 - i\omega\tau} \tag{6.161}$$

for the second viscosity in relaxation processes.[34] Also, we can write a generalized sound velocity, given by

$$c^2 = \frac{c_0^2 - c_\infty^2}{1 - i\omega\tau} + c_\infty^2 \ , \tag{6.162}$$

which shows indeed that c_∞ is a parameter corresponding to the sound velocity in slow processes. From this equation we get the wavevector

$$k = \omega\sqrt{(1 - i\omega\tau)/(c_0^2 - c_\infty^2 i\omega\tau)} \ , \tag{6.163}$$

which gives the attenuation coefficient in such processes.

We note that c_∞ should be larger than c_0, since ζ must be positive. Indeed, according to Le Chatelier's principle, the pressure should decrease in the out-of-equilibrium state, to its value in the equilibrium state, so $c_\infty > c_0$.

[34] L. I. Mandelstam and M. A. Leontovich, ZhETF **7** 438 (1937) (in Russian), quoted by L. Landau and E. Lifshitz, *Course of Theoretical Physics*, vol. 6 (*Fluid Mechanics*), Elsevier, Oxford (1987).

7 Thermal Conduction

7.1 Temperature distribution

Let us consider the equation of thermal transfer in fluids

$$\rho c_p \left(\frac{\partial T}{\partial t} + \boldsymbol{v} gradT \right) = div(\kappa gradT) + \sigma'_{ij}\partial_j v_i + S \ , \tag{7.1}$$

where ρ is the density, c_p is the specific heat per unit mass at constant pressure, $\boldsymbol{v}$ is the velocity, T is the temperature, κ is the thermoconductivity, σ'_{ij} is the viscosity tensor and S is an external source. The external source is $S = div\boldsymbol{q}$, where $\boldsymbol{q}$ is the external heat flux density. The term on the left is the variation of the entropy per unit volume and per unit time (multiplied by temperature). The viscosity tensor is a source of heat. We assume that the thermoconductivity is constant, such that we may replace $div(\kappa gradT)$ by $\kappa\Delta T$. We may assume that all the heat is conducted, *i.e.* the term on the right in the above equation is zero. Then, the entropy is constant, therefore the flow is adiabatic. Moreover, we may leave aside the viscosity term. Then, we get the equation of thermoconduction

$$\kappa\Delta T = -S \ , \tag{7.2}$$

and the flow is decoupled from the thermal conduction. This equation can also be written as $\kappa gradT = -\boldsymbol{q}$, which is Fourier's law.

Let us consider a solid wall in the yz-plane with a source of heat, localized on the wall, such that the temperature of the wall is a constant T_0. The temperature varies only along the x-distance, so we may write $S = \frac{1}{l}q$, $q = q_0 l\delta(x)$, where l is a small distance of the order of the spatial extension of the thickness of the wall; we have to solve the equation in one dimension

$$\frac{d^2T}{dx^2} = -\frac{q_0}{\kappa}\delta(x) \ ; \tag{7.3}$$

the solution is $T = -\frac{q_0}{2\kappa} \mid x \mid + T_0$ (the derivative is solely determined by the source). The temperature decreases linearly up tp $\mid x \mid = 2\kappa T_0/q_0$; beyond that point the solution becomes unphysical, because the plane is assumed infinite; its transverse extension limits the validity of equation (7.3).

It is worth noting that the solution is defined not only by T_0, but also by the thermal coupling of the source to the medium (the coefficient $q_0/2\kappa$). We note that $dT/dx = -\frac{q_0}{2\kappa} sgn(x)$, such that the solution can be written as $T = (dT/dx)x + T_0$, or $T = (dT/dx)_0 x + T_0$.

A similar problem can be formulated in two and three dimensions, where we need the solution of the Poisson equation with a singular source (or Laplace equation with the singular source treated as a boundary condition). In two dimensions the solution is again restricted by the finite length of the source (which extends uniformly along the coordinate which does not enter the equation). In both cases the singularity of the function $\delta(\boldsymbol{r})$ must be viewed as a physical singularity. Indeed, let us consider the equation

$$\Delta T = -\frac{q_0}{\kappa} r_0^2 \delta(\boldsymbol{r}) \tag{7.4}$$

in three dimensions, where r_0 is a small cutoff length. The solution is

$$T = \frac{q_0 r_0^2}{4\pi\kappa} \frac{1}{r} + T_\infty \ , \tag{7.5}$$

where T_∞ is the temperature at infinity. We must have $T = T_0$ for $r = r_0$, such that we get

$$T = (T_0 - T_\infty)\frac{r_0}{r} + T_\infty \ . \tag{7.6}$$

Since the parameter r_0 is small, the temperature difference is small.

7.2 Diffusion equation

In usual situations the temperature variations are small. Also, the velocities and their spatial variations are small. Consequently, we may view a constant density in equation (7.1) and neglect the viscosity and transport terms. Equation (7.1) becomes

$$\frac{\partial T}{\partial t} = \chi \Delta T + S \ , \tag{7.7}$$

where $\chi = \kappa/\rho c_p$ is called thermometric conductivity[1] and S is an external source. The equation is defined for $t > 0$ (and the source acts also for $t > 0$). This is the equation of heat (or of diffusion). The entropy is not constant anymore. It is given by $\delta s = \frac{c_p}{T_0}\delta T$ (per unit mass), where δT is given by equation (7.7) and T_0 is the equilibrium temperature.

For a finite spatial domain we expand the temperature and the source in equation (7.7) in series of eigenfunctions of the laplacian with boundary conditions. Then we get a first-order time equation for the expansion coefficients, which is solved immediately with initial condition. For an infinite spatial domain we use a Fourier expansion; the Fourier expansion and its derivatives fall off to zero at infinity. For a semi-infinite spatial domain we use a boundary condition at the border of the domain, another one being the vanishing of the function at infinity.

The solution is the sum of a particular solution and the solution of the homogeneous equation. For the particular solution it is convenient to use

$$T = \int dt' d\boldsymbol{r}' G(t - t', \boldsymbol{r} - \boldsymbol{r}') S(t', \boldsymbol{r}') \ , \tag{7.8}$$

where G is the (heat) Green function, which satisfies the equation

$$\frac{\partial G}{\partial t} - \chi \Delta G = \delta(t)\delta(\boldsymbol{r}) \ . \tag{7.9}$$

In one dimension we get immediately, by using Fourier transforms,

$$G = \theta(t)\frac{e^{-\frac{x^2}{4\chi t}}}{\sqrt{4\pi\chi t}} \ . \tag{7.10}$$

This is a gaussian with height $1/\sqrt{4\pi\chi \mid t \mid}$ and width $\Delta x = \sqrt{2\chi \mid t \mid}$. Its height "propagates" according to the diffusion law $x = \sqrt{4\chi t}$, while

[1]For air $\chi \simeq 10^{-1} cm^2/s$, for water $\chi \simeq 10^{-3} cm^2/s$.

becoming flatter and flatter. For $t \to 0$ the gaussian is zero for any $x \neq 0$ and ∞ for $x = 0$; in addition

$$\int dxG = 1 \ , \tag{7.11}$$

which means that the gaussian is similar to a δ-function.

In n dimensions we get, by Fourier transforming equation (7.9),

$$G = \frac{1}{(2\pi)^n} \int d\boldsymbol{k} e^{-\chi t k^2 + i\boldsymbol{kr}} = \theta(t) \frac{e^{-\frac{r^2}{4\chi t}}}{(4\pi\chi t)^{n/2}} \ , \tag{7.12}$$

with the same properties as above.

We may include the boundary conditions in the diffusion equation. For instance, let us assume a half-space $x > 0$, and a solution which depends only on the coordinate x. We multiply equation (7.7) by $\theta(x)$ and look for the solution $\theta(x)T$, which we denote by T, and retain it for $x > 0$. Equation (7.7) becomes

$$\frac{\partial T}{\partial t} - \chi \frac{\partial^2 T}{\partial x^2} = -\chi T_0 \delta'(x) - \chi T_1 \delta(x) \ , \tag{7.13}$$

where $T_0 = T_{x=0}$ and $T_1 = T'_{x=0}$ (without other sources). The solution is

$$\begin{aligned} T = & -\chi \textstyle\int_t dt' T_0(t') \frac{\partial}{\partial x} G(t-t', x) - \\ & -\chi \textstyle\int_t dt' T_1(t') G(t-t', x) \ , \end{aligned} \tag{7.14}$$

whence

$$\begin{aligned} T_0 &= -\chi \textstyle\int_t dt' T_1(t') G(t-t', 0) \ , \\ T_1 &= -\chi \textstyle\int_t dt' T_0(t') G(t-t', 0) \ . \end{aligned} \tag{7.15}$$

These equalities cannot be satisfied together, which shows that, indeed, the conditions T_0 and T_1 are too many; we must limit ourselves to one condition, for instance T_0.

The solution given above needs also the solution for the homogeneous equation, in order to satisfy the initial condition. However, we may include the initial condition in equation by the same procedure, such that, finally, we can write the solution as

$$T = \int dx' T_i(x') G(t, x - x') - \\ -\chi \int_t dt' T_0(t') \frac{\partial}{\partial x} G(t - t', x) \ , \tag{7.16}$$

where T_i is the initial condition.

Let us consider an yz-plane wall with a temperature $\sim e^{-i\omega t}$ and be interested in the temperature for $x > 0$ (and $t > 0$). Obviously, the wall is a localized source or a boundary condition. The solution will be proportional to $e^{-i\omega t}$. Therefore, we have to solve the equation

$$-i\omega T - \chi \frac{\partial^2 T}{\partial x^2} = -\chi 2 T_0 \delta'(x) \tag{7.17}$$

(equation (7.13)). We note that we put $2T_0$ for the source, because this equation is written for the whole x-axis, such that only half of the strength of the δ-function acts for $x > 0$. The equation is solved immediately by a Fourier transformation. We get

$$T = T_0 e^{-i\omega(t - x/\sqrt{2\omega\chi})} e^{-\sqrt{\omega/2\chi}x} \ . \tag{7.18}$$

The solution is a damped wave,[2] which propagates with velocity $\sqrt{2\omega\chi}$ and decays with a decaying length $\sqrt{2\chi/\omega}$ (χ has dimensions of squared length/time). Of course, we should take the real part of this solution.

Let us consider the same problem for a pointlike source (equation (7.9)). The equation is

$$\Delta T + \frac{i\omega}{\chi} T = -\frac{T_0 v}{\tau\chi} \delta(\boldsymbol{r}) \ , \tag{7.19}$$

where v is the small volume of the source and τ is a characteristic time (of the order of $1/\omega$). By using a Fourier transformation we get

$$T = \frac{T_0 v}{4\pi\tau\chi} \frac{e^{-i\omega(t - r/\sqrt{2\omega\chi})}}{r} e^{-\sqrt{\omega/2\chi}r} \ , \tag{7.20}$$

which is a damped spherical wave.

---

[2] A damped wave is rather a vibration.

7.3 Temperature in a half-space

Let us consider a half space $x > 0$, with a yz-plane surface at $x = 0$, and the heat equation

$$\frac{\partial T}{\partial t} = \chi \frac{\partial^2 T}{\partial x^2} \tag{7.21}$$

for $t > 0$. Let us consider a zero heat flux at $x = 0$, *i.e.* the boundary condition $\frac{\partial T}{\partial x} |_{x=0} = 0$. Then, the temperature has the form

$$T = \int_0 dk T(k,t) \cos kx \ , \tag{7.22}$$

where $T(k,t)$ is real. The equation gives $\dot{T}(k,t) = -\chi k^2 T(k,t)$, such that $T(k,t) = T(k) e^{-\chi k^2 t}$ and

$$T = \int_0 dk T(k) e^{-\chi k^2 t} \cos kx \ , \tag{7.23}$$

for $t > 0$. The coefficients $T(k)$ are determined by the initial condition, which we choose $T |_{t=0} = 2T_0 l \delta(x)$, l being a small distance over which there exists the temperature T_0 at $x = 0$. This choice corresponds to an initial temperature T_0 at the surface. We note the prefactor 2, which accounts for the integration of the function $\delta(x)$ from $x = 0$. We multiply the equation

$$T |_{t=0} = \int_0 dk T(k) \cos kx = 2T_0 l \delta(x) \tag{7.24}$$

by $\cos k'x$ and integrate from $x = 0$ to a large distance L. In order to determine the integral of the product $\cos kx \cos k'x$, apart from an exponential factor with a small exponent at large distance, we use the discretization $k = \frac{2\pi}{L} \times integer$ (the so-called cyclic condition), which leads to the well-known formula

$$\int_0^L dx \cos kx \cos k'x = \pi \delta(k - k') \ ; \tag{7.25}$$

we get $T(k) = T_0 l / \pi$ and

$$T = \frac{T_0 l}{\pi} \int_0 dk e^{-\chi k^2 t} \cos kx \ . \tag{7.26}$$

Finally, this integral gives

$$T = \frac{T_0 l}{\sqrt{\pi\chi t}} e^{-\frac{x^2}{4\chi t}} \ . \tag{7.27}$$

We can see that the temperature decreases in time, including the temperature at the surface. In order to keep a constant temperature T_0 at the surface we need a source of heat, which corresponds to

$$\frac{\partial T}{\partial t} = \chi \frac{\partial^2 T}{\partial x^2} + \frac{q_0}{\rho c_p}\delta(x) \ , \tag{7.28}$$

as in equation (7.3). The solution includes the static particular solution $-\frac{q_0}{2\kappa} \mid \kappa \mid + T_0$, as shown in a preceding section.

Let us keep constant the temperature T_1 at the surface $x = 0$ of the half-space, and assume an initial (constant) temperature T_0 for equation (7.21). It is convenient to set $T_1 = 0$, *i.e.* to reckon the temperature from T_1; for instance, T_0 becomes $T_0 - T_1$. The boundary condition is $T \mid_{x=0} = 0$, such that we seek a solution of the form

$$T = \int_0 dk T(k) e^{-\chi k^2 t} \sin kx \ , \tag{7.29}$$

where we use the discretization $k = \frac{\pi}{L} \times integer$. The calculations are straightforward, and we get

$$\begin{aligned} T &= \frac{T_0}{\sqrt{4\pi\chi t}} \int_0^\infty dx' \left[e^{-(x'-x)^2/4\chi t} - e^{-(x'+x)^2/4\chi t} \right] = \\ &= \frac{T_0}{\sqrt{\pi\chi t}} \int_0^x d\xi e^{-\xi^2/4\chi t} = T_0 erf(x/2\sqrt{\chi t}) \ , \end{aligned} \tag{7.30}$$

where erf is the error function; Actually,

$$T = T_1 + (T_0 - T_1) erf(x/2\sqrt{\chi t}) \ . \tag{7.31}$$

For $x/2\sqrt{\chi t} \simeq 1$ the error function is very close to unity; the temperature T_1 is quickly attained. We note the difference in the discretization of the eigenfunctions $\cos kx$, $\sin kx$ of the laplacian, which should preserve its self-adjoint character.[3] If a source is present, we use its decomposition in the eigenfunctions of the laplacian.

[3] See A. Corciovei, "On the "cyclic condition" in the study of the secular problem for finite one dimensional bodies", Rev. Roum. Phys. **10** 3 (1965).

7.4 Thermal waves

Let us consider the simplifed equations of fluid mechanics, described in the Sound chapter. The velocity is small and has small spatial variations. The continuity equation can be written as

$$\frac{\partial \delta\rho}{\partial t} + \rho_0 div\boldsymbol{v} = 0 \ , \tag{7.32}$$

where ρ_0 is the equilibrium density and $\delta\rho$ is the density variation. We may neglect the inertial term and the viscosity term in the Navier-Stokes equation, which becomes

$$\rho_0 \frac{\partial \boldsymbol{v}}{\partial t} = -grad\delta p \ , \tag{7.33}$$

where δp is the pressure variation. From these equations we get

$$\frac{\partial^2 \delta\rho}{\partial t^2} - \Delta\delta p = 0 \ . \tag{7.34}$$

In the heat transfer equation the temperature variation is small, so we may neglect the transport term and the viscosity term,

$$\rho_0 c_p \frac{\partial \delta T}{\partial t} \doteq \kappa \Delta \delta T \ . \tag{7.35}$$

Henceforth, we drop out the suffix 0 of the density.
The variation of the pressure is given by

$$\delta p = \frac{K}{\rho}\delta\rho + \alpha\delta T \ , \tag{7.36}$$

where $K = -V\left(\frac{\partial p}{\partial V}\right)_T$ is the isothermal modulus of compression, $\alpha = \left(\frac{\partial p}{\partial T}\right)_V$ is the thermal pressure coefficient and $V = 1/\rho$ is the volume of the unit mass. The variation of the entropy is

$$\delta s = -\frac{\alpha}{\rho^2}\delta\rho + \frac{c_v}{T}\delta T \ , \tag{7.37}$$

where c_v is the specific heat per unit mass at constant volume. As shown in the Ideal Fluids chapter, part of the temperature variation is compensated by density variations, due to the fact that the local equilibrium processes are much more rapid than the sound propagation (over the same molecular distance). This is an adiabatic process. Therefore, we write equation (7.37) as

$$\delta s = -\frac{\alpha}{\rho^2}\delta\rho + \frac{c_v}{T}\delta T_{ad} + \frac{c_v}{T}\delta T \tag{7.38}$$

and put

$$-\frac{\alpha}{\rho^2}\delta\rho + \frac{c_v}{T}\delta T_{ad} = 0 \ . \tag{7.39}$$

In the remaining part of equation (7.37) we replace c_v by c_p, as if the difference is a second-order effect: indeed,

$$c_p - c_v = \frac{K\beta^2 T}{\rho} \ , \tag{7.40}$$

where $\beta = \frac{1}{V}\left(\frac{\partial V}{\partial T}\right)_p$ is the dilatation coefficient (see Appendix, and the Ideal Fluids chapter). Consequently, the remaining part in equation (7.38) leads to the heat transfer equation (7.35). A similar decomposition is made in equation (7.36), which reads

$$\delta p = \frac{K}{\rho}\delta\rho + \alpha\delta T_{ad} + \alpha\delta T \ . \tag{7.41}$$

From equation (7.39) we get

$$\delta T_{ad} = \frac{\alpha T}{\rho^2 c_v}\delta\rho \ , \tag{7.42}$$

which, inserted in equation (7.41), gives

$$\delta p = \frac{K}{\rho}\delta\rho + \frac{\alpha^2 T}{\rho^2 c_v} + \alpha\delta T \ . \tag{7.43}$$

Here we use the thermodynamic relation $\alpha = K\beta$ (see Appendix), and get

$$\delta p = \frac{K}{\rho}\left(1 + \frac{K\beta^2 T}{\rho c_v}\right)\delta\rho + \alpha\delta T \ . \tag{7.44}$$

By another thermodynamic relation, we have

$$K_{ad} = K\left(1 + \frac{K\beta^2 T}{\rho c_v}\right) \tag{7.45}$$

(see Appendix; and $K_{ad} = K\frac{c_p}{c_v}$), such that, we have finally

$$\delta p = \frac{K_{ad}}{\rho}\delta\rho + \alpha\delta T \tag{7.46}$$

and equation (7.33) becomes

$$\rho\frac{\partial \boldsymbol{v}}{\partial t} = -\frac{K_{ad}}{\rho} grad\delta\rho - \alpha grad\delta T \ . \tag{7.47}$$

We can use the variables p and T, such that we get $\frac{c_p}{c_v}\alpha grad\delta T$ in equation (7.47), instead of $\alpha grad\delta T$ (see the Ideal Fluids chapter). By making use of the continuity equation (7.32), we get the equation of sound

$$\frac{\partial^2 \delta\rho}{\partial t^2} - c^2\Delta\delta\rho = \alpha\Delta\delta T \ , \tag{7.48}$$

where $c = \sqrt{K_{ad}/\rho}$ is the velocity of sound. Equation (7.35)

$$\rho c_p \frac{\partial \delta T}{\partial t} = \kappa\Delta\delta T \ . \tag{7.49}$$

must be added, which is independent of equation (7.48); in equation (7.48) δT plays the role of source of sound.

In the heat transfer equation we can have a source. Let us assume that the source is given by equation (7.18), written as

$$\delta T = T_0 e^{-i\omega t + ik_0 x} \ , \tag{7.50}$$

where $k_0 = \sqrt{\omega/2\chi}(1+i)$. The solution of equation (7.48) is

$$\delta\rho = -\frac{\alpha T_0}{c^2 + i\omega\chi} e^{-i\omega(t - x/\sqrt{2\omega\chi})} e^{-\sqrt{\omega/2\chi}x} \ . \tag{7.51}$$

This is a "sound" produced by thermal waves, propagating with velocity $\sqrt{2\omega\chi}$ and damped over distances of the order of $\sqrt{2\chi/\omega}$; it implies density and pressure variations, produced by a thermal wave. A similar result

$$\begin{aligned} \delta\rho = &-\frac{\alpha T_0 v}{4\pi\tau\chi(c^2+i\omega\chi)} \frac{e^{-i\omega(t-r/\sqrt{2\omega\chi})}}{r} e^{-\sqrt{\omega/2\chi}r} - \\ &-\frac{\alpha T_0 v}{2\pi c^2\tau}\frac{\omega}{\omega\chi - ic^2}\frac{\cos\omega r/c}{r} e^{-i\omega t} \end{aligned} \tag{7.52}$$

is obtained from equations (7.19) and (7.48). We note that apart from the "sound" produced by the thermal waves we have also genuine sound vibrations, with a small amplitude (since $\omega\chi \ll c^2$); indeed, in solving these equations with a spatial Fourier transform we should place the poles given by $k = \pm\omega/c$ in the upper half-plane ($k = \pm\omega/c + i\varepsilon$, $\varepsilon \to 0^+$), in order to have a vanishing solution at infinity $e^{\pm i\omega r/c - \varepsilon r}$. Also we note that the time dependence in the above equations of thermal waves implies a pole $\omega = ic^2/\chi$, which indicates an instability.

7.5 Entropic sound

The velocity of propagation of the thermal waves is of the order of $\sqrt{\omega\chi}$. It may happen that the thermal conduction is so high that the local thermodynamic equilibrium is not attained. Consequently, we cannot make the decomposition $\delta T/\delta T_{ad}$ in equation (7.38). In this case we use the continuity equation (7.32),

$$\frac{\partial \delta\rho}{\partial t} + \rho div\boldsymbol{v} = 0 \ , \tag{7.53}$$

the Euler equation (7.47)

$$\rho\frac{\partial \boldsymbol{v}}{\partial t} = -\frac{K}{\rho} grad\delta\rho - \alpha grad\delta T \ . \tag{7.54}$$

and the heat transfer equation

$$\rho T\frac{d\delta s}{dt} \simeq \rho T\frac{\partial \delta s}{\partial t} = \kappa\Delta\delta T \ , \tag{7.55}$$

where

$$\delta s = -\frac{\alpha}{\rho^2}\delta\rho + \frac{c_v}{T}\delta T \tag{7.56}$$

(equation (7.38), for small velocities). From the continuity equation and the Euler equation we get

$$\frac{\partial^2 \delta\rho}{\partial t^2} = c_s^2\Delta\delta\rho + \alpha\Delta\delta T \ , \tag{7.57}$$

where $c_s = \sqrt{K/\rho}$ is the velocity of the thermal sound (this is the velocity of the entropic sound according to the Ideal Fluids chapter). Equations (7.55) and (7.56) give

$$-\frac{\alpha}{\rho^2}\frac{\partial \delta\rho}{\partial t} + \frac{c_v}{T}\frac{\partial \delta T}{\partial t} = \frac{\kappa}{\rho T}\Delta\delta T \ . \tag{7.58}$$

We seek solutions of the form $e^{-i\omega t + i\boldsymbol{kr}}$ of the equations (7.57) and (7.58). In the condition of compatibility we use equation (7.45) and $K_{ad}/K = c_p/c_v$ (see Appendix). We get

$$k^4 - \left(\frac{\omega^2}{c_s^2} + \frac{i\omega}{\chi}\right)k^2 + \frac{i\omega^3}{\chi c^2} = 0 \ , \tag{7.59}$$

where $c = \sqrt{K_{ad}/\rho}$ is the velocity of sound. For $\omega\chi \ll c^2$ the dispersion relation is $\omega^2 = c^2k^2$, which corresponds to the adiabatic sound. The condition $\omega\chi \ll c^2$ is equivalent with $\chi/\omega \ll \lambda^2$, where $\sqrt{\chi/\omega}$ is the attenuation length of heat conduction and λ is the wavelength; the sound is propagated with a very small damping. In the opposite limit $\omega\chi \gg c^2$ the dispersion relation is $\omega^2 = c_s^2k^2$ and the sound is entropic; the attenuation is again very small.

7.6 Free convection

Let us write down the variation of the pressure

$$\delta p = \frac{K}{\rho}\delta\rho + \alpha\delta T \tag{7.60}$$

(equation (7.36)). For the same change in pressure we have the relative changes in density and temperature

$$\delta\rho/\rho = \frac{\delta p}{K} \ , \ \delta T/T = \frac{\delta p}{\alpha T} \ , \tag{7.61}$$

such that their ratio is

$$\frac{\delta\rho/\rho}{\delta T/T} = \frac{\alpha T}{K} = \beta T \ . \tag{7.62}$$

For a monoatomic ideal gas $\beta T = 1$. In general, the relative variations $\delta\rho/\rho$ and $\delta T/T$ are of comparable orders of magnitude. The identity $\alpha = K\beta$ used in equation (7.62) is valid in thermodynamic equilibrium.

Let us assume that a temperature change δT occurs, which produces a dilatation at constant pressure. The change in density is

$$\delta\rho = \left(\frac{\partial\rho}{\partial T}\right)_p \delta T = -\rho\beta\delta T \; ; \tag{7.63}$$

there is no compression at constant temperature in this case.

In the continuity equation

$$\frac{\partial\rho}{\partial t} + div(\rho\boldsymbol{v}) = 0 \tag{7.64}$$

we may view a significant variation of the velocity, such that $\delta\rho$ may be neglected. We get the continuity equation for incompressible fluids

$$div\boldsymbol{v} = 0 \; . \tag{7.65}$$

We expect the production of entropy caused by viscosity to be small, in comparison with the temperature change; so, we write the heat transfer equation as

$$\rho T\frac{ds}{dt} = \kappa\Delta T \; . \tag{7.66}$$

At constant pressure for an incompressible fluid the entropy change is $\delta s = \frac{c_p}{T}\delta T$, such that the heat transfer equation becomes

$$\rho c_p\left(\frac{\partial T}{\partial t} + \boldsymbol{v}gradT\right) = \kappa\Delta T \; . \tag{7.67}$$

Let us assume that the fluid is placed in uniform gravitational field; the Navier-Stokes equation reads

$$\frac{\partial\boldsymbol{v}}{\partial t} + (\boldsymbol{v}grad)\boldsymbol{v} = -\frac{1}{\rho}gradp + \boldsymbol{g} + \nu\Delta\boldsymbol{v} \; ; \tag{7.68}$$

here the viscosity does contribute. In this equation we write

$$-\tfrac{1}{\rho}gradp + \boldsymbol{g} \Rightarrow$$

$$\Rightarrow -\tfrac{1}{\rho}gradp + \tfrac{1}{\rho^2}gradp \cdot \delta\rho - \tfrac{1}{\rho}grad\delta p + \boldsymbol{g} = \tag{7.69}$$

$$= \tfrac{\boldsymbol{g}}{\rho}\delta\rho - \tfrac{1}{\rho}grad\delta p = -\beta\boldsymbol{g}\delta T - \tfrac{1}{\rho}grad\delta p \; .$$

The Navier-Stokes equation becomes

$$\frac{\partial \boldsymbol{v}}{\partial t} + (\boldsymbol{v}grad)\boldsymbol{v} = \beta\boldsymbol{g}\delta T - \frac{1}{\rho}grad\delta p + \nu\Delta\boldsymbol{v} \; . \tag{7.70}$$

We note the change of sign in the buoyancy force $-\boldsymbol{g}\delta\rho$ (according to the discussion in the Ideal Fluids chapter). In equation (7.67) T may be replaced by δT. We view δT, δp and $\boldsymbol{v}$ as unknowns, for the three equations

$$div\boldsymbol{v} = 0 \; ,$$

$$\tfrac{\partial T}{\partial t} + \boldsymbol{v}gradT = \chi\Delta T \; , \tag{7.71}$$

$$\tfrac{\partial \boldsymbol{v}}{\partial t} + (\boldsymbol{v}grad)\boldsymbol{v} = \beta\boldsymbol{g}T - \tfrac{1}{\rho}gradp + \nu\Delta\boldsymbol{v} \; ,$$

where δT and δp are re-denoted by T and p. These are the Oberbeck-Boussinesq equations of free convection.[4]

The first two equations (7.71) are determined. We note the inconsistency between the assumptions underlying these two equations and the third equation (7.71), which, however, may look like a reasonable approximation. The variations of the quantities should be taken with respect to the thermal equilibrium in gravitational field (see the Ideal Fluids chapter).

The continuity equation and the heat transfer equation lead to

$$\boldsymbol{v} = \chi grad \ln T \; , \; \Delta \ln T = 0 \tag{7.72}$$

[4] A. Oberbeck, "Ueber die Waermeleitung der Flussigkeiten bei Beruecksichtigung der Stroemungen infolge von Temperaturdifferenzen", Ann. Physik **243** 271 (1879); J. Boussinesq, *Theorie Analytique de la Chaleur*, Gauthiers-Villars, Paris (1903).

(where we assume $T > 0$).

For an infinite horizontal extension the solution does not depend on the horizontal coordinates; it is given by $\ln(T/T_0) = v_0 z/\chi$, where T_0 is the temperature on the boundary $z = 0$ and $v_0 < 0$ is the vertical velocity. The warmer fluid raises, once the temperature change is established, and its place is taken by the colder fluid. For negative T the velocities change sign. This is the convection phenomenon. The pressure change follows immediately from the Navier-Stokes equation (7.71). The solution has a limited validity for finite z, due to the unphysical infinite horizontal extension of the boundary $z = 0$. For a finite horizontal extension, as, for instance, a column of fluid with a transverse radius R, the solution is of the form $J_0(kr)e^{-kz}$. (Compatible) boundary conditions are necessary, both for $z = 0$ and $r = R$.[5]

Convection is also called convective instability. In general, the stability of the solution can be investigated by giving small perturbations to the unknowns and linearizing equations (7.71). Positive imaginary parts in the frequency in the time factor $e^{-i\omega t}$ indicates an instability (sometimes it is called an absolute instability). Such instabilities signal the occurrence of turbulence.[6] Within our approximation the solution is stable, but, in general, it may exhibit instabilities.[7] However, we should note that the solution given by equation (7.72) does not fulfil the stability condition of convection, given in the Ideal Fluids chapter, for any reasonable temperature T_0 and velocity v_0 (*i.e.*, we have convection).

[5] See, for instance, G. A. Ostroumov, "Free convection in closed cavities: a review of work carried out at Perm, USSR", Int. J. Heat&Mass Transfer **8** 259 (1965).

[6] C. C. Lin, *The Theory of Hydrodyamic Stability*, Cambridge University Press, Cambridge (1955); P. G. Drazin and W. H. Reid, *Hydrodynamic Stability*, Cambridge University Press, Cambridge (1981).

[7] S. Chandrasekhar, *Hydrodynamic and Hydromagnetic Stability*, Clarendon Press, Oxford (1961) and References therein.

8 Diffusion

8.1 Binary mixture

Let us consider $N_{1,2}$ particles with mass $M_{1,2}$ in the same volume, such that

$$N_1 M_1 + N_2 M_2 = m \ , \tag{8.1}$$

where m is the total mass; or

$$m_1 + m_2 = m \ , \tag{8.2}$$

where $m_{1,2} = N_{1,2} M_{1,2}$. This is a mixture. We define the mass concentrations $c_{1,2} = m_{1,2}/m$, such that the densities are $\rho_{1,2} = c_{1,2}\rho$, where ρ is the density of the mixture. Since $c_1 + c_2 = 1$, we use the notation $c = c_1$ and $c_2 = 1 - c$. We call c the mass concentration of the mixture.

It is also convenient to introduce the number of particles per unit mass $n_{1,2} = N_{1,2}/m$, such that equation (8.1) becomes

$$n_1 M_1 + n_2 M_2 = 1 \ ; \tag{8.3}$$

also, we have $c_{1,2} = n_{1,2} M_{1,2}$.

The total mass is conserved, so we have

$$\frac{\partial \rho}{\partial t} + div(\boldsymbol{\rho v}) = 0 \ , \tag{8.4}$$

where $\boldsymbol{v}$ is the velocity of the mixture. If we multiply this equation by $\boldsymbol{r}$ and integrate, we get

$$\frac{d}{dt}\int dV \boldsymbol{r}\rho = \int dV \rho \boldsymbol{v} \ , \tag{8.5}$$

where on the left is the velocity of the centre of mass. Therefore, $\rho\boldsymbol{v}$ is the momentum of unit volume and $\boldsymbol{v}$ is the momentum of unit mass. Consequently, the Navier-Stokes equation remains unchanged (with different viscosity coefficients).

The mass of each mixture component must also be conserved; for instance, we would have an equation (8.4) for $c\rho$. This mass density changes not only through the velocity $\boldsymbol{v}$ but also by a mass flux density $\boldsymbol{i}$ (diffusion current), which accounts for the molecular mixing of the two kinds of particles. Therefore, we have an additional equation of continuity

$$\frac{\partial(c\rho)}{\partial t}+div(c\rho\boldsymbol{v})+div\boldsymbol{i}=0\ ; \tag{8.6}$$

the mass flux density is $c\rho\boldsymbol{v}+\boldsymbol{i}$, while for the other component it is $(1-c)\rho\boldsymbol{v}-\boldsymbol{i}$. Making use of equation (8.4), we can write the one-component continuity equation (8.6) as

$$\rho\left(\frac{\partial c}{\partial t}+\boldsymbol{v}gradc\right)+div\boldsymbol{i}=0\ . \tag{8.7}$$

In order to write down the energy conservation we start with the fundamental law of Thermodynamics

$$d\varepsilon=-pdV+Tds+\mu_1dn_1+\mu_2dn_2\ , \tag{8.8}$$

where $\mu_{1,2}$ are the chemical potentials; p denotes pressure, V is volume per unit mass, T is temperature and s is entropy per unit mass. From equation (8.3) we have $M_1dn_1+M_2dn_2=0$ and $M_1dn_1=dc$, such that

$$d\varepsilon=-pdV+Tds+\left(\frac{\mu_1}{M_1}-\frac{\mu_2}{M_2}\right)dc\ . \tag{8.9}$$

We denote by μ the bracket in equation (8.9), such that we can write

$$d\varepsilon=\frac{p}{\rho^2}d\rho+Tds+\mu dc \tag{8.10}$$

for energy and

$$dw=\frac{dp}{\rho}+Tds+\mu dc \tag{8.11}$$

for the enthalpy $w=\varepsilon+p/\rho$. We call μ reduced chemical potential.

The energy conservation is obtained as in the Viscous Fluids chapter, with additional contributions; it reads

$$\frac{\partial}{\partial t}\left(\frac{1}{2}\rho v^2+\rho\varepsilon\right)=-\partial_j\left[\rho v_j\left(\frac{1}{2}v^2+w\right)-v_i\sigma'_{ij}+q_j\right]+$$
$$+\rho T\left(\frac{\partial s}{\partial t}+\boldsymbol{v}grads\right)-\sigma'_{ij}\partial_j v_i+div\boldsymbol{q}-\mu div\boldsymbol{i}\ , \tag{8.12}$$

where the heat flux density $\boldsymbol{q}$ is added and subtracted (heat current); obviously, now it does not depend only on temperature. The term in square brackets in this equation is the energy flux density: the first two contributions arise from the mechanical macroscopic motion, the other two contributions $(v_i\sigma'_{ij}+q_j)$ correspond to the dissipated energy. The equation of energy conservation is

$$\frac{\partial}{\partial t}\left(\frac{1}{2}\rho v^2+\rho\varepsilon\right)=-\partial_j\left[\rho v_j\left(\frac{1}{2}v^2+w\right)-v_i\sigma'_{ij}+q_j\right] \tag{8.13}$$

and the heat transfer equation is

$$\rho T\left(\frac{\partial s}{\partial t}+\boldsymbol{v}grads\right)=\sigma'_{ij}\partial_j v_i-div(\boldsymbol{q}-\mu\boldsymbol{i})-\boldsymbol{i}grad\mu\ . \tag{8.14}$$

The additional unknown c is compensated by the additional one-component continuity equation (8.7). The two additional terms in this equation bring a new contribution to the variation of the entropy

$$\frac{d}{dt}\int dV\rho s=-\int dV\frac{(\boldsymbol{q}-\mu\boldsymbol{i})gradT}{T^2}-\int dV\frac{\boldsymbol{i}grad\mu}{T}+\dots\ . \tag{8.15}$$

8.2 Mass transfer and thermal diffusion

Small heat flux density $\boldsymbol{q}$ and mass flux density $\boldsymbol{i}$ (of the order of the mean freepath to the dimension of the fluid) depend linearly on the temperature gradient and the chemical-potential gradient,

$$\boldsymbol{i}=-\alpha grad\mu-\beta gradT\ ,$$
$$\boldsymbol{q}=-\delta grad\mu-\gamma gradT+\mu\boldsymbol{i}\ . \tag{8.16}$$

The flux densities imply molecular velocities, so they may be viewed as proportional to the time derivatives of some parameters x_a, which define the thermodynamic state. In equilibrium

$$X_a = -\frac{\partial S}{\partial x_a} \tag{8.17}$$

should be zero, where S is the entropy. It follows that

$$\dot{x}_a = -\sum_b \gamma_{ab} X_b \ , \tag{8.18}$$

for small deviations from equilibrium, where γ_{ab} is called the matrix of the kinetic coefficients. On the other side,

$$\dot{S} = -\sum_a X_a \dot{x}_a = \sum_{a,b} \gamma_{ab} X_a X_b \ ; \tag{8.19}$$

therefore, γ_{ab} is a symmetric matrix. These are Onsager's reciprocity relations of the kinetic coefficients.[1] If x_a depend on position, we must introduce a volume integral in the expression of entropy.

In our case $\dot{x}_a$ correspond to $\boldsymbol{i}$ and $\boldsymbol{q} - \mu\boldsymbol{i}$. By comparing equations (8.15) and (8.19) we get X_a as $\frac{grad\mu}{T}$ and $\frac{gradT}{T^2}$; it follows that the matrix of the kinetic coefficients is

$$\gamma_{ab} = \begin{pmatrix} \alpha T & \beta T^2 \\ \delta T & \gamma T^2 \end{pmatrix} \ , \tag{8.20}$$

so we must have $\delta = \beta T$. Equations (8.16) become

$$\begin{aligned} \boldsymbol{i} &= -\alpha grad\mu - \beta gradT \ , \\ \boldsymbol{q} &= -\beta T grad\mu - \gamma gradT + \mu\boldsymbol{i} \ , \end{aligned} \tag{8.21}$$

or

$$\begin{aligned} \boldsymbol{i} &= -\alpha grad\mu - \beta gradT \ , \\ \boldsymbol{q} &= (\mu + \beta T/\alpha)\boldsymbol{i} - (\gamma - \beta^2 T/\alpha) gradT \ . \end{aligned} \tag{8.22}$$

[1] L. Onsager, "Reciprocal relations in irreversible processes, I.", Phys. Rev. **37** 405 (1931); II., **38** 2265 (1931); H. B. Callen, "The application of Onsager's reciprocal relations to thermodynamic, thermomagnetic and galvanomagnetic effects", Phys. Rev. **73** 1349 (1948).

We denote $\gamma - \beta^2 T/\alpha = \kappa$. If there is no diffusion, $\boldsymbol{i} = 0$ and κ is the thermoconductivity. In addition, $\alpha\delta\mu + \beta\delta T = 0$ for $\boldsymbol{i} = 0$, which leads to a relation between c,p and T (in equilibrium $p = 0$). Making use of these equations in equation (8.15) we get

$$\frac{d}{dt}\int dV \rho s = \int dV \frac{\kappa (gradT)^2}{T^2} + \int dV \frac{\boldsymbol{i}^2}{\alpha T} \ . \tag{8.23}$$

such that κ and α must be positive.

Equations (8.22) can be put in another form,[2]

$$\begin{gathered} \boldsymbol{i} = -\rho D\left[gradc + \tfrac{k_T}{T} gradT + \tfrac{k_p}{p} gradp\right] \ , \\ \boldsymbol{q} = \left[k_T(\partial\mu/\partial c)_{p,T} - T(\partial\mu/\partial T)_{p,c} + \mu\right]\boldsymbol{i} - \kappa gradT \ , \end{gathered} \tag{8.24}$$

where

$$\begin{gathered} D = \tfrac{\alpha}{\rho}(\partial\mu/\partial c)_{p,T} \ , \\ \rho k_T D/T = \alpha(\partial\mu/\partial T)_{c,p} + \beta \ , \\ k_p = p(\partial V/\partial c)_{p,T}/(\partial\mu/\partial c)_{p,T} \ . \end{gathered} \tag{8.25}$$

D is called the diffusion coefficient (or mass transfer coefficient); since $\alpha > 0$ and $(\partial\mu/\partial c)_{p,T} > 0$ (see Appendix), $D > 0$. $k_T D$ is the thermal diffusion coefficient (k_T is called thermal diffusion ratio); $k_p D$ is called the barodiffusion coefficient. Equations (8.24) are obtained by eliminating $grad\mu$ in favour of $grad$ of c , p and T. The Gibbs free energy (per unit mass) reads

$$d\Phi = Vdp - sdT + \mu dc \ , \tag{8.26}$$

where V is the specific volume (per unit mass). The chemical potential μ is a function of p, T and c. The derivative $(\partial\mu/\partial p)_{c,T}$ is $\partial^2\Phi/\partial p\partial c = (\partial V/\partial c)_{p,T}$. In a single fluid $\boldsymbol{i} = 0$, such that k_T and k_p should be zero.

[2]L. Landau and E. Lifshitz, *Course of Theoretical Physics*, vol. 6 (*Fluid Mechanics*), Elsevier, Oxford (1987).

In a simplified approach we may assume that $grad p = 0$ and $\boldsymbol{v} = 0$. Equations (8.7) and (8.14) become

$$\rho\frac{\partial c}{\partial t} + div\boldsymbol{i} = 0 \ , \qquad \rho T\frac{\partial s}{\partial t} = -div(\boldsymbol{q} - \mu\boldsymbol{i}) \tag{8.27}$$

(the term $\boldsymbol{i} grad\mu$ is of the second order); in addition, the coefficients in equations (8.24) are assumed constant. We insert these equations in equations (8.27). Also, we use

$$\frac{\partial s}{\partial t} = \left(\frac{\partial s}{\partial T}\right)_{c,p}\frac{\partial T}{\partial t} + \left(\frac{\partial s}{\partial c}\right)_{T,p}\frac{\partial c}{\partial t} = \frac{c_p}{T}\frac{\partial T}{\partial t} - \left(\frac{\partial \mu}{\partial T}\right)_{c,p}\frac{\partial c}{\partial t} \ , \tag{8.28}$$

since $(\partial s/\partial c)_{T,p} = -\partial^2\Phi/\partial c\partial T - (\partial\mu/\partial T)_{c,p}$, where c_p is the specific heat per unit mass at constant pressure. We get

$$\frac{\partial c}{\partial t} = D\left[\Delta c + \frac{k_T}{T}\Delta T\right] \ , \qquad \frac{\partial T}{\partial t} - \frac{k_T}{c_p}\left(\frac{\partial \mu}{\partial c}\right)_{p,T}\frac{\partial c}{\partial t} = \chi\Delta T \ , \tag{8.29}$$

where $\chi = \kappa/\rho c_p$ is the thermometric conductivity. These are the concentration (mass) and heat diffusion equations. For small concentration $k_T \rightarrow 0$.

The boundary conditions for the mass diffusion are various. If the solid boundary is insoluble in the fluid we should have $\boldsymbol{i} = -\rho D grad c = 0$ on the boundary (*i.e.* $\partial c/\partial \boldsymbol{n} = 0$, where $\boldsymbol{n}$ is the normal to the surface). If the solid boundary is soluble in the fluid, we should have $c = c_0$, where c_0 is the saturation concentration. If the solid boundary absorbs the diffusing substance, then $c = 0$ on the surface.

In general, in addition to the terms included in the expansions (8.16) we should take into account terms arising from the spatial derivatives of the velocity ($\partial_i v_j$); these are the formations $\Delta\boldsymbol{v}$ and $grad\, div\boldsymbol{v}$ arising in the viscosity tensor.

8.3 Brownian motion

Let us assume that we put a concentration c_0 of particles at the initial moment of time at the origin in a fluid. By collisions with the molecules of the fluid they tend to form a thermodynamic ensemble in equilibrium; therefore, they tend to increase their entropy. Consequently, the particles diffuse in the fluid. This is the Brownian motion of particles suspended in a fluid. For small concentration the diffusion equation in this case is

$$\frac{\partial c}{\partial t} - D\Delta c = c_0 v_0 \delta(\boldsymbol{r})\delta(t) \ , \tag{8.30}$$

where v_0 is the initial volume (equation (8.29)). The solution is

$$c = \frac{c_0 v_0}{(4\pi Dt)^{3/2}} e^{-\frac{r^2}{4Dt}} \ . \tag{8.31}$$

The probability $dw = \Delta c/c_0$ of finding a particle in the volume $\Delta v_0 = 4\pi r^2 dr$ is

$$dw = \frac{1}{(4\pi Dt)^{3/2}} e^{-\frac{r^2}{4Dt}} \cdot 4\pi r^2 dr \ . \tag{8.32}$$

The mean distance travelled by a particle in time t is given by

$$\overline{r^2} = 6Dt \ . \tag{8.33}$$

Let us assume that a constant external force acts upon the particles. In equilibrium this force should be balanced by the drag force exerted by the fluid upon the particles. The drag force is proportional to the velocity $\boldsymbol{v}$ (for small velocities), such that

$$\boldsymbol{v} = b\boldsymbol{f} \ , \tag{8.34}$$

where the constant b is called mobility. For instance, for spherical particles with radius R the force is the Stokes force $6\pi R\eta v$, such that the mobility is $b = 1/6\pi R\eta$, where η is the viscosity.

The diffusion current $\boldsymbol{i} = -\rho D grad c$ must include the velocity contribution $\rho c\boldsymbol{v}$; it reads

$$\boldsymbol{i} = -\rho D grad c + \rho c b\boldsymbol{f} \tag{8.35}$$

(for constant temperature); or,

$$\boldsymbol{i} = -\frac{\rho D}{(\partial\mu/\partial c)_{p,T}} grad\mu + \rho c b\boldsymbol{f} \ , \tag{8.36}$$

where μ is the chemical potential of the suspended particles. For a weak solution we have $\mu = T \ln c$ up to a function of p and T,[3] such that we get

$$\boldsymbol{i} = -\frac{\rho D c}{T} grad\mu + \rho c b \boldsymbol{f} \ . \tag{8.37}$$

In equilibrium $\boldsymbol{i} = 0$ and $\mu + U$ should be constant, where U is the potential energy of the external field. Therefore, $grad\mu = \boldsymbol{f}$, and equation (8.37) gives

$$D = bT \ . \tag{8.38}$$

This is known as Einstein's relation between diffusion coefficient and mobility.[4]

A similar treatment can be done for the diffusion of rotatory Brownian motion, where the moment of the force appears instead of the force.

8.4 A non-linear diffusion equation

We derive here a non-linear diffusion equation, by taking into account the local variations of solvent pressure or volume, within a mechanism of diffusion driven by collisions between the fluid particles.[5] A particular class of radially-symmetric solutions is discussed, which are localized over finite ranges, localized either on the origin or on finite values of the radius. In the case of a thin layer of solute they look like disks or rings, such that they might be appropriate for describing the shapes of clouds or wreaths of smoke. These patterns exhibit diffusion fronts propagating slower and slower, and grow in time at a constant rate of the total number of solute particles.

The ascent of warm, moist air in Earth's atmosphere has been mod-

---

[3] L. Landau and E. Lifshitz, *Course of Theoretical Physics*, vol. 5 (*Statistical Physics*), Pergamon, Oxford (1980).

[4] A. Einstein, "Ueber die von der molekularkinetischen Theorie der Waerme geforderte Bewegung von in ruhenden Fluessigkeiten suspendierten Teilchen", Ann. Phys. **17** 549 (1905); see also M. Apostol, *Physical Kinetics*, Cambridge Scholars Publishing, Newcastle upon Tyne (2020) and References therein.

[5] B. F. Apostol, "On a non-linear diffusion equation describing clouds or wreaths of smoke", Physics Letters **A235** 363 (1997).

elled[6] with the Kardar-Parisi-Zhang equation,[7] known from the crystal growth on atomic surfaces. This equation has been solved numerically,[8] the fractal aspects of its solutions being emphasized. Leaving aside the randomness term responsible for the fractal behaviour, the remaining part of this equation looks like a non-linear diffusion equation, though the non-linear term has originally been interpreted as describing growth. By taking into account the local variations of the solvent pressure within a diffusion mechanism driven by the collisions between the fluid particles we show here that this non-linear term may indeed arise in a diffusion process in non-equilibrium conditions; while in equilibrium, the local variations of the solvent volume lead to an identical term, only with opposite sign. The non-linear diffusion equation obtained thereby is solved for a particular class of radially-symmetric solutions. In the first case it is found that the solute concentrations extend over finite ranges in space, having the shape of a disk in two dimensions. In the second case solutions of annular shape are obtained in two dimensions. These patterns of the solute concentration exhibit diffusion fronts propagating slower and slower, and grow in time at a constant rate of the total number of solute particles. We suggest that they might be interpreted as describing clouds or wreaths of smoke. Specifically, we discuss the two-dimensional case, but relevant features, wherever appropriate, are pointed out in one and three dimensions.

Suppose that we have a thin layer of fluid and that its particles may jump from the position (x, y) at time t to the neighbouring positions $(x \pm a, y)$, $(x, y \pm a)$, where a is a mean, atomic scale length; and suppose further on that these jumps proceed with a mean frequency $\nu(x, y)$. Then, the time variation of the fluid concentration $n(x, y, t)$

---

[6] J. D. Pelletier, "Kardar-Parisi-Zhang model for the fractal structure of cumulus cloud fields", arXiv: ao-sci/9605001v1 (1996).

[7] M. Kardar, G. Parisi and Y. C. Zhang, "Dynamic scaling of growing interfaces", Phys. Rev. Lett **56** 889 (1986).

[8] J. G. Amar and F. Family, "Numerical solutions of a continuum equation for interface growth in 2+1 dimensions", Phys. Rev. **A41** 3399 (1990); K. Moser, D. E. Wolf and J. Kertesz, "Numerical solution of the Kardar-Parisi-Zhang equation in one, two and three dimensions", Physica **A178** 215 (1991).

is given by[9]

$$\begin{aligned}\frac{\partial n}{\partial t} = \nu(x+a,y)n(x+a,y)+\\ +\nu(x-a,y)n(x-a,y)-2\nu n+\\ +\nu(x,y+a)n(x,y+a)+\\ +\nu(x,y-a)n(x,y-a)-2\nu n\ .\end{aligned} \tag{8.39}$$

Here we have assumed that n varies slowly in time over lapses of time much longer than the time scale ν^{-1}; and we assume further on that n and ν vary slowly in space too, over distances much larger than a. Then, by series expansion equation (8.39) becomes

$$\begin{aligned}\frac{\partial n}{\partial t} = a^2\nu\left(\frac{\partial^2 n}{\partial x^2}+\frac{\partial^2 n}{\partial y^2}\right)+\\ +2a^2\left(\frac{\partial \nu}{\partial x}\frac{\partial n}{\partial x}+\frac{\partial \nu}{\partial y}\frac{\partial n}{\partial y}\right)+a^2 n\left(\frac{\partial^2 \nu}{\partial x^2}+\frac{\partial^2 \nu}{\partial y^2}\right)\ .\end{aligned} \tag{8.40}$$

We are interested in small values of the solute mean concentration n, so that we may neglect the last term in (8.40); we obtain therefore

$$\frac{\partial n}{\partial t} = a^2\nu \triangle n + 2a^2 grad\nu \cdot gradn\ . \tag{8.41}$$

We adopt now a kinetic model of fluid, wherein the collision frequency ν is given by $\nu = (1/3)\,\sigma u N$, σ being the particle cross-section, u- the particle mean velocity and N- the solvent concentration. We assume that by adding n solute particles per unit volume (of similar nature as those of the solvent) the non-equilibrium local pressure of the solvent changes from $p = (1/3)\,mu^2N$ to $p' = p/(1-n/N) = (1/3)mu'^2N$, where m is the mass of the fluid particles; whence, for $n/N \ll 1$, we obtain $\nu' = b(N+n/2)$, with $b = (1/3)\sigma u$. Similarly, when equilibrium is established by local variations of the solvent volume, the pressure becomes $p' = p/(1+n/N)$ and $\nu' = b(N-n/2)$. The first process is expected to occur more frequently in dilute gases, while the latter is more appropriate for dense gases or liquids. Introducing these n-dependences of the collision frequency ν in equation (8.41) we get in the former case

$$\frac{1}{S}\frac{\partial n}{\partial t} = \triangle n + A(gradn)^2\ , \tag{8.42}$$

[9]F. Despa and M. Apostol, "A note on off-centre diffusion", Solid State Commun. **94** 153 (1995); "On the geometrical factor in off-centre diffusion", J. Phys. Chem. Solids **57** 1231 (1996).

and

$$\frac{1}{S}\frac{\partial n}{\partial t} = \triangle n - A(gradn)^2 \ , \tag{8.43}$$

in the latter, where $S = a^2bN$ and $A = 1/N$. These are non-linear diffusion equations, the first one (equation (8.42)) being the non-stochastic part of the Kardar-Parisi-Zhang equation. We note that the non-linear term in (8.42) and (8.43) occurs as a result of taking into account the local variations of the solvent pressure or volume within a mechanism of diffusion driven by the collisions between the fluid particles. This is in contrast with the diffusion by tunneling in an external potential, as, for example, in a (homogeneous) solid, where the jump frequency ν does not depend on the solute concentration n. Of course, if N is very large then we may neglect the non-linear term in (8.42) and (8.43), and we get the usual, linear diffusion equation.

We note, first, that the non-linear diffusion equations do not conserve the total number of solute particles, in contrast with the linear diffusion equation. Indeed, we have

$$\frac{\partial}{\partial t}\int n \cdot dV = \pm SA \int (gradn)^2 \cdot dV \ , \tag{8.44}$$

and requiring the conservation of the particle number would imply the trivial solution $n = const$. We shall look for radially-symmetric solutions (as those corresponding to radially-symmetric initial conditions), in which case equation (8.42) reads

$$\frac{1}{S}\frac{\partial n}{\partial t} = \frac{\partial^2 n}{\partial r^2} + \frac{1}{r}\frac{\partial n}{\partial r} + A\left(\frac{\partial n}{\partial r}\right)^2 \ ; \tag{8.45}$$

further on, we assume that n depends on $\xi = r/\sqrt{t}$ only, which represents a particular class of solutions, corresponding to an increase of the total number of solute particles at a constant rate,

$$\int n(\xi) \cdot dV = const \cdot t \tag{8.46}$$

($\sqrt{t}$ in one dimension, $t^{3/2}$ in three dimensions). Then, equation (8.45) reads

$$n'' + \left(\frac{1}{\xi} + \frac{\xi}{2S}\right) n' + An'^2 = 0 \ , \tag{8.47}$$

which is a Bernoulli-type equation. Introducing $f = 1/n'$ we get an equation of the type

$$f' + af + b = 0 \ , \tag{8.48}$$

whose solution is[10]

$$f = e^{-F}\left(const - \int^{\xi} be^{F} d\xi\right) \ , \tag{8.49}$$

where $F = \int^{\xi} a d\xi$. In this way we obtain for equation (8.47)

$$n' = \frac{\exp\left(-\xi^2/2\right)}{\xi\left[const + \frac{1}{2}A\sqrt{S/2}\cdot Ei\left(-\xi^2/2\right)\right]} \ , \tag{8.50}$$

where Ei is the exponential integral ($Ei(x) = -\int_{-x}^{\infty}\left(e^{-t}/t\right)dt$). The exponential integral $Ei\left(-\xi^2/2\right)$ is a monotonously increasing function which has the following asymptotic behaviours:

$$\begin{aligned} Ei\left(-\xi^2/2\right) &\sim \ln\left(\xi^2/2\right) \ , \quad \xi^2/2 \ll 1 \ , \\ Ei\left(-\xi^2/2\right) &\sim -\frac{1}{\left(\xi^2/2\right)}\cdot e^{-\xi^2/2} \ , \quad \xi^2/2 \gg 1 \ . \end{aligned} \tag{8.51}$$

Now, it is easy to see that for *const* ≥ 0 in equation (8.50) we would get a negative, unphysical solution n, while for *const* < 0 we get a monotonously decreasing solution n whose asymptotics are

$$n \sim \frac{1}{A\sqrt{2S}} \ln\left|\ln\left(\xi^2/2\right)\right| \ , \quad \xi^2/2 \ll 1 \ , \tag{8.52}$$

$$n \sim \frac{1}{\xi^2/2}\cdot e^{-\xi^2/2} \ , \quad \xi^2/2 \gg 1 \ . \tag{8.53}$$

Therefore the small solute density is given by the smooth function n in equation (8.53) for $\xi^2/2 \gg 1$, while it increases rapidly for $\xi^2/2 \ll 1$, as suggested by equation (8.52). Consequently, there exists a front of diffusion placed at $\xi^2/2 \sim 1$, *i.e.* at $r \sim 2\sqrt{St}$, which propagates slower and slower ($dr/dt \sim 1/\sqrt{t} \longrightarrow 0$ for $t \longrightarrow \infty$), encompassing an area which increases linearly with time, exactly as the total number of solute particles does. We may say that the solute particles, placed

[10]See, for example, E. Kamke, *Differential Equations*, Leipzig (1959).

initially over a certain, small area, propagate rather compactly (with a diffusion front), slower and slower, while being continuously fed at a constant rate. This suggests that we may speak in this case of a disk-like, cloud pattern, as an atmospheric, or a smoke cloud.

We note that a similar solution exists also for the linear diffusion equation ($A = 0$), $n = const \cdot Ei(-\xi^2/2)$, but not for the number-conserving solution of the linear diffusion equation

$$n \sim \frac{1}{t} \cdot \exp(-\xi^2/2) \ , \ \ \xi^2/2 \gg 1 \ , \tag{8.54}$$

which decreases uniformly to zero for $t \to \infty$.

Similar solutions exist for the non-linear equation in three dimensions and in one dimension, but in the later case the existence of the diffusion front is doubtful, since the solution n and its derivative are finite at the origin.

A similar analysis can be carried out straightforwardly for equation (8.43), which amounts to changing the sign of A in equations (8.47) and (8.50). It is easy to see, in this case, that we would get an unphysical solution for $const \geq 0$ in equation (8.50), but a physical one for $const < 0$ (and $\xi^2/2 \gg 1$). However, this solution has a logarithmic singularity at certain, finite values ξ_0 (depending on this integration constant), where it looks like $n \sim -\ln|(\xi - \xi_0)/2|$. For large values of the variable $\xi^2/2$ the solution behaves like the previous one, given by equation (8.53). One may say, therefore, that there exists in this case a finite value ξ_0 (and a finite value of the radius r_0), where the solute density is concentrated; it extends over a finite range of the order ξ_0, so that we may identify two diffusion fronts, one by each side of the singularity, propagating (in opposite directions) slower and slower, and encompassing an area which increases again linearly in time, in the same manner as the total number of solute particles. The pattern has an annular shape, suggesting again atmospheric clouds or wreaths of smoke. We remark that such a solution might be expected for the equilibrium diffusion process, corresponding to equation (8.43). Indeed, a spot of solute particles cannot proceed to diffuse at equilibrium by removing the solvent particles only outwardly, they have to displace them inwardly too, thereby acquiring the ring-like shape of an annulus.

Similar conclusions may also be reached in one and three dimensions, though in the former case the solution may be peaked at the origin, too ($\xi_0 = 0$).

Finally, we mention that the smoothness conditions required for deriving equation (8.41) are met by our solutions given by equation (8.53) for distances $\triangle r$ and lapses of time $\triangle t$ much smaller than the distance r and the time t of observation (but certainly much larger than a and ν^{-1}); we note also that a low diffusion coefficient S favours the fulfillment of the asymptotic condition $\xi^2/2 \gg 1$ where our solutions hold.

9 Surface Phenomena

9.1 Laplace's formula

Let us assume two continuous media in contact. A thin superficial layer appears, such that the surface is not perfectly defined geometrically. However, the thickness of this superficial layer is very small, so we may neglect it. The molecular forces in the two media are different, so they will curve the surface; until, the pressure difference establishes equilibrium. The mechanical work done by these forces is $\alpha\delta f$, where α is the surface-tension coefficient and δf is the change in the area of the surface.[1] On the other hand, the infinitesimal work done by the pressure is $(p_2 - p_1)d\zeta df$, where $p_{1,2}$ are the pressures in the two media and $d\zeta$ is the element of displacement along the normal to the surface from medium 1 to medium 2. Therefore, we must have the equilibrium condition

$$-\int (p_1 - p_2)d\zeta df + \alpha\delta f = 0 \; . \tag{9.1}$$

Let $dl_{1,2}$ be the two arc elements along the principal circles with radii $R_{1,2}$ of the surface. They become $dl_{1,2}\left(1 + \frac{d\zeta}{R_{1,2}}\right)$ during the displacement. The change of the area is

$$\begin{aligned} d\delta f &= dl_1(1 + d\zeta/R_1)dl_2(1 + d\zeta/R_2) - dl_1 dl_2 = \\ &= \left(\frac{1}{R_1} + \frac{1}{R_2}\right) d\zeta df \; . \end{aligned} \tag{9.2}$$

[1] α may vary with position, which leads to a tangential force $grad\alpha$ per unit area. This force is meaningful only for viscous fluids, and it puts a bound upon the Helmholtz-Kelvin instability condition.

Therefore, from equations (9.1) and (9.2) we obtain

$$p_1 - p_2 = \alpha\left(\frac{1}{R_1} + \frac{1}{R_2}\right) . \tag{9.3}$$

This is Laplace's formula.[2] We can see that the pressure is higher ($p_1 > p_2$) in the convex medium ($R_{1,2} > 0$). For a plane surface ($R_{1,2} \to \infty$) the pressure is the same. If the surface is free, the pressures are constant, and $1/R_1 + 1/R_2 = const$. If the whole surface is free $R_1 = R_2$, *i.e.* the surface is spherical. Thin films have a convex surface on one side and a concave surface on the other, such that the equilibrium condition is $1/R_1 + 1/R_2 = 0$. In uniform gravitational field $p_2 = const$, $p_1 = const - \rho g z$, such that equation (9.3) becomes

$$\frac{1}{R_1} + \frac{1}{R_2} + \frac{\rho g z}{\alpha} = const . \tag{9.4}$$

Let $\zeta(x, y)$ be the equation of a surface. The arc element is $dl = \sqrt{dx^2 + (\partial\zeta/\partial x)^2 dx^2}$, such that the area is

$$\begin{aligned} f &= \int \sqrt{1 + (\partial\zeta/\partial x)^2 + (\partial\zeta/\partial y)^2}\, dxdy \simeq \\ &\simeq \int \left[1 + \tfrac{1}{2}(\partial\zeta/\partial x)^2 + \tfrac{1}{2}(\partial\zeta/\partial y)^2\right] dxdy \end{aligned} \tag{9.5}$$

for small deviations from a plane. The variation of the area is

$$\delta f = -\int \left(\frac{\partial^2\zeta}{\partial x^2} + \frac{\partial^2\zeta}{\partial y^2}\right) \delta\zeta dxdy ; \tag{9.6}$$

comparing this result with equation (9.2) we get

$$\begin{aligned} \tfrac{1}{R_1} + \tfrac{1}{R_2} &= -\left(\tfrac{\partial^2\zeta}{\partial x^2} + \tfrac{\partial^2\zeta}{\partial y^2}\right) , \\ p_1 - p_2 = \alpha\left(\tfrac{1}{R_1} + \tfrac{1}{R_2}\right) &= -\alpha\left(\tfrac{\partial^2\zeta}{\partial x^2} + \tfrac{\partial^2\zeta}{\partial y^2}\right) . \end{aligned} \tag{9.7}$$

Inserting this equation in equations (9.3) or (9.4), we get the equation of the equilibrium surface.

[2]P. S. de Laplace, Supplement au *Traite de Mecanique Celeste*, vol. 4, Courcier, Paris (1805); C. F. Gauss, *Principia Generalia Theoriae Figurae Fluidorum in statu Aequilibrii*, Dieterichs, Goettingen (1830).

The infinitesimal tangent to the surface is proportional to $(dx, dy, \zeta_1 dx + \zeta_2 dy)$, where $\zeta_1 = \partial\zeta/\partial x$, $\zeta_2 = \partial\zeta/\partial y$. Therefore, the normal is proportional to $(-\zeta_1, -\zeta_2, 1)$, and, for $\mid \zeta_{1,2} \mid \ll 1$ the unit normal is precisely $\boldsymbol{n} = (-\zeta_1, -\zeta_2, 1)$; we can see that it is $grad(z - \zeta)$. It follows $div\boldsymbol{n} = -\Delta\zeta = 1/R_1 + 1/R_2$. Therefore, the Laplace equation is written as $\delta p = p_2 - p_1 = \alpha\Delta\zeta = -\alpha div\boldsymbol{n}$.

Let us suppose a solid wall placed at $x = 0$ and a fluid with the free surface placed at $z = 0$ in uniform gravitational field. Equations (9.4) and (9.7) lead to

$$-z'' + \frac{\rho g z}{\alpha} = 0 \ , \tag{9.8}$$

or $z = he^{-\sqrt{\frac{\rho g}{\alpha}}x}$, where h is the height of the fluid on the wall. This is the capillarity phenomenon. $a = \sqrt{2\alpha/\rho g}$ is called the capillary constant. The result is valid for $h \ll a$ (for water $a = 0.4cm$ and $\alpha \simeq 72dyn/cm$). The angle of contact is given by $\tan\theta = (dx/dz)_{x=0} = a/\sqrt{2}h$. For a capillary tube with radius a equations (9.7) give $\rho gh = 2\alpha/R$, where $a = R\cos\theta$ (Jurin's law[3]). Indeed, we assume a spherical free surface in the capillary, with radius R; its equation is $\zeta = R\cos\varphi - z$. The Laplace equation gives $\Delta\zeta = \frac{2}{R}\cos\varphi$, such that $\rho gh = \frac{2\alpha}{R}\cos\varphi$, where $\varphi = \pi/2 - \theta$; since $\cos\varphi = \sqrt{1 - a^2/R^2}$, we may set $\cos\varphi = 1$ for $a \ll R$; and $\cos\theta = \sin\varphi = a/R$.

The equilibrium surface can also be determined by minimizing the (relevant part of the) free energy

$$\alpha \int df + \rho g \int z dV \ , \tag{9.9}$$

with the condition of constant volume ($\int dV = const$).

For two fluids in motion the force acting upon their joint boundary per unit area is $n_j(\sigma_{2,ij} - \sigma_{1,ij})$, where $\sigma_{1,2,ij}$ are the stress tensors and $\boldsymbol{n}$ is the normal to the surface; since σ includes $-p$, from equation

[3] J. Jurin, "An account of some experiments shown before the Royal Society; with an enquiry into the cause of some ascent and suspension of water in capillary tubes", Trans. Roy. Soc. London **30** 739 (1718); "An account of some new experiments, relating to the action of glass tubes upon water and quicksilver", *ibid.* **30** 1083 (1719).

(9.3) we get the condition of equilibrium with surface tension

$$n_j(\sigma_{2,ij} - \sigma_{1,ij}) = \alpha \left(\frac{1}{R_1} + \frac{1}{R_2} \right) n_i \ ; \tag{9.10}$$

introducing the viscosity tensor, we get

$$(p_1 - p_2)n_i = n_j(\sigma'_{1,ij} - \sigma'_{2,ij}) + \alpha \left(\frac{1}{R_1} + \frac{1}{R_2} \right) n_i \ . \tag{9.11}$$

9.2 Capillary waves

For oscillations small in comparison with wavelength the flow is potential, *i.e.* the velocity is given by $\boldsymbol{v} = grad\Phi$, where the potential Φ satisfies the equation

$$\frac{\partial \Phi}{\partial t} + \frac{p}{\rho} + g\zeta = 0 \tag{9.12}$$

(see the Ideal Fluids chapter). This equation, with the small displacement ζ, can be viewed as the boundary condition for the potential Φ, which, for incompressible fluids, satisfies the equation $\Delta\Phi = 0$. From equations (9.3) and (9.7) we have

$$p - p_0 = -\alpha \left(\frac{\partial^2 \zeta}{\partial x^2} + \frac{\partial^2 \zeta}{\partial y^2} \right) \ , \tag{9.13}$$

where p is the pressure in the fluid, near the surface, and p_0 is the constant external pressure. Inserting this equation in equation (9.12) we get

$$\rho g \zeta + \rho \frac{\partial \Phi}{\partial t} - \alpha \left(\frac{\partial^2 \zeta}{\partial x^2} + \frac{\partial^2 \zeta}{\partial y^2} \right) = 0 \ , \tag{9.14}$$

where p_0 is absorbed in Φ. Now we differentiate this expression with respect to t and write $\partial\zeta/\partial t = \partial\Phi/\partial z$:

$$\rho g \frac{\partial \Phi}{\partial z} + \rho \frac{\partial^2 \Phi}{\partial t^2} - \alpha \frac{\partial}{\partial z} \left(\frac{\partial^2 \Phi}{\partial x^2} + \frac{\partial^2 \Phi}{\partial y^2} \right) = 0 \ . \tag{9.15}$$

This is the boundary condition at $z = 0$ for vibrations given by $\Delta\Phi = 0$. As it is customary we seek a solution $\Phi \sim e^{kz}\cos(kx - \omega t)$; equation (9.15) gives the dispersion relation

$$\omega^2 = gk + \alpha k^3/\rho \ . \tag{9.16}$$

For long wavelengths the effect of capillarity may be neglected, and we have gravity waves. For short wavelengths ($\lambda \ll a$) we have capillary waves. They are known also as ripples.[4]

9.3 Capillary vibrations

In spherical polar coordinates the area of a surface is

$$f = \int \sqrt{r^2 + \left(\frac{\partial r}{\partial \theta}\right)^2 + \frac{1}{\sin^2\theta}\left(\frac{\partial r}{\partial \varphi}\right)^2}\, r\sin\theta d\theta d\varphi \ . \tag{9.17}$$

Let ζ be a small deviation from the constant radius R of a sphere, $r = R + \zeta$. The variation of the area is

$$\delta f = \int \left[2(R+\zeta) - \frac{1}{\sin\theta}\frac{\partial}{\partial\theta}\left(\sin\theta\frac{\partial\zeta}{\partial\theta}\right) - \frac{1}{\sin^2\theta}\frac{\partial^2\zeta}{\partial\varphi^2}\right]\cdot$$
$$\cdot r\sin\theta d\theta d\varphi \ . \tag{9.18}$$

In order to find the sum of the curvature radii from equation (9.2) we must divide by $(R+\zeta)^2$; we get

$$\frac{1}{R_1} + \frac{1}{R_2} = \frac{2}{R} - \frac{2\zeta}{R^2} - \frac{1}{R^2}\left[\frac{1}{\sin\theta}\frac{\partial}{\partial\theta}\left(\sin\theta\frac{\partial\zeta}{\partial\theta}\right) + \frac{1}{\sin^2\theta}\frac{\partial^2\zeta}{\partial\varphi^2}\right] \ . \tag{9.19}$$

Therefore, the boundary condition for the velocity potential is

$$\rho\frac{\partial\Phi}{\partial t} + \alpha\{\frac{2}{R} - \frac{2\zeta}{R^2} - \frac{1}{R^2}\cdot$$
$$\cdot[\frac{1}{\sin\theta}\frac{\partial}{\partial\theta}(\sin\theta\frac{\partial\zeta}{\partial\theta}) + \frac{1}{\sin^2\theta}\frac{\partial^2\zeta}{\partial\varphi^2}]\} + p_0 = 0 \ , \tag{9.20}$$

[4] W. Thomson, "Ripples and waves", Nature **5** 1 (1871).

where the constant $2/R + p_0$ may be omitted. By differentiating with respect to time, this equation becomes

$$\rho\frac{\partial^2\Phi}{\partial t^2} - \frac{\alpha}{R^2}\{2\frac{\partial\Phi}{\partial r} + $$
$$+\frac{\partial}{\partial r}[\frac{1}{\sin\theta}\frac{\partial}{\partial\theta}(\sin\theta\frac{\partial\Phi}{\partial\theta}) + \frac{1}{\sin^2\theta}\frac{\partial^2\Phi}{\partial\varphi^2}]\} = 0 \ . \tag{9.21}$$

This is the boundary condition at $r = R$ for the Laplace equation $\Delta\Phi = 0$. The solution is of the form $e^{-i\omega t}r^l Y_{lm}$, where Y_{lm} are the spherical harmonics. Equation (9.21) gives

$$\omega^2 = \alpha l(l-1)(l+2)/\rho R^3 \ . \tag{9.22}$$

This is the frequency spectrum of the vibrations of a sphere with radius R, due to the surface tension.[5] For $l = 0$ we have radial oscillations, which are impossible for an incompressible fluid. For $l = 1$ we have a translatory motion. The lowest vibration frequency is given by $\omega^2 = 8\alpha/\rho R^3$ for $l = 2$.

[5]Lord Rayleigh, "On the capillary phenomena of jets", Proc. Roy. Soc. London **29** 71 (1879).

10 Surface Waves

10.1 Gravity waves

We consider an incompressible ideal fluid in an external potential $\varphi(\boldsymbol{r})$. Since the (internal) pressure is constant, the Euler equation reads

$$\frac{\partial \boldsymbol{v}}{\partial t} + (\boldsymbol{v}grad)\boldsymbol{v} = -grad\varphi \ , \tag{10.1}$$

where $\boldsymbol{v}$ is the velocity and ρ is the density. In time Δt the fluid particle moves with distance a, so $v = a/\Delta t$; in the same duration of time the velocity changes over a distance λ. The two terms on the left in equation (10.1) are of the order of $a/\Delta t^2$ and $a^2/\lambda\Delta t^2$. We assume $a \ll \lambda$, such that the inertial term may be disregarded. The equation becomes

$$\frac{\partial \boldsymbol{v}}{\partial t} = -grad\varphi \ . \tag{10.2}$$

In this equation the potential φ must be taken at the position of the fluid particle. The continuity equation $div\boldsymbol{v} = 0$ must be added.

From equation (10.2) it follows $\boldsymbol{v} = grad\Phi$, and

$$grad\left[\frac{\partial \Phi}{\partial t} + \varphi(\boldsymbol{r})\right] = 0 \ ; \tag{10.3}$$

the flow is potential, and Φ is the velocity potential. The variation of the external potential is

$$\delta\varphi = \boldsymbol{u}grad\varphi \ , \tag{10.4}$$

where $\boldsymbol{u}$ is the small displacement at position $\boldsymbol{r}$. Therefore,

$$grad\left(\frac{\partial \Phi}{\partial t} + \boldsymbol{u}grad\varphi\right) = 0 \tag{10.5}$$

(where $grad\varphi$ is not affected by the derivatives). The term $\boldsymbol{u}grad\varphi$ is the mechanical work done by the external field, while $\dot{\Phi}$ is the kinetic energy (per unit mass). Since the energy is conserved, we must have

$$\frac{\partial\Phi}{\partial t} + \boldsymbol{u}grad\varphi = 0\ . \tag{10.6}$$

This is an integral of motion; it can be written as

$$\frac{\partial^2\Phi}{\partial t^2} + grad\Phi \cdot grad\varphi = 0\ . \tag{10.7}$$

On the other hand, the incompressibility condition $div\boldsymbol{v} = 0$ gives

$$\Delta\Phi = 0\ . \tag{10.8}$$

Let us assume that the fluid is placed in uniform gravitational field, $\varphi = p_0/\rho + gz$, where p_0 is a constant pressure, g is the gravitational acceleration and z is the vertical coordinate. The pressure generated by the motion is

$$\delta p = \rho g u_z\ . \tag{10.9}$$

Equation (10.7) becomes

$$\frac{\partial^2\Phi}{\partial t^2} + g\frac{\partial\Phi}{\partial z} = 0\ . \tag{10.10}$$

Equations (10.8) and (10.10) should be solved with initial and boundary conditions. They describe vibrations.

We consider the half-space $z < 0$ occupied by the fluid. The boundary conditions are the vanishing of Φ and its derivatives at infinity. The elementary solutions of equation (10.10) are $\Phi = \Phi_k(\boldsymbol{r})e^{-i\omega t+kz}$, where

$$\omega^2 = gk \tag{10.11}$$

($k > 0$) and $\boldsymbol{r}$ is the in-plane position (surface waves).[1] These are called gravity "waves". Equation (10.8) becomes

$$\Delta_2\Phi_k + k^2\Phi_k = 0\ . \tag{10.12}$$

[1] J. W. Strutt (Lord Rayleigh), *The Theory of Sound*, Cambridge University Press, Cambridge (1877).

The solutions of this equation, which are regular in the origin, are the Bessel functions $J_m(kr)$.[2] Let us assume that we are interested in solutions with axial symmetry; then, $m = 0$. Therefore, the elementary solutions are $\Phi = J_0(kr)e^{-i\sqrt{gk}t+kz}$. We can see that these solutions are vibrations, not waves, since the time dependence is separated from the spatial dependence.

We can include an external force, generating vibrations, in Euler's equation (10.1), which appears in equation (10.10); this equation becomes the equation of a harmonic oscillator. We may formulate a Green function equation, with causal conditions. We prefer to keep the free equation, preserve the dependence $e^{-i\sqrt{gk}t}$, and impose an initial condition.

In order to satisfy the initial condition we form the superposition

$$\Phi = \int_0^\infty dkC(k)J_0(kr)e^{-i\sqrt{gk}t+kz} \ . \tag{10.13}$$

The vertical displacement is

$$u_z = \int_0^\infty dk\frac{ik}{\omega}C(k)J_0(kr)e^{-i\sqrt{gk}t+kz} \ . \tag{10.14}$$

We take as initial condition the function

$$u_z(t=0) = h(r)e^{kz} \ , \tag{10.15}$$

where $h(r) = a$ for $r < r_0$ and zero otherwise. Both a and r_0 should satisfy the conditions $ak \ll 1$, $r_0k \ll 1$. This function can be expanded in a Fourier-Bessel series as

$$h(r) = a\int_0^\infty dkkJ_0(kr)\int_0^{r_0} dr'r'J_0(kr') \ . \tag{10.16}$$

The inner integral is

$$\int_0^{r_0} dr'r'J_0(kr') = \frac{1}{k^2}\int_0^{kr_0} d\rho\rho J_0(\rho) \ . \tag{10.17}$$

[2] M. Apostol, *Equations of Mathematical Physics*, Cambridge Scholars Publishing, Newastle upon Tyne (2018).

From the Bessel equation we have

$$\frac{d}{d\rho}(\rho J_0^{'}) + \rho J_0 = 0 \ , \tag{10.18}$$

where $J_0^{'} = -J_1$; therefore, $\rho J_0 = \frac{d}{d\rho}(\rho J_1)$, and the integral (10.17) is

$$\int_0^{r_0} dr' r' J_0(kr') = \frac{r_0}{k} J_1(kr_0) \ . \tag{10.19}$$

Consequently, equation (10.16) becomes

$$h(r) = ar_0 \int_0^{\infty} dk J_0(kr) J_1(kr_0) \ . \tag{10.20}$$

By comparing this equation to equation (10.14), we get the coefficients

$$C(k) = \frac{ar_0\sqrt{g}}{i\sqrt{k}} J_1(kr_0) \ , \tag{10.21}$$

such that the velocity potential (equation (10.13)) becomes

$$\Phi = -iar_0\sqrt{g} \int_0^{\infty} dk \frac{J_1(kr_0)}{\sqrt{k}} J_0(kr) e^{-i\sqrt{gk}t + kz} \tag{10.22}$$

and the vertical displacement (equation (10.14))

$$u_z = ar_0 \int_0^{\infty} dk J_1(kr_0) J_0(kr) e^{-i\sqrt{gk}t + kz} \ . \tag{10.23}$$

Since the displacements must be much smaller than the characteristic length k^{-1}, we may use the approximation $J_1(kr_0) \simeq \frac{1}{2}kr_0$ for $kr_0 \ll 1$, and get

$$\begin{gathered} \Phi \simeq -\tfrac{1}{2}iar_0^2\sqrt{g} \int_0^{\infty} dk\sqrt{k} J_0(kr) e^{-i\sqrt{gk}t + kz} \ , \\ u_z = \tfrac{1}{2}ar_0^2 \int_0^{\infty} dk k J_0(kr) e^{-i\sqrt{gk}t + kz} \ . \end{gathered} \tag{10.24}$$

This approximation can be obtained directly from $h(r) = ar_0\delta(r)$, by noticing that

$$\delta(\boldsymbol{r}) = \frac{1}{2\pi} \int_0^{\infty} dk k J_0(kr) \tag{10.25}$$

and

$$\int d\boldsymbol{r}\delta(\boldsymbol{r}) = 2\pi\int_0^\infty dr r\delta(\boldsymbol{r}) = 1 \ , \tag{10.26}$$

whence $2\pi r\delta(\boldsymbol{r}) = 2\delta(r)$; therefore,

$$\delta(r) = \pi r\delta(\boldsymbol{r}) \simeq \frac{r_0}{2}\int_0^\infty dkkJ_0(kr) \ , \tag{10.27}$$

whence $h(r)$ and u_z given by equations (10.24).

We write the vertical surface displacement given by equations (10.24) as

$$u_{z=0} = \frac{1}{2\pi}ar_0^2\int_0^\pi d\alpha\int_0^\infty dkke^{i(kr\cos\alpha-\sqrt{gk}t)} \ . \tag{10.28}$$

Both integrals can be effected by the method of the stationary phase (the method of steepest descent).[3] The stationary points are $\sqrt{k_0} = \sqrt{g}t/2r\cos\alpha$ and $\alpha = 0$. The result is

$$u_{z=0} = \frac{ar_0^2 g}{\sqrt{2}}\frac{t^2}{r^3}e^{-i\frac{gt^2}{4r}} \tag{10.29}$$

(valid for $k_0 r_0 \gg 1$, i.e. $gt^2/r \gg r/r_0 \gg 1$).

10.2 Ship waves

The above result is derived by assuming that the initial perturbation is placed at $t = 0$ and $r = 0$. Let us assume now that the perturbation moves with constant velocity $\boldsymbol{v}$, such that it appears at times t_a at positions $\boldsymbol{r}_a = \boldsymbol{v}t_a$, from $t_a = -T$, where T is some long time in the past. We are interested in the vertical surface displacement given by equation (10.29) at time $t = 0$; obviously, it is given by

$$u = \frac{ar_0^2 g}{\sqrt{2}}\frac{v}{r_0}\int_{-T}^0 dt_a\frac{t_a^2}{\mid \boldsymbol{r}+\boldsymbol{v}t_a\mid^3}e^{-i\frac{gt_a^2}{4\mid \boldsymbol{r}+\boldsymbol{v}t_a\mid}} \ . \tag{10.30}$$

[3] W. Thomson (Lord Kelvin), "On the waves produced by a single impulse in water of any depth, or in a dispersive medium", Proc. Roy. Soc. London **42** 80 (1887); A. Sommerfeld, *Vorlesungen ueber Theoretische Physik*, Band II (*Mechanik der Deformierbaren Medien*), Akad. Verlagsgesellschaft, Leipzig (1970); Band VI (*Partielle Differentialgleichungen der Physik*), Akad. Verlagsgesellschaft, Leipzig (1966).

Since $gt_a^2/\boldsymbol{r} + \boldsymbol{v}t_a \mid\gg 1$ the phase factor in this integral has a rapid variation, so we can estimate this integral by the method of the stationary phase. The equation

$$\frac{d}{dt_a}\frac{gt_a^2}{4 \mid \boldsymbol{r} + \boldsymbol{v}t_a \mid} = 0 \tag{10.31}$$

is

$$v^2 t_a^2 + 3vr\cos\theta \cdot t_a + 2r^2 = 0 \ , \tag{10.32}$$

where θ is the angle made by $\boldsymbol{v}$ and $\boldsymbol{r}$; its roots must be real, such that we have the condition

$$\cos^2\theta > \frac{8}{9} \ . \tag{10.33}$$

The ship waves (on the surface) are confined to a triangle, left behind the moving pointlike perturbation, and determined by the angle with $\cos\theta_0 = -2\sqrt{2}/3$.[4]

10.3 Plane gravity waves

The elementary plane-wave solutions of equation (10.12) are $\Phi = \Phi_0 e^{i\boldsymbol{kr}} e^{-i\omega t + kz}$, where $\omega^2 = gk$; the in-plane velocity is

$$\boldsymbol{v}_{\parallel} = i\boldsymbol{k}\Phi_0 e^{i\boldsymbol{kr}} e^{-i\omega t + kz} \tag{10.34}$$

and the transverse velocity is

$$v_z = k\Phi_0 e^{i\boldsymbol{kr}} e^{-i\omega t + kz} \ . \tag{10.35}$$

The displacement is given by

$$\boldsymbol{u}_{\parallel} = -\frac{\boldsymbol{k}}{\omega}\Phi_0 e^{i\boldsymbol{kr}} e^{-i\omega t + kz} \ , \ u_z = i\frac{k}{\omega}\Phi_0 e^{i\boldsymbol{kr}} e^{-i\omega t + kz} \ . \tag{10.36}$$

The trajectory of the particles are circles with a decreasing radius with increasing the depth,

$$u_{\parallel}^2 + u_z^2 = \frac{k}{g}\Phi_0 e^{2kz} \ . \tag{10.37}$$

[4] W. Thomson (Lord Kelvin), "On ship waves", Proc. Inst. Mech. Eng. **38** 409, 641 (1887).

The stream function ψ is given by $\frac{\partial\psi}{\partial r} = -v_z$, $\frac{\partial\psi}{\partial z} = v_\parallel$ in the rz-plane, up to the time dependence. We get $\psi = \Phi_0 \sin kre^{kz}$. Both the particle path and the streamlines can be photographed by using metal fillings mixed up in a liquid, with a long and short time exposure, respectively.

For a fluid with a finite depth h the z-dependence is given by $Ae^{kz} + Be^{-kz}$, with the boundary condition $v_z \mid_{z=h} = 0$; hence, $\Phi = \Phi_0(t)\cosh k(z+h)e^{i\mathbf{kr}}$. Equation (10.10) gives

$$\ddot{\Phi}_0 + gk \tanh k(z+h)\Phi_0 = 0 \ , \tag{10.38}$$

which shows that ω depends on z:

$$\omega^2 = gk \tanh k(z+h) \tag{10.39}$$

(for a constant ω the wavelength depends on the depth). For $z \to -h$ the frequency tends to zero; for $z = 0$ the frequency of the surface waves is given by $\omega^2 = gk \tanh kh$. For shallow waters $\omega = \sqrt{ghk}$, and the phase velocity is $\sqrt{gh}$. It follows that the wave crest, with larger h, propagates faster than the wave trough, with a smaller h. This is the surf phenomenon.[5]

[5] A. Sommerfed, *Vorlesungen ueber Theoretische Physik*, vol. II (*Mechanik der Deformierbaren Medien*), Akademische Verlag, Leipzig (1970).

11 Explosion Waves

11.1 Explosions

We consider a classical gas in equilibrium, consisting of identical molecules with mass m. We identify a small region with dimension l in the gas, where, suddenly, a large number N of identical molecules appear, with mass M and uniform density, in general distinct from the gas molecules, with a very high energy E_0. Their apparition may be caused by a chemical reaction, which produces a strong explosion (a detonation), or a nuclear explosion. We assume that each molecule appearing in this region has a radial high velocity v_0 (a non-zero angular momentum is less likely), much higher than their mean thermal velocity, such that the N molecules acquire a non-equilibrium state, which is a one mechanical state, characterized by velocity v_0. This means that their entropy is appreciably reduced. We assume that each molecule in this state has an energy $\varepsilon_0 = \frac{1}{2}Mv_0^2$, and their total energy is $E_0 = N\varepsilon_0$.

According to our hypothesis of radial high velocity those N molecules go away from the small region with dimension l and fill a spherical shell with a small thickness l and radius r at some moment of time t. Their (uniform) density decreases according to $n = N/4\pi r^2 l = n_0 l^2/r^2$, where n_0 is their initial (very large) density.

During their motion through the gas the M-molecules lose energy by collisions with the gas molecules. Let us consider an M-molecule moving with velocity v to the right, and colliding to an m-molecule moving along the direction of v with velocity u. Since $v \gg |u|$ these molecules collide. We need not consider the m-molecules placed at the left of the M-molecules, since they never collide with the M-molecule. Also, we need not consider collisions with velocities u perpendicular to the direction of v, since their effect is null on the average. The conservation of the momentum and the energy reads

$$\begin{aligned} Mv' + mu' &= Mv + mu \, , \\ Mv'^2 + mu'^2 &= Mv^2 + mu^2 \, , \end{aligned} \tag{11.1}$$

where v' and u' are the velocities after collision. From these equations we get[1]

$$\begin{aligned} v' &= \tfrac{M-m}{M+m} v + \tfrac{2m}{M+m} u \, , \\ u' &= \tfrac{2M}{M+m} v - \tfrac{M-m}{M+m} u \, . \end{aligned} \tag{11.2}$$

Since $v \gg \mid u \mid$ we may write

$$v' \simeq \frac{M-m}{M+m} v \; ; \tag{11.3}$$

the momentum variation of the M-molecule is

$$\Delta P = -\frac{2mM}{M+m} v \; . \tag{11.4}$$

The mean time between two successive collisions is $\Delta t = \Lambda / v$, where Λ is the mean freepath of the gas molecules. It is given by $\Lambda = a^3/d^2$, where a is the mean separation distance between the gas molecules and d is the molecular diameter (taken the same for both kinds of molecules). It follows that the Newton law for the motion of an M-molecule can be written as[2]

$$\frac{dv}{dt} = -\frac{2m}{\Lambda(M+m)} v^2 \; ; \tag{11.5}$$

hence,

$$\begin{aligned} v &= \frac{v_0}{1 + \frac{2mv_0}{\Lambda(M+m)} t} \, , \\ r &= \tfrac{\Lambda(M+m)}{2m} \ln \left(1 + \tfrac{2mv_0}{\Lambda(m+M)} t \right) \, . \end{aligned} \tag{11.6}$$

We can see that the velocity decreases with a high rate $1/\tau = 2mv_0/\Lambda(M + m)$, while the distance r has a slower logarithmic increase.

[1] M. Apostol, "Drag force through gases and plasma", Ann. Math. Phys. **5** doi.org/10.17352/amp. 000031 (2022).

[2] Lord Rayleigh, "On the resistance of fluids", Phil. Mag. **2** 430 (1876).

The velocity v reaches the mean thermal velocity u of the gas molecules after a time lapse Δt given by

$$\Delta t = \tau(v_0/u - 1) \simeq \frac{v_0 \tau}{u} . \tag{11.7}$$

During this time the peripheral molecular layer of the spherical-shell wave is thermalized. Therefore, the total thermalization time of the wave is estimated by

$$T' \simeq \frac{v_0 \tau}{u} N^{1/3} . \tag{11.8}$$

During this time the wave covers the distance

$$r \simeq v_0 \tau \ln(v_0 N^{1/3}/u) . \tag{11.9}$$

In normal conditions the mean freepath of the air molecules is $\Lambda = 10^{-5}cm$; the molecular mass of the air is $m = 5 \times 10^{-23}g$ and the mean thermal velocity is $u = 4 \times 10^4 cm/s$.[3] For the explosion we take the uranium bomb dropped at Hiroshima. This bomb contained $64kg$ uranium with molecular mass $M = 4 \times 10^{-22}g$; therefore, $N = 1.6 \times 10^{26}$ (approximately $300mol$). The density of uranium is $19g/cm^3$, so the dimension of the focus was $l = 10cm$. The released energy was $E_0 = 1.5 \times 10^{20} erg$,[4] whence we estimate an initial velocity $v_0 = 10^8 cm/s$ (though a reduction factor of 10^{-5} - the nucleon mass to the atom mass - may account for the radiation energy, see equations (11.1) and (11.2)). From these data we get a characteristic time $\tau \simeq 4 \times 10^{-13}s$, a thermalization time $T' \simeq 0.5s$ and a very small distance $r \simeq 10^{-3}cm$. The thermalization occurs in a very short time, due to the numerous collisions with the gas molecules; the number of collisions is appreciably increased precisely due to the high explosion velocity. Actually, these estimations correspond in fact to the explosion.

The remnant explosion molecules diffuse in the gas. According to the diffusion law the distance covered in time t is $d = \sqrt{Dt}$, where D is the coefficient of diffusion. An estimate of this coefficient is $D = n\Lambda^4 u$,

---

[3] $u = \sqrt{3T/m}$, where T is the temperature.

[4] This would correspond to an earthquake with magnitude $M_w = 3$, see B. F. Apostol and L. C. Cune, *A Guide to Practical Seismology*, Cambridge Scholars Publishing, Newcastle upon Tyne (2023).

where n is the gas density.[5] In normal conditions $n \simeq 10^{19}cm^{-3}$ (and $u \simeq 10^4 cm/s$ estimated above). By using $\Lambda = 10^{-5}cm$, we get $D = 10^3 cm^2/s$, which is a high diffusion coefficient. Variations of the diffusion coefficient may be expected, since the fourth power of Λ brings a strong dependence on numerical factors.

11.2 Explosion waves

The effect of the explosion consists in the energy transferred to the gas molecules (a similar situation holds explosions in solids). Indeed, from equations (11.2) we can see that after collisions the velocity of the gas molecule is $u' \simeq 2Mv/(M+m)$, *i.e.* it is comparable to the velocity v of the explosion molecules. At the moment of the explosion a pressure $p_0 = E_0/l^3$ occurs in the small volume l^3. Therefore, a force $-E_0 grad\delta(\boldsymbol{R})$ occurs per unit volume, where the δ-function is localized in the small volume l^3. The time dependence of this force is written as $-E_0 T\delta(t) grad\delta(\boldsymbol{R})$, where the function $\delta(t)$ is localized in some very short time interval T. The energy E_0 corresponds to the seismic moment M.[6]

The equation of motion for the displacement $\boldsymbol{u}$ of the gas molecules reads

$$\rho\frac{d^2\boldsymbol{u}}{dt^2} = -E_0 T\delta(t) grad\delta(\boldsymbol{R}) - gradp \ , \tag{11.10}$$

where ρ is the density of the gas and p is its internal pressure (we leave aside the viscosity and the thermoconduction). The solution $\boldsymbol{u}$ is determined by the external force. The external force should be treated as a boundary condition, due to its singular nature. Consequently, outside its domain the solution is a smooth function, so we may neglect the transport term in equation (11.10), and write this equation as

$$\rho\frac{\partial^2\boldsymbol{u}}{\partial t^2} = -E_0 T\delta(t) grad\delta(\boldsymbol{R}) - gradp \ . \tag{11.11}$$

[5] M. Apostol, *Physical Kinetics*, Cambridge Scholars Publishing, Newcastle upon Tyne (2020).

[6] B. F. Apostol, *Seismology*, Nova, NY (2020).

The variation of the internal pressure is given by

$$\delta p = -\frac{K}{V}\delta V + \alpha\delta T \ , \qquad (11.12)$$

where V is the volume, T is the temperature, $K = -V\left(\partial p/\partial V\right)_T$ is the isothermal modulus of compression and $\alpha = \left(\partial p/\partial T\right)_V$ is the thermal expansion coefficient (see the Ideal Fluids chapter). The thermodynamical quantities vary slowly in comparison to the molecular collision which realize the equilibrium, such that the motion is adiabatic. The variation of the entropy per unit mass is given by

$$\delta s = \alpha\delta V + \frac{c_v}{T}\delta T = 0 \ , \qquad (11.13)$$

where c_v is the specific heat per unit mass at constant volume and $V = 1/\rho$ is the specific volume per unit mass (we use the variation of the free energy per unit mass $dF = -pdV - sdT$). We get

$$\delta T = -\frac{\alpha T}{c_v}\delta V \qquad (11.14)$$

and

$$\delta p = -\frac{K}{V}\left(1 + \frac{\alpha^2 T}{\rho c_v K}\right)\delta V \ . \qquad (11.15)$$

The bracket in this equation gives rise to the adiabatic modulus of compressibility ($K_{ad} = (c_p/c_v)K$, where c_p is the specific heat per unit mass at constant pressure; also $\alpha = K\beta$, where $\beta = \frac{1}{V}\left(\partial V/\partial T\right)_p$ is the expansion (dilatation) coefficient); therefore,

$$\delta p = -\frac{K_{ad}}{V}\delta V \ . \qquad (11.16)$$

It is easy to see that the variations of the density are given by $\delta\rho = -\rho div\boldsymbol{u}$, such that the variations of the volume are $\delta V = V div\boldsymbol{u}$. It follows that the variations of the pressure are

$$\delta p = -K_{ad} div\boldsymbol{u} \qquad (11.17)$$

and the variations of the temperature are

$$\delta T = -\frac{\alpha T}{\rho c_v} div\boldsymbol{u} \ . \qquad (11.18)$$

Since $\alpha = K\beta$, where $\beta = \frac{1}{V}\left(\partial V/\partial T\right)_p$ is the expansion (dilatation) coefficient, and $K_{ad} = (c_p/c_v)K$ (see the Ideal Fluids chapter), the temperature variation can also be written as

$$\delta T = -\frac{K_{ad}\beta T}{\rho c_p} div\boldsymbol{u} = -\frac{c^2\beta T}{c_p} div\boldsymbol{u} \ , \tag{11.19}$$

where $c = \sqrt{K_{ad}/\rho}$ is the sound velocity.
It follows that equation (11.11) becomes

$$\frac{\partial^2 \boldsymbol{u}}{\partial t^2} - c^2 graddiv\boldsymbol{u} = -\frac{E_0 T}{\rho}\delta(t) grad\delta(\boldsymbol{R}) \ . \tag{11.20}$$

Obviously, $\boldsymbol{u}$ is given by a gradient, $\boldsymbol{u} = grad\psi$, where the potential ψ satisfies the wave equation

$$\ddot{\psi} - c^2\Delta\psi = -\frac{E_0 T}{\rho}\delta(t)\delta(\boldsymbol{R}) \ . \tag{11.21}$$

The solution is

$$\psi = -\frac{E_0 T}{4\pi\rho c^2}\frac{\delta(t - R/c)}{R} = -\frac{E_0 T}{4\pi\rho c}\frac{\delta(R - ct)}{R} \ . \tag{11.22}$$

We can see that the explosion wave is a spherical shell of thickness l and time duration $T = l/c$, which propagates with the sound velocity. In the far field region ($R \gg l$) the displacement is

$$\boldsymbol{u} = -\frac{E_0 T}{4\pi\rho c}\frac{\delta'(R - ct)}{R^2}\boldsymbol{R} \ . \tag{11.23}$$

It is worthwhile estimating the pressure and the temperature in this explosion wave (equations (11.17) and (11.19)). We can see that these variations can be written as

$$\delta p = -\rho c^2\Delta\psi \ , \ \delta T = -\tfrac{c^2\beta T}{c_p}\Delta\psi \ , \qquad \delta T = \tfrac{\beta T}{\rho c_p}\delta p \ , \tag{11.24}$$

where we estimate $\Delta\psi$ as $E_0 T/\rho c l^3 R = E_0/\rho c^2 l^2 R$. We get

$$\delta p = \frac{E_0}{l^2 R} = p_0\frac{l}{R} \ , \tag{11.25}$$

which can be an enormous pressure. For $E_0 = 10^{20} erg$ and $l = 10cm$ we get $p_0 = 10^{17} dyn/cm^2$. The temperature variation is given by

$$\delta T = \frac{\beta T p_0}{\rho c_p} \frac{l}{R} = T_0 \frac{l}{R} \ , \tag{11.26}$$

where $T_0 = \beta T p_0 / \rho c_p$ is the explosion temperature. For air $\rho = 10^{-3} g/cm^3$, $\beta = 10^{13}/erg$ and $c_{v,p} \simeq 10^{23}/g$ (see the Ideal Fluids chapter), such that $T_0 = 10^{-7} p_0 T$; for $p_0 = 10^{17} dyn/cm^2$ we get an enormous increase in temperature. If we use the reduction factor 10^{-5} for energy, we get T_0 of the order of $10^6 K$ (with $T = 300K$), much higher than Sun's temperature $\simeq 6 \times 10^3 K$.[7]

11.3 On the mean freepath

The mean freepath Λ is the mean distance covered by a particle between two successive collisions in a statistical ensemble in equilibrium.[8] Let us assume a classical gas consisting of identical molecules, with the mean separation distance between the molecules a (density $n = 1/a^3$) and the molecular diameter d. A molecule has a^2/d^2 chances to pass between the other molecules, and once passed through, it encounters a molecule at distance a. It follows $\Lambda = a^3/d^2$. The quantity $\sigma = d^2$ can be viewed as a cross-section. The mean collision time is $\tau = \Lambda / v_{th}$, where v_{th} is the mean thermal velocity.

In a binary gaseous mixture in equilibrium, a molecule suffers competitive collisions, with alike molecules and different molecules. If the densities of the two components are $n_{1,2} = 1/a_{1,2}^3$, the density of the mixture is $n = n_1 + n_2$, so we can define a mean separation distance $a = 1/n^{1/3}$ given by

$$\frac{1}{a^3} = \frac{1}{a_1^3} + \frac{1}{a_2^3} \ . \tag{11.27}$$

For the cross-section of a molecule in the mixture we may take $\sigma = \sigma_1 + \sigma_2$, so we get

$$\frac{\sigma}{a^3} = \frac{\sigma}{a_1^3} + \frac{\sigma}{a_2^3} \ , \tag{11.28}$$

[7] "Brighter than a thousand Suns", J. R. Oppenheimer.

[8] R. Clausius, "Ueber die Art der Bewegung, welche wir Waerme nennen", Ann. Phys. **176** 353 (1857).

whence the mean freepath $\Lambda_{12} = a^3/\sigma$. We can see that we may write the above equation as

$$\frac{1}{\Lambda_{12}} = \frac{1}{\Lambda_1} + \frac{1}{\Lambda_2} , \qquad (11.29)$$

where $\Lambda_{1,2} = a_{1,2}^3/\sigma$. For a common thermal velocity this equation gives

$$\frac{1}{\tau_{12}} = \frac{1}{\tau_1} + \frac{1}{\tau_2} \qquad (11.30)$$

for the collision times. This relation is known as Matthiessen's rule.[9]

In some cases (as in an ionized gas, see the Viscous Fluids chapter) there may exist correlations between the two species of compoennts, *i.e.* a large mutual cross-section; then, the collisions are governed mainly by the smaller cross–secton, and the total cross-section is given by $1/\sigma_{12} = 1/\sigma_1 + 1/\sigma_2$.

[9] A. Matthiessen, Rep. Brit. Ass. Adv. Sci. **32** 144 (1862).

12 Superfluidity

12.1 Superfluid hydrodynamics

At very low temperatures the atoms with integral spins, called bosons, occupy the lowest, ground-state quantum-mechanical state in a large, macroscopic proportion. This is the Bose-Einstein condensation.[1] Part of them remain in excited states, which, for long wavelengths are sound waves, for shorter wavelengths are single atoms. In these conditions the ensemble of atoms is formed by two parts, a condensed phase and a normal phase.[2] The two phases have distinct properties. While the normal part behaves as any other fluid, with collisions, friction, viscosity and thermoconduction, the condensed phase consists of entangled atoms, without collisions, friction, viscosity or thermoconduction.[3] Consequently, this phase may flow upwards on solid walls. This is why it is termed the superfluid part of the ensemble.[4] Moreover, it behaves as a single entity, with a potential velocity, *i.e.* its

[1] A. Einstein, "Quantentheorie des einatomigen idealen Gases. 2. Abhandlung", Sitzungsber. Preuss. Akad. Wiss. **3** 18 (1925); S. Bose, "Plancks Gesetz und Lichtquantenhypothese", Z. Phys. **26** 178 (1924).

[2] L. Tisza, "Transport phenomena in helium II", Nature **141** 913 (1938); "Sur la theorie des liquides quantiques. Application a l'helium liquid. I, II.", J. Phys. Radium **1** 164, 350 (1940); "The theory of liquid helium", Phys. Rev. **72** 838 (1947); see also, F. London, *Superfluids*, Dover, NY (1964).

[3] F. London, "On the Bose-Einstein condensation", Phys. Rev. **54** 947 (1938).

[4] P. Kapitsa, "Viscosity of liquid helium below the λ-point", Nature **141** 3558 (1938); J. F. Allen and A. D. Misener, "Flow of liquid helium II", Nature **142** 3597 (1938).

velocity v_s satisfies the equation[5]

$$curl\boldsymbol{v}_s = 0 \ . \tag{12.1}$$

Indeed, the velocity $\boldsymbol{v} = -\frac{i\hbar}{2m}[\psi^* grad\psi - (grad\psi^*)\psi]$ for a quantum-mechanical condensate with wavefunction $\psi \sim e^{i\Phi}$ and mass m is $v = \frac{\hbar}{m}grad\Phi$, and $curl\boldsymbol{v} = 0$ (where $\hbar$ is Planck's constant).[6]
This happens for He_4 below $2.19K$, where it suffers a second-order phase transition. The critical temperature $2.19K$ is called the λ-point, and below this temperature the liquid helium is called *He II*. Any other atomic bosons suffer such a superfluid transition, but the temperature range of their liquid state is rather narrow, and they solidify quickly; in contrast with He_4, which remains liquid down to zero temperature.
The density of the two-fluid helium is $\rho = \rho_s + \rho_n$, where $\rho_{s,n}$ are the densities of the superfluid and normal parts. The mass flux density is $\boldsymbol{j} = \rho_s\boldsymbol{v}_s + \rho_n\boldsymbol{v}_n$, where $\boldsymbol{v}_{s,n}$ are the velocities of the two parts. We must have the equation of mass conservation (continuity equation)

$$\frac{\partial\rho}{\partial t} + div\boldsymbol{j} = 0 \ . \tag{12.2}$$

We assume now that the velocities are small, and have a slight spatial variation. Therefore, the equation of motion reads

$$\frac{\partial\boldsymbol{j}}{\partial t} + gradp = 0 \ , \tag{12.3}$$

where p is the pressure. We note that we leave aside the viscosity. Also, in the same conditions, the equation of heat transfer is

$$\frac{\partial(\rho s)}{\partial t} + \rho s div\boldsymbol{v}_n = 0 \ , \tag{12.4}$$

[5]L. Landau, "The theory of superfluidity of helium II", ZhETF **11** 592 (1941) (J. Phys. USSR **5** 71 (1941)) (in Russian); "On the hydrodynamics of helium II", ZhETF **14** 112 (1944) (J. Phys. USSR **8** 1 (1944)) (in Russian); "On the theory of superfluidity of helium II", J. Phys. USSR **11** 91 (1947) (in Russian); "Theory of the superfluidity of helium II", Phys. Rev, **60** 356 (1941); "On the theory of superfluidity", Phys. Rev. **75** 884 (1949).

[6]M. Apostol, "Bose-Einstein condensate and superfluidity", Roum. J. Phys. **48**, Suppl. I, 281 (2003); *Physical Kinetics*, Cambridge Scholars Publishing, Newcastle upon Tyne (2020) and References therein.

since only the normal part transfer heat. We note that we may leave aside the thermoconduction, as for an ideal normal fluid.

The pressure and the entropy in these equations are functions of density ρ and temperature T. Therefore, we have nine unknowns: $\rho_{s,n}$, $\boldsymbol{v}_{s,n}$ and T, and eigth equations (12.1) to (12.4). Actually, equation (12.1) is satisfied identically by the velocity gradient, such that we have seven unknowns and five equations. An additional equation was provided by Landau, who gave also the basic equation (12.1) and the excitation spectrum of liquid helium (References quoted in footnote #1). This is why we owe the theory of superfluidity to Landau.

The chemical potential μ is the energy of a particle in the thermodynamic ensemble, at constant volume and entropy. $-grad\mu$ is a force (per unit mass). This force does not act upon the particle, if that particle is an independent entity. However, when the particle is entangled, it does act. Therefore,

$$\frac{\partial \boldsymbol{v}_s}{\partial t} + grad\mu = 0 \ , \tag{12.5}$$

where μ is a function of p and T. Equation (12.5) is in fact only one equation for the velocity gradient. The problem remains single undetermined. Actually, since the equations are homogeneous, compatibility conditions may occur.

12.2 Second sound

We take the derivative with respect to time in equation (12.2) and use equation (12.3); we get

$$\frac{\partial^2 \rho}{\partial t^2} - \Delta p = 0 \ . \tag{12.6}$$

Since $d\mu = -sdT + dp/\rho$, we have

$$gradp = \rho grad\mu + \rho s gradT \tag{12.7}$$

and

$$\rho s gradT = -\frac{\partial}{\partial t}(\rho_s \boldsymbol{v}_s + \rho_n \boldsymbol{v}_n) + (\rho_s + \rho_n)\frac{\partial \boldsymbol{v}_s}{\partial t} \ , \tag{12.8}$$

or

$$\rho s grad T = \rho_n \frac{\partial}{\partial t}(\boldsymbol{v}_s - \boldsymbol{v}_n) \ , \tag{12.9}$$

where equations (12.3) and (12.5) are used. Now, from equations (12.4) and (12.2) we have

$$\frac{\partial s}{\partial t} = \frac{s\rho_s}{\rho} div(\boldsymbol{v}_s - \boldsymbol{v}_n) \ , \tag{12.10}$$

such that, by making use of equation (12.9), we get

$$\frac{\partial^2 s}{\partial t^2} - \frac{\rho_s s^2}{\rho_n}\Delta T = 0 \ . \tag{12.11}$$

Now we take small variations of ρ, p, s and T in equations (12.6) and (12.11), with respect to their equilibrium values, and use

$$\begin{aligned} \delta\rho &= \tfrac{\partial\rho}{\partial p}\delta p + \tfrac{\partial\rho}{\partial T}\delta T \ , \\ \delta s &= \tfrac{\partial s}{\partial p}\delta p + \tfrac{\partial s}{\partial T}\delta T \ . \end{aligned} \tag{12.12}$$

Equations (12.6) and (12.11) become

$$\begin{aligned} \tfrac{\partial\rho}{\partial p}\tfrac{\partial^2\delta p}{\partial t^2} + \tfrac{\partial\rho}{\partial p}\tfrac{\partial^2\delta T}{\partial t^2} - \Delta\delta p = 0 \ , \\ \tfrac{\partial s}{\partial p}\tfrac{\partial^2\delta p}{\partial t^2} + \tfrac{\partial s}{\partial T}\tfrac{\partial^2\delta T}{\partial t^2} - \tfrac{\rho_s s^2}{\rho_n}\Delta\delta T = 0 \ . \end{aligned} \tag{12.13}$$

A compatibility condition appears for these two equations. For a wave-like solution $\sim e^{-i\omega(t-x/u)}$, the compatibility condition is

$$u^4\frac{\partial(s,\rho)}{\partial(T,p)} - u^2\left(\frac{\partial s}{\partial T} + \frac{\rho_s s^2}{\rho_n}\frac{\partial\rho}{\partial p}\right) + \frac{\rho_s s^2}{\rho_n} = 0 \ . \tag{12.14}$$

This equation can be transformed into[7]

$$u^4 - u^2\left[\left(\frac{\partial p}{\partial\rho}\right)_s + \frac{\rho_s T s^2}{\rho_n c_v}\right] + \frac{\rho_s T s^2}{\rho_n c_v}\left(\frac{\partial p}{\partial\rho}\right)_T = 0 \ , \tag{12.15}$$

[7] L. Landau and E. Lifshitz, *Course of Theoretical Physics*, vol. 6 (*Fluid Mechanics*), Elsevier, Oxford (1987).

where c_v is the specific heat per unit mass at constant volume. For $\rho_s = 0$ the solution is

$$u = \sqrt{(\partial p/\partial \rho)_s} \ , \tag{12.16}$$

i.e. the velocity of the usual adiabatic sound (first sound).

For helium $c_v \simeq c_p$ denoted by c (where c_p is the specific heat per unit mass at constant pressure); also, $(\partial p/\partial \rho)_s = \frac{c_p}{c_v}(\partial p/\partial \rho)_T \simeq (\partial p/\partial \rho)_T$, denoted by $\partial p/\partial \rho$. The solutions are

$$u_1 = \sqrt{\partial p/\partial \rho} \ , \ u_2 = \sqrt{Ts^2\rho_s/c\rho_n} \ ; \tag{12.17}$$

u_2 is the velocity of the "superfluid" sound, called the second sound.[8]

At very low temperatures $\rho_s \simeq \rho$, $\rho_n = cT\rho/3u_1^2$ and $c = 3s$, such that $u_2 = u_1/\sqrt{3}$.[9]

In equations (12.2) to (12.6) and (12.11) all the quantities are proportional to each other; we can eliminate them as functions of one, say v_s. By doing so, we get $\boldsymbol{v}_n \simeq \boldsymbol{v}_s$ for the first sound and $\boldsymbol{v}_n \simeq -\rho_s\boldsymbol{v}_s/\rho_n$ for the second sound. Therefore, in the former case the atoms move in phase in the two fluids, while in the latter the atoms move in anti-phase, such that $\boldsymbol{j} = 0$: the normal phase and the superfluid phase move in opposition.[10]

The general equations of superfluid hydrodynamics, not restricted to small velocities, include $(\boldsymbol{v}_s - \boldsymbol{v}_n)^2$ as a thermodynamic variable (besides ρ and T, for instance), together with viscosity and thermocoduction.[11] Close to the critical λ-point the thermodynamic state has a long relaxation time τ. This generate dispersion and absorption of sound. The above calculations for sound are made for $\omega\tau \ll 1$.

[8]Third and fourth sounds in superfluid helium on solid surfaces or in capillaries are known.

[9]L. Landau and E. Lifshitz, *Course of Theoretical Physics*, vol. 9 (*Statistical Physics, Part 2*), Butterworth-Heinemann, Oxford (1991).

[10]E. Lifshitz, "Radiation of sound in helium II", J. Phys. USSR **8** 110 (1944).

[11]L. Landau and E. Lifshitz, *Course of Theoretical Physics*, vol. 6 (*Fluid Mechanics*), Elsevier, Oxford (1987); I. M. Khalatnikov, *An Introduction to the Theory of Superfluidity*, Benjamin, NY (1965).

12.3 Thermomechanical effect

Let us suppose two vessels filled with superfluid helium, and connected by a capillary. Helium flows from one vessel at pressure p_1 and temperature T_1 to the other vessel at pressure p_2 and temperature T_2. The superfluid helium carries no entropy, so, the mass transport decreases the mass and increases the entropy. Consequently, the temperature should rise in the vessel which loses helium. This is the thermomechanical effect.[12] For equilibrium the chemical potentials (per unit mass) should be equal,

$$\mu(p_1, T_1) = \mu(p_2, T_2) \; ; \tag{12.18}$$

for small variations, we have $\Delta p/\rho = s\Delta T$, since $(\partial\mu/\partial p)_s = 1/\rho$ and $(\partial\mu/\partial T)_p = s$, where ρ is the density and s is the entropy per unit mass. Therefore,

$$\frac{\partial p}{\partial T} = \rho s > 0 \; , \tag{12.19}$$

which means that the temperature increases in the vessel where the pressure is higher.[13]

[12] A similar, much smaller effect occurs in rarefied gases, due to the so-called temperature slip (L. Landau and E. Lifshitz, *Course of Theoretical Physics*, vol. 10 (*Physical Kinetics*), Elsevier, Oxford (1981)).

[13] H. London, "Thermodynamics of the thermomechanical effect of liquid He II", Proc. Roy. Soc. **A171** 484 (1939).

13 Appendix

13.1 Cylindrical coordinates

The cylindrical coordinates r and θ are given by

$$x = r\cos\theta \ , \ y = r\sin\theta \ , \ z \ , \tag{13.1}$$

where x, y and z are the cartesian coordinates. Since $\boldsymbol{r} = x\boldsymbol{e}_x + y\boldsymbol{e}_y + z\boldsymbol{e}_z$, where $\boldsymbol{e}_{x,y,z}$ are the cartesian unit vectors, and

$$d\boldsymbol{r} = dx\boldsymbol{e}_x + dy\boldsymbol{e}_y + dz\boldsymbol{e}_z \ , \tag{13.2}$$

we have

$$d\boldsymbol{r} = (\cos\theta\boldsymbol{e}_x + \sin\theta\boldsymbol{e}_y)dr + (-\sin\theta\boldsymbol{e}_x + \cos\theta\boldsymbol{e}_y)rd\theta + dz\boldsymbol{e}_z \tag{13.3}$$

and the unit vectors in cylindrical coordinates

$$\boldsymbol{e}_r = \cos\theta\boldsymbol{e}_x + \sin\theta\boldsymbol{e}_y \ , \ \boldsymbol{e}_\theta = -\sin\theta\boldsymbol{e}_x + \cos\theta\boldsymbol{e}_y \ ; \tag{13.4}$$

and

$$\boldsymbol{e}_x = \cos\theta\boldsymbol{e}_r - \sin\theta\boldsymbol{e}_\theta \ , \ \boldsymbol{e}_y = \sin\theta\boldsymbol{e}_r + \cos\theta\boldsymbol{e}_\theta \ . \tag{13.5}$$

For any vector $\boldsymbol{A}$ we have

$$\boldsymbol{A} = A_x\boldsymbol{e}_x + A_y\boldsymbol{e}_y + A_z\boldsymbol{e}_z = A_r\boldsymbol{e}_r + A_\theta\boldsymbol{e}_\theta + A_z\boldsymbol{e}_z \ , \tag{13.6}$$

$$A_x = A_r\cos\theta - A_\theta\sin\theta \ , \ A_y = A_r\sin\theta + A_\theta\cos\theta \tag{13.7}$$

and

$$A_r = A_x\cos\theta + A_y\sin\theta \ , \ A_\theta = -A_x\sin\theta + A_y\cos\theta \ . \tag{13.8}$$

The differential operators are computed by using

$$\frac{\partial}{\partial x} = \cos\theta\frac{\partial}{\partial r} - \frac{\sin\theta}{r}\frac{\partial}{\partial\theta} \ , \ \frac{\partial}{\partial y} = \sin\theta\frac{\partial}{\partial r} + \frac{\cos\theta}{r}\frac{\partial}{\partial\theta} \ . \tag{13.9}$$

For instance,

$$\begin{aligned} grad f &= \frac{\partial f}{\partial x}\boldsymbol{e}_x + \frac{\partial f}{\partial y}\boldsymbol{e}_y + \frac{\partial f}{\partial z}\boldsymbol{e}_z = \\ &= \frac{\partial f}{\partial r}\boldsymbol{e}_r + \frac{\partial f}{r\partial\theta}\boldsymbol{e}_\theta + \frac{\partial f}{\partial z}\boldsymbol{e}_z \ , \end{aligned} \tag{13.10}$$

where equations (13.5) and (13.9) are used.
Similarly,

$$div\boldsymbol{A} = \frac{1}{r}\frac{\partial}{\partial r}(rA_r) + \frac{1}{r}\frac{\partial A_\theta}{\partial\theta} + \frac{\partial A_z}{\partial z} \ , \tag{13.11}$$

$$\Delta f = \frac{1}{r}\frac{\partial}{\partial r}(r\frac{\partial f}{\partial r}) + \frac{1}{r^2}\frac{\partial^2 f}{\partial\theta^2} + \frac{\partial^2 f}{\partial z^2} \ , \tag{13.12}$$

$$\begin{aligned} (\Delta\boldsymbol{A})_r &= \Delta A_r - \frac{A_r}{r^2} - \frac{2}{r^2}\frac{\partial A_\theta}{\partial\theta} \ , \\ (\Delta\boldsymbol{A})_\theta &= \Delta A_\theta - \frac{A_\theta}{r^2} + \frac{2}{r^2}\frac{\partial A_r}{\partial\theta} \ , \\ (\Delta\boldsymbol{A})_z &= \Delta A_z \end{aligned} \tag{13.13}$$

and

$$\begin{aligned} (curl\boldsymbol{A})_r &= \frac{1}{r}\frac{\partial A_z}{\partial\theta} - \frac{\partial A_\theta}{\partial z} \ , \\ (curl\boldsymbol{A})_\theta &= \frac{\partial A_r}{\partial z} - \frac{\partial A_z}{\partial r} \ , \\ (curl\boldsymbol{A})_z &= \frac{1}{r}\frac{\partial}{\partial r}(rA_\theta) - \frac{1}{r}\frac{\partial A_r}{\partial\theta} \ . \end{aligned} \tag{13.14}$$

For a tensor σ_{ij}, $i, j = x, y, z$, we use the form $\sigma_{ij}\boldsymbol{e}_i\boldsymbol{e}_j$, which is invariant under rotations, and identify the coefficients of $\boldsymbol{e}_\alpha\boldsymbol{e}_\beta$, $\alpha, \beta = r, \theta, z$. For instance, for the stress tensor $\sigma_{ij} = -p\delta_{ij} + \eta(\partial_i v_j + \partial_j v_i)$ of an incompressible fluid we get[1]

$$\begin{aligned} \sigma_{rr} &= -p + 2\eta\frac{\partial v_r}{\partial r} \ , \\ \sigma_{\theta\theta} &= -p + 2\eta\left(\frac{1}{r}\frac{\partial v_\theta}{\partial\theta} + \frac{v_r}{r}\right) \ , \\ \sigma_{zz} &= -p + 2\eta\frac{\partial v_z}{\partial z} \end{aligned} \tag{13.15}$$

and

$$\begin{aligned} \sigma_{r\theta} &= \eta\left(\frac{1}{r}\frac{\partial v_r}{\partial\theta} + \frac{\partial v_\theta}{\partial r} - \frac{v_\theta}{r}\right) \ , \\ \sigma_{\theta z} &= \eta\left(\frac{\partial v_\theta}{\partial z} + \frac{1}{r}\frac{\partial v_z}{\partial\theta}\right) \ , \\ \sigma_{zr} &= \eta\left(\frac{\partial v_z}{\partial r} + \frac{\partial v_r}{\partial z}\right) \ . \end{aligned} \tag{13.16}$$

[1]The η-part of this tensor is similar to the strain tensor of Elasticity.

Also, the term $(\boldsymbol{v}grad)\boldsymbol{v}$ in the Navier-Stokes equation is written as

$$\left(v_r\frac{\partial}{\partial r}+v_\theta\frac{1}{r}\frac{\partial}{\partial\theta}+v_z\frac{\partial}{\partial z}\right)(v_r\boldsymbol{e}_r+v_\theta\boldsymbol{e}_\theta+v_z\boldsymbol{e}_z)\ , \tag{13.17}$$

where the derivatives act also upon the unit vectors. For instance, by making use of equation (13.4), $\frac{\partial}{\partial\theta}\boldsymbol{e}_\theta=-\boldsymbol{e}_r$.

13.2 Spherical coordinates

The spherical polar coordinates r, θ and φ are given by

$$x=r\sin\theta\cos\varphi\ ,\ y=r\sin\theta\sin\varphi\ ,\ z=r\cos\theta\ , \tag{13.18}$$

where x, y and z are the cartesian coordinates. Since $\boldsymbol{r}=x\boldsymbol{e}_x+y\boldsymbol{e}_y+z\boldsymbol{e}_z$, where $\boldsymbol{e}_{x,y,z}$ are the cartesian unit vectors, and

$$d\boldsymbol{r}=dx\boldsymbol{e}_x+dy\boldsymbol{e}_y+dz\boldsymbol{e}_z\ , \tag{13.19}$$

we have

$$\begin{gathered}d\boldsymbol{r}=(\sin\theta\cos\varphi\boldsymbol{e}_x+\sin\theta\sin\varphi\boldsymbol{e}_y+\cos\theta\boldsymbol{e}_z)dr+\\+(\cos\theta\cos\varphi\boldsymbol{e}_x+\cos\theta\sin\varphi\boldsymbol{e}_y-\sin\theta\boldsymbol{e}_z)rd\theta+\\+(-\sin\varphi\boldsymbol{e}_x+\cos\varphi\boldsymbol{e}_y)r\sin\theta d\varphi\end{gathered} \tag{13.20}$$

and the unit vectors are

$$\begin{gathered}\boldsymbol{e}_r=\sin\theta\cos\varphi\boldsymbol{e}_x+\sin\theta\sin\varphi\boldsymbol{e}_y+\cos\theta\boldsymbol{e}_z\ ,\\ \boldsymbol{e}_\theta=\cos\theta\cos\varphi\boldsymbol{e}_x+\cos\theta\sin\varphi\boldsymbol{e}_y-\sin\theta\boldsymbol{e}_z\ ,\\ \boldsymbol{e}_\varphi=-\sin\varphi\boldsymbol{e}_x+\cos\varphi\boldsymbol{e}_y\end{gathered} \tag{13.21}$$

and

$$\begin{gathered}\boldsymbol{e}_x=\sin\theta\cos\varphi\boldsymbol{e}_r+\cos\theta\cos\varphi\boldsymbol{e}_\theta-\sin\varphi\boldsymbol{e}_\varphi\ ,\\ \boldsymbol{e}_y=\sin\theta\sin\varphi\boldsymbol{e}_r+\cos\theta\sin\varphi\boldsymbol{e}_\theta+\cos\varphi\boldsymbol{e}_\varphi\ ,\\ \boldsymbol{e}_z=\cos\theta\boldsymbol{e}_r-\sin\theta\boldsymbol{e}_\theta\ .\end{gathered} \tag{13.22}$$

For any vector $\boldsymbol{A}$ we have

$$\boldsymbol{A}=A_x\boldsymbol{e}_x+A_y\boldsymbol{e}_y+A_z\boldsymbol{e}_z=A_r\boldsymbol{e}_r+A_\theta\boldsymbol{e}_\theta+A_\varphi\boldsymbol{e}_\varphi\ , \tag{13.23}$$

$$
\begin{aligned}
A_x &= \sin\theta\cos\varphi A_r + \cos\theta\cos\varphi A_\theta - \sin\varphi A_\varphi \ , \\
A_y &= \sin\theta\sin\varphi A_r + \cos\theta\sin\varphi A_\theta + \cos\varphi A_\varphi \ , \\
A_z &= \cos\theta \boldsymbol{A}_r - \sin\theta A_\theta
\end{aligned}
\tag{13.24}
$$

and

$$
\begin{aligned}
A_r &= \sin\theta\cos\varphi A_x + \sin\theta\sin\varphi A_y + \cos\theta A_z \ , \\
A_\theta &= \cos\theta\cos\varphi A_x + \cos\theta\sin\varphi A_y - \sin\theta A_z \ , \\
A_\varphi &= -\sin\varphi A_x + \cos\varphi A_y \ .
\end{aligned}
\tag{13.25}
$$

The differential operators are computed by using

$$
\begin{aligned}
\frac{\partial}{\partial x} &= \sin\theta\cos\varphi\frac{\partial}{\partial r} + \frac{\cos\theta\cos\varphi}{r}\frac{\partial}{\partial\theta} - \frac{\sin\varphi}{r\sin\theta}\frac{\partial}{\partial\varphi} \ , \\
\frac{\partial}{\partial y} &= \sin\theta\sin\varphi\frac{\partial}{\partial r} + \frac{\cos\theta\sin\varphi}{r}\frac{\partial}{\partial\theta} + \frac{\cos\varphi}{r\sin\theta}\frac{\partial}{\partial\varphi} \ , \\
\frac{\partial}{\partial z} &= \cos\theta\frac{\partial}{\partial r} - \frac{\sin\theta}{r}\frac{\partial}{\partial\theta} \ .
\end{aligned}
\tag{13.26}
$$

For instance,

$$
\begin{aligned}
grad f &= \frac{\partial f}{\partial x}\boldsymbol{e}_x + \frac{\partial f}{\partial y}\boldsymbol{e}_y + \frac{\partial f}{\partial z}\boldsymbol{e}_z = \\
&= \frac{\partial f}{\partial r}\boldsymbol{e}_r + \frac{\partial f}{r\partial\theta}\boldsymbol{e}_\theta + \frac{\partial f}{r\sin\theta\partial\varphi}\boldsymbol{e}_\varphi \ .
\end{aligned}
\tag{13.27}
$$

Similarly,

$$
div\boldsymbol{A} = \frac{1}{r^2}\frac{\partial}{\partial r}(r^2 A_r) + \frac{1}{r\sin\theta}\frac{\partial}{\partial\theta}(\sin\theta A_\theta) + \frac{\partial A_\varphi}{r\sin\theta\partial\varphi} \ , \tag{13.28}
$$

$$
\Delta f = \frac{1}{r^2}\frac{\partial}{\partial r}(r^2\frac{\partial f}{\partial r}) + \frac{1}{r^2\sin\theta}\frac{\partial}{\partial\theta}(\sin\theta\frac{\partial f}{\partial\theta}) + \frac{1}{r^2\sin^2\theta}\frac{\partial^2 f}{\partial\varphi^2} \ , \tag{13.29}
$$

$$
\begin{aligned}
(\Delta\boldsymbol{A})_r &= \Delta A_r - \frac{2}{r^2}\left(A_r + \frac{1}{\sin\theta}\frac{\partial}{\partial\theta}(\sin\theta A_\theta) + \frac{1}{\sin\theta}\frac{\partial A_\varphi}{\partial\varphi}\right) \ , \\
(\Delta\boldsymbol{A})_\theta &= \Delta A_\theta + \frac{2}{r^2}\left(\frac{\partial A_r}{\partial\theta} - \frac{A_\theta}{2\sin^2\theta} - \frac{\cos\theta}{\sin^2\theta}\frac{\partial A_\varphi}{\partial\varphi}\right) \ , \\
(\Delta\boldsymbol{A})_\varphi &= \Delta A_\varphi + \frac{2}{r^2\sin\theta}\left(\frac{\partial A_r}{\partial\varphi} + \cot\theta\frac{\partial A_\theta}{\partial\varphi} - \frac{1}{2\sin\theta}A_\varphi\right)
\end{aligned}
\tag{13.30}
$$

and

$$
\begin{aligned}
(curl\boldsymbol{A})_r &= \frac{1}{r\sin\theta}\left(\frac{\partial}{\partial\theta}(\sin\theta A_\varphi) - \frac{\partial A_\theta}{\partial\varphi}\right) \ , \\
(curl\boldsymbol{A})_\theta &= \frac{1}{r\sin\theta}\frac{\partial A_r}{\partial\varphi} - \frac{1}{r}\frac{\partial}{\partial r}(rA_\varphi) \ , \\
(curl\boldsymbol{A})_\varphi &= \frac{1}{r}\left(\frac{\partial}{\partial r}(rA_\theta) - \frac{\partial A_r}{\partial\theta}\right) \ .
\end{aligned}
\tag{13.31}
$$

The stress tensor $\sigma_{ij} = -p\delta_{ij} + \eta(\partial_i v_j + \partial_j v_i)$ of an incompressible fluid is

$$\begin{aligned}\sigma_{rr} &= -p + 2\eta\frac{\partial v_r}{\partial r} \ , \\ \sigma_{\varphi\varphi} &= -p + 2\eta\left(\frac{1}{r\sin\theta}\frac{\partial v_\varphi}{\partial\varphi} + \frac{v_r}{r} + \frac{v_\theta\cot\theta}{r}\right) \ , \\ \sigma_{\theta\theta} &= -p + 2\eta\left(\frac{1}{r}\frac{\partial v_\theta}{\partial\theta} + \frac{v_r}{r}\right)\end{aligned} \tag{13.32}$$

and

$$\begin{aligned}\sigma_{r\theta} &= \eta\left(\frac{1}{r}\frac{\partial v_r}{\partial\theta} + \frac{\partial v_\theta}{\partial r} - \frac{v_\theta}{r}\right) \ , \\ \sigma_{\theta\varphi} &= \eta\left(\frac{1}{r\sin\theta}\frac{\partial v_\theta}{\partial\varphi} + \frac{1}{r}\frac{\partial v_\varphi}{\partial\theta} - \frac{v_\varphi\cot\theta}{r}\right) \ , \\ \sigma_{\varphi r} &= \eta\left(\frac{\partial v_\varphi}{\partial r} + \frac{1}{r\sin\theta}\frac{\partial v_r}{\partial\varphi} - \frac{v_\varphi}{r}\right) \ .\end{aligned} \tag{13.33}$$

13.3 Thermodynamic relations

These are relations between the derivatives of the thermodynamic quantities, in terms of the variables in the equation of state (pressure p, volume V and temperature T).[2]

1) The free energy is given by

$$dF = -pdV - SdT \ , \tag{13.34}$$

where S is the entropy. For the specific heat C_v at constant volume We have

$$\left(\frac{\partial C_v}{\partial V}\right)_T = T\frac{\partial^2 S}{\partial T\partial V} = -T\frac{\partial^3 F}{\partial V\partial^2 T} = T\left(\frac{\partial^2 p}{\partial T^2}\right)_V \ . \tag{13.35}$$

Similarly, for the specific heat at constant pressure,

$$\left(\frac{\partial C_p}{\partial p}\right)_T = T\frac{\partial^2 S}{\partial T\partial V} = -T\frac{\partial^3 F}{\partial V\partial^2 T} = -T\left(\frac{\partial^2 V}{\partial T^2}\right)_p \ , \tag{13.36}$$

by using the Gibbs free energy

$$d\Phi = V dp - SdT \ . \tag{13.37}$$

[2] L. Landau and E. Lifshitz, *Course of Theoretical Physics*, vol. 5 (*Statistical Physics*), Peragamon, Oxford (1980).

2)

$$\left(\frac{\partial S}{\partial V}\right)_T = -\frac{\partial^2 F}{\partial T \partial V} = \left(\frac{\partial p}{\partial T}\right)_V \tag{13.38}$$

and

$$\left(\frac{\partial S}{\partial p}\right)_T = -\left(\frac{\partial V}{\partial T}\right)_p . \tag{13.39}$$

3)

$$\left(\frac{\partial E}{\partial V}\right)_T = -p + T\left(\frac{\partial S}{\partial V}\right)_T = -p + T\left(\frac{\partial p}{\partial T}\right)_V , \tag{13.40}$$

$$\begin{aligned} \left(\frac{\partial E}{\partial p}\right)_T &= -T\left(\frac{\partial V}{\partial T}\right)_p - p\left(\frac{\partial V}{\partial p}\right)_T , \\ \left(\frac{\partial E}{\partial T}\right)_p &= C_p - p\left(\frac{\partial V}{\partial T}\right)_p \end{aligned} \tag{13.41}$$

and

$$\begin{aligned} \left(\frac{\partial W}{\partial V}\right)_T &= T\left(\frac{\partial p}{\partial T}\right)_V + V\left(\frac{\partial p}{\partial V}\right)_T , \\ \left(\frac{\partial W}{\partial p}\right)_T &= V - T\left(\frac{\partial V}{\partial T}\right)_p , \\ \left(\frac{\partial W}{\partial T}\right)_p &= C_V + V\left(\frac{\partial p}{\partial T}\right)_V \end{aligned} \tag{13.42}$$

for energy ($dE = -pdV + TdS$) and enthalpy ($dW = Vdp + TdS$).

4)

$$\begin{aligned} C_v &= T\left(\frac{\partial S}{\partial T}\right)_V = T\frac{\partial(S,V)}{\partial(T,V)} = T\frac{\partial(S,V)/\partial(T,p)}{\partial(T,V)/\partial(T,p)} = \\ &= T\frac{(\partial S/\partial T)_p(\partial V/\partial p)_T - (\partial S/\partial p)_T(\partial V/\partial T)_p}{(\partial V/\partial p)_T} = \\ &= C_p - T\frac{(\partial S/\partial p)_T(\partial V/\partial T)_p}{(\partial V/\partial p)_T} . \end{aligned} \tag{13.43}$$

Inserting equation (13.39),

$$C_p - C_v = -T\frac{(\partial V/\partial T)_p^2}{(\partial V/\partial p)_T} . \tag{13.44}$$

i.e. the relation between the specific heats. Starting with $C_p = T\left(\frac{\partial S}{\partial T}\right)_p$, we get

$$C_p - C_v = -T\frac{(\partial p/\partial T)_V^2}{(\partial p/\partial V)_T} . \tag{13.45}$$

Since $(\partial p/\partial V)_T < 0$, we have $C_p > C_v$.

5)

$$\left(\frac{\partial T}{\partial V}\right)_S = \frac{\partial(T,S)}{\partial(V,S)} = \frac{\partial(T,S)/\partial(V,T)}{\partial(V,S)/\partial(V,T)} = \\ = -\frac{(\partial S/\partial p)_T}{(\partial S/\partial T)_V} = -\frac{T}{C_v}\left(\frac{\partial S}{\partial V}\right)_T , \tag{13.46}$$

or, by making use of equation (13.38),

$$\left(\frac{\partial T}{\partial V}\right)_S = -\frac{T}{C_v}\left(\frac{\partial p}{\partial T}\right)_V . \tag{13.47}$$

Similarly, we get

$$\left(\frac{\partial T}{\partial p}\right)_S = \frac{T}{C_p}\left(\frac{\partial V}{\partial T}\right)_p . \tag{13.48}$$

6)

$$\left(\frac{\partial V}{\partial p}\right)_S = \frac{\partial(V,S)}{\partial(p,S)} = \frac{\partial(V,S)/\partial(V,T)}{\partial(p,S)/\partial(p,T)} \cdot \frac{\partial(V,T)}{\partial(p,T)} = \\ = \frac{(\partial S/\partial T)_V}{(\partial S/\partial T)_p}\left(\frac{\partial V}{\partial p}\right)_T , \tag{13.49}$$

i.e. the relation between the two compressibilities (adiabatic and isothermal)

$$\left(\frac{\partial V}{\partial p}\right)_S = \frac{C_v}{C_p}\left(\frac{\partial V}{\partial p}\right)_T . \tag{13.50}$$

7) From equation (13.50) and equations (13.44) and (13.45) we get

$$\left(\frac{\partial V}{\partial p}\right)_S = \left(\frac{\partial V}{\partial p}\right)_T + \frac{T}{C_p}\left(\frac{\partial V}{\partial T}\right)_p^2 , \\ \left(\frac{\partial p}{\partial V}\right)_S = \left(\frac{\partial p}{\partial V}\right)_T - \frac{T}{C_v}\left(\frac{\partial p}{\partial T}\right)_V^2 . \tag{13.51}$$

For thermodynamic stability (maximum of the entropy) the inequalities $C_v > 0$ and $(\partial p/\partial V)_T < 0$ are valid; also, $(\partial \mu/\partial c)_{p,T} > 0$, where μ is the chemical potential and c is the concentration.

Index

T

V

W